6/21/93

D1272304

Fish Energetics

Fish

Energetics

New Perspectives

Edited by

Peter Tytler & Peter Calow

THE JOHNS HOPKINS UNIVERSITY PRESS
Baltimore, Maryland

© Peter Tytler and Peter Calow
All rights reserved

First published in the United States of America in 1985 by
The Johns Hopkins University Press, 701 West 40th Street,
Baltimore, Maryland 21211

Library of Congress Cataloging in Publication Data
Main entry under title:

Fish Energetics.

1. Fishes—Physiology. 2. Bioenergetics. I. Tytler, Peter.
II. Calow, Peter.
QL639.1.F554 1985   597'.019'121   85-26
ISBN 0-8018-2792-2

Printed and bound in Great Britain

CONTENTS

PREFACE

It is almost thirty years since Professor G.G. Winberg established the basis for experimental studies in fish energetics with the publication of his monograph, *Rate of Metabolism and Food Requirements of Fishes*. His ultimate aim was to develop a scientific approach to fish culture and management, and the immense volume of literature generated in the ensuing years has been mainly in response to the demand for information from a rapidly expanding, world-wide aquaculture industry and to the shortcomings of contemporary practices in fisheries management.

The purpose of this book is not to review this literature comprehensively, but, assuming an informed readership, to focus attention on topics in which new knowledge and theory are beginning to be applied in practice.

Most emphasis has been placed on food; feeding; production (growth and reproduction) and energy budgeting, as these have most influence on the development of fish culture. Some chapters offer practical advice for the selection of methods, and warn of pitfalls in previous approaches. In others the influence of new theory on the interpretation of studies in fish energetics is discussed in the context of resource allocation and adaptation. We hope that the scope of material presented here will have sufficient interest and value to help significantly to fulfil Winberg's original objectives.

The book was conceived during a scientific meeting of the Fisheries Society of the British Isles in December 1982. Most of the authors were participants in this meeting. The book is, however, in no sense a proceedings of the meeting and all chapters have been written since the meeting specifically for the book.

P.T.

LIST OF CONTRIBUTORS

A.E. Brafield, Department of Biology, Queen Elizabeth College, Campden Hill Road, London W8 7AH, England

P. Calow, Department of Zoology, University of Sheffield, Sheffield S10 2TN, England

C.B. Cowey, NERC, Institute of Marine Biochemistry, St Fittick's Road, Aberdeen AB1 3RA, Scotland

A.D. Hawkins, Department of Agriculture and Fisheries for Scotland, Marine Laboratory, Aberdeen AB9 8D8, Scotland

M. Jobling, Institutt for Fiskerifag, Universitetet I Tromso, 9000 Tromsø, Norway

B. Knights, Applied Ecology Research Group, The Polytechnic of Central London, London W1M 8JS, England

K.P. Lone, Department of Zoology, University of the Punjab, New Campus, Lahore, Pakistan

A.J. Matty, Department of Biological Sciences, University of Aston, Gosta Green, Birmingham B4 7ET, England

T.J. Pandian, School of Biology, Madurai Kamaraj University, Madurai, 625 021 India

I.G. Priede, Department of Zoology, University of Aberdeen, Tillydrone Avenue, Aberdeen AB9 2TN, Scotland

N. Soofiani, Department of Agriculture and Fisheries for Scotland, Marine Laboratory, Aberdeen AB9 8D8, Scotland

A.G.J. Tacon, Aquaculture Development Coordination Programme, FAO, Via delle Termi di Caracalla, 00100 Rome, Italy

C. Talbot, Department of Agriculture and Fisheries for Scotland, Freshwater Fisheries Laboratory, Perthshire PH16 5LB, Scotland

C.R. Townsend, Department of Biology, University of East Anglia, Norwich NR4 7TJ, England

E. Vivekanandan, School of Biological Sciences, Madurai Kamaraj University, Madurai, 625 021 India

I.J. Winfield, Department of Biology, University of East Anglia, Norwich NR4 7TJ, England

R.J. Wootton, Department of Zoology, The University College of Wales, Penglais, Aberystwyth, Dyfed SY23 2AX

PART ONE:
EVOLUTIONARY ASPECTS OF ENERGY BUDGETS

1 ADAPTIVE ASPECTS OF ENERGY ALLOCATION

Peter Calow

1.1 Introduction

Fishes, like all organisms, use ingested food resources (C) as building blocks in the synthesis of tissues (production, P) and as fuel in the metabolic processes that power this synthesis and other physico-chemical work (R). Some of the resources are lost as waste products (E). All these aspects of metabolism can be represented in energy units (joules; 4.2 J/cal) and, since all biological systems obey the laws of thermodynamics, can be combined as follows:

$$C = P + R + E \qquad 1.1$$

Total metabolism (R) comprises a number of subcomponents: standard metabolism (R_S), which is recorded when the organism is at rest; routine metabolism (R_R), which is recorded in 'routinely' active animals; feeding metabolism (R_F), associated with animals that have just fed (sometimes known as specific dynamic action or effect), and active metabolism (R_A), which is recorded in animals undergoing sustained activity (i.e. swimming in fishes). If the energy demands additional to routine metabolism are additive, then:

$$R = R_S + aR_{R-S} + bR_{F-S} + cR_{A-S} \qquad 1.2$$

where a, b and c are constants expressing the fraction of time that each type of metabolism is used. P can consist of both somatic (Pg) and/or reproductive (Pr) components. Finally, E can be decomposed to faeces (F), the excretory products such as urea and ammonia (U), and miscellaneous secretions such as mucus (Muc).

When these separate terms are assembled, we have:

$$C = (R_S + aR_{R-S} + bR_{F-S} + cR_{A-S}) +$$

$$(P_g + P_r) + (F + U + Muc)$$

This version of the energy budget follows Brett & Groves (1979). Energy acquisition patterns (C and $A = C-F$) are considered further in Chapters 3 to 5, and energy utilization patterns (R and P) further in Chapters 8 to 10. The methodology used to measure the various components of the energy budget and some specimen results on whole budgets from both laboratory and field populations are given in Chapters 10 to 11.

The way energy is partitioned between metabolic demands is likely to have a profound effect on form and function and hence on time to and between breeding (t), survivorship (s) and fecundity (n). Thus investment in activity, by facilitating escape from predators, will influence survival (s); investment in Pg will influence the time it takes for somatic structures to become sufficiently developed to allow reproduction (t); and investment in Pr will influence fecundity (n) (Figure 1.1). So the energy budget must be sensitive to natural selection (Alexander, 1967). That is to say, the allocation of energy between the components of the budget is controlled by enzymes and hence ultimately by genes, and whether or not specific genes increase or recede in frequency in a population depends on how they influence s, n and t. This capacity to spread, fitness, is conveniently expressed as F, the rate of increase per individual with a particular trait, and this can be defined in terms of s, n and t using the classical Euler–Lotka equation (see Charlesworth, 1980):

$$1 = \sum_t e^{-Ft} s_t . n_t \qquad 1.3$$

F is sometimes considered equivalent to r, the intrinsic rate of increase, but this can lead to confusion (Sibly & Calow, 1985).

Clearly those genes that maximize s and n and minimize t should be most successful and so, in practice, natural selection should favour maximum investment in all the aspects of metabolism illustrated in Figure 1.1. Yet the resources available for allocation are finite and limited, because the feeding processes and structures are themselves limited. Hence resources used in one aspect of metabolism will not be available for use in others; for example energy used to support locomotion, which might improve ability to escape from predators and hence enhance s, will not be available for production and this will have a negative effect on n and a positive one on t. This means that there will be trade-offs between the various components of metabolism, and hence the components of fitness influenced by them. As will be made clear below, the form of these trade-offs puts important constraints on the way natural selection can work and on the optimum solutions likely to evolve (Calow, 1984a,b).

Figure 1.1: Metabolic and Possible Fitness Effects of Energy Allocation.

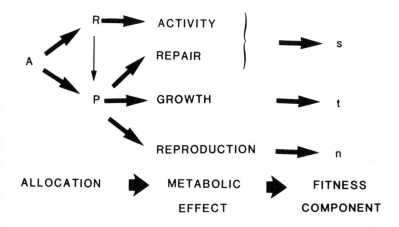

This book is concerned with new developments in the general principles and practices of fish energetics. In this introduction, special attention is given to the adaptational aspects of energetics and in particular to a definition of the trade-offs that limit the action of selection. The theme, therefore, is that physiology has as much to contribute to an understanding of fitness as the theory of natural selection has to an understanding of physiology.

Throughout, physiology is equated with energetics. This is only partially legitimate because the food supplies various kinds of resource, including energy, and the one that is most limited is likely to be most influenced by natural selection. This need not be energy (Chapters 6 and 12). Nevertheless, energy is used in all the processes of the body and in the building of all structures. Moreover, it is easily measured and modelled. Hence, though it need not always be the most appropriate currency, it is likely to be the one most generally appropriate and that is the justification for emphasizing it here (Calow, 1984a).

1.2 Interaction Between Production and Metabolism

One very obvious metabolic conflict in fishes is between using resources to produce biomass (both somatic and reproductive) and paying the costs of overt activity. Biomass production ought to be positively correlated with fitness since increases in it should reduce the time between birth and reaching a size capable of reproducing, and should ultimately enhance reproductive output. However, as much as 25% of ingested energy may be invested in active metabolism by carnivorous fishes (Brett & Groves, 1979; Chapters 2 and 10). Herbivorous fishes probably invest considerably less in this aspect of metabolism (Chapter 4). Swimming is metabolically costly but it can bring both survival gains by allowing escape from predators and metabolic gains by facilitating the capture of prey. Ware (1975) has modelled the conflict between the metabolic costs of swimming and the feeding gains (summarized in Figure 1.2) and found that young bleak (*Alburnus alburnus*), which are planktivorous, swim at speeds that maximize the energy available for production. Different speeds would have been required to maximize growth efficiency (P_g/C) and growth efficiency per unit food intake $(P_g/C)/C$, both of which have been advocated as phenotypic, physiological measures of fitness. *A priori*, it would be expected that selection should maximize production rate, since what matters in the end is the numbers of viable progeny that are produced, not necessarily the efficiency at which they are produced. This argument is considered further in Chapter 2.

If it is production rate which is maximized, and swimming is costly, then it is to be expected that morphological and behavioural characteristics should have evolved that economize on these costs (economization principle; Calow, 1984a). A power reduction in swimming might also reduce the possibility of fishes exceeding their metabolic scope and so cause a reduction in the chances of death (Chapter 2). Economization of this kind has been documented by Ware (1981). It involves streamlining, the capacity to rest when food is unavailable, the adoption of efficient searching strategies and an ability to migrate into waters that provide increased food abundance and/or allow reduced costs (e.g. colder water). Stream-dwelling salmonids 'rest' in areas of low water velocity to minimize the energy expended in swimming, yet close to swift currents to maximize access to invertebrate drift; i.e. positions that maximize net energy returns (Fausch, 1984).

Figure 1.2: Increased Swimming Speed (S) Increases the Rate at Which Food is Ingested (C) in the Way Illustrated but Causes Increased Metabolic Output (O) in the Way Illustrated. The balance between C and O determines production, P, and so the influence of S on P, P/C and (P/C)/C can be calculated. Argument from Ware (1975).

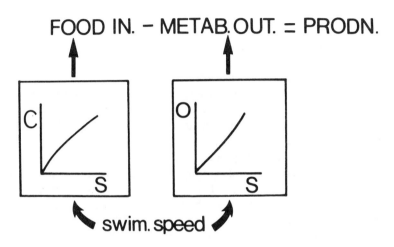

Another less obvious cost is that associated with maintaining biomass in an organized state some distance from thermodynamic equilibrium. Biological organization is continuously being disorganized by thermal noise, autoxidation, mistakes in synthesis, etc., and to remain viable the disorganized structures have to be replaced and repaired. The continuous turnover of molecules and cells even in adults is a manifestation of this (Calow, 1978), and these dynamic processes are expensive both in building blocks and energy; for example, as much as 25% basal metabolic rate may be involved in paying these costs (Waterlow, 1980). At least part of the so-called specific dynamic action (above) might be paying the costs of a surge of synthesis, possibly involved in this process of repair and replacement, following the intake of new resources in a meal (Parry, 1983; Vahl, 1984; and see Chapter 8 but cf. Chapter 12). The costs are metabolic and so put constraints on production, but the gains are in terms of viability and hence survival. However, the investment in this process need not be fixed and there is some evidence, from animals other than fishes, that when the post-reproductive life-span (due to extrinsic mortality) is short, the investment in reproductive production may be at the expense of these maintenance processes, causing accelerated senescence (Calow, 1978). In some fishes there is enhanced loss of scales during the breeding season, possibly due to reduced replacement and the production of translucent otoliths indicative of reduced protein synthesis in somatic structures (reviewed in Miller, 1984). The mortality associated with massive investment in reproduction in species such as some lampreys and salmon might be due to accelerated senescence and deserves some further investigation (Larsen, 1973).

One possible conclusion from this section is that biomass production should be maximized within the physiological limits imposed by costs of living. However, growth rates can be enhanced above normal by endocrine manipulation (Chapter 7), implying that under normal circumstances production is constrained below the physiological capacity of the organism by active endocrine control. This implies, further, that sub-maximum production (i.e. growth) rates are optimum and suggests that there may be non-physiological, perhaps ecological, restraints at work; for example, fishes that grow more rapidly than the average for their size class might be more conspicuous to predators. These possibilities are reviewed in Calow (1982) and mean that maximization principles, as applied in models of food acquisition (foraging theory — Chapter 3) and utilization, have to be modulated, at least in certain circumstances, by sub-maximal controls. The existence of endocrine-controlled, not necessarily maximum, growth rates has been advocated for fishes by Iles (1974) from the program-like nature observed in growth patterns in herring (see also Miller, 1979). Chapter 7 provides a detailed review of the general features of endocrine control of metabolism.

1.3 Interaction Between Reproduction and Somatic Maintenance

With the initiation of reproduction, interactions become possible between reproductive production, somatic production and somatic maintenance. This section concentrates on interactions between reproduction and maintenance since these entail trade-offs between reproductive output and the survival of parents. The complicating interaction between reproduction, growth and subsequent reproductive success will be considered in the next section.

With limited resources for use in metabolism, an allocation to the formation of gametes is likely to be at the expense of the maintenance of somatic processes and structures. This could lead to a quantitative reduction in somatic biomass (Shevchenko, 1972; Roff, 1983) and a qualitative deterioration of the soma (above). It can also lead to a reduction in swimming activity (Koch & Wieser, 1983) and hence ability to escape predators. In turn these effects are all likely to increase the mortality risks to which the parent is subjected (Calow, 1979; see also Chapter 2). Negative correlations have been recorded between fecundity and subsequent survival of the parent in a variety of species including fishes (Calow & Sibly, 1983; but cf. Bell, 1984a,b) and it has been possible to extend the longevities of some fishes by artificially preventing a reproductive drain, in particular for salmon and lampreys (reviewed in Larsen, 1973).

Equation 1.3 indicates that F, fitness, should be a positive function of both n, fecundity (or if gamete size is approximately constant, investment in reproduction, z; since $n = z/k$, where k is gamete size) and survivorship. With certain simplifying assumptions, it can be shown from the algebra (Sibly & Calow, 1983, and legend to Figure 1.3) that if z is plotted against adult survival chances (s_a), the combinations of these parameters which give the same F lie on straight lines with negative slopes proportional to juveniles' survival chances (s_j) — i.e. increasing in slope with s_j. These are fitness contours which represent increasing values of F as they shift away from the origin (Figure 1.3). Hence the optimal combination of s_a and n is that *feasible combination* which lies on the highest F-contour, and feasibility is determined in a large part by the trade-offs discussed above.

It therefore becomes necessary to define these feasible options, but this turns out to be experimentally difficult, since natural selection, by definition, will have favoured restriction of s_a and z to only a subset of all feasible combinations — i.e. the optimal subset. One possible way of defining full trade-off sets, however, is from an understanding of their physiological basis. For example, a possible hypothesis is that resources are diverted away from somatic maintenance to reproduction according to a set of physiological priorities — the first few eggs being formed from resources in excess of parental requirements (e.g. from stores laid down specifically to support reproduction), and this has no effect on adult survival; the next few eggs from resources required by but not essential to maintenance and hence having a

Figure 1.3: Fitness Contours in the s_a vs n (or z) Plane are Straight Lines with a Negative Slope Proportional to s_j. If it is assumed that the fecundity and mortality of adults is independent of age, equation 1.3 reduces to $1 = e^{-Ft_j} s_j n + e^{-Ft_a} s_a$. If $t_j = t_a = 1$, then after rearrangements $F = \log_e (s_j n + s_a)$. Hence to keep F constant, an increase in s_a must be compensated by a *proportional* reduction in n, with the proportionality being *directly dependent* on s_j. In the figure F increases in the direction shown, i.e. with increase in s_a and n, as is intuitively reasonable.

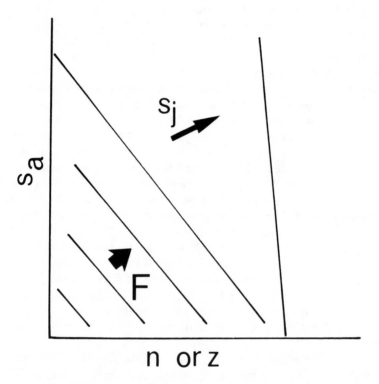

depressing effect on s_a, and the next few eggs from resources essential for maintenance and hence driving s_a to zero. This yields the trade-off curve in Figure 1.4A&B. An alternative is illustrated in Figure 1.4C&D. Here the same physiological priorities operate but there is never a point in the process where adult mortality becomes a certainty, so s_a never completely reduces to zero and the curve is pulled out to the left to yield the sigmoid relationship.

A possible way of distinguishing between these alternatives is to assume that resources are used in gamete formation according to the same set of physiological priorities that operate under conditions of

Figure 1.4: Relationship of Post-reproductive Mortality Rate (μ) and Survivorship (s_a) to Fecundity (n or Z). It is assumed that mortality of adults is random so $s_a = e^{-\mu t}$; A & B and C & D are different models. See text for further explanation. The broken and dotted lines in B are fitness contours and are explained in the text. After Calow (1984).

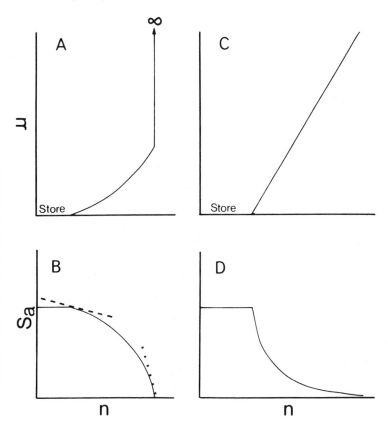

starvation (Calow, 1984b). If this is true, survivorship curves for cohorts of starving fishes should give some clue as to the form of the trade-off curve. Figure 1.5 shows data on this for the fry of 12 species (taken from Ivlev, 1961). In some species there is an initial period of high mortality, but this probably represents the death of 'weaklings'. If this is ignored, the curves are rather ambiguous. In some, survival reduces continuously to zero whereas in others there is some 'slowing-off' as starvation continues so that curves are pulled to the right. However, visual inspection suggests that this is never very marked. More work is needed on survival under starvation, particularly on older fishes.

Figure 1.5: Survivorship Curves for Starving Fry; days = time starved. After Ivlev (1961).

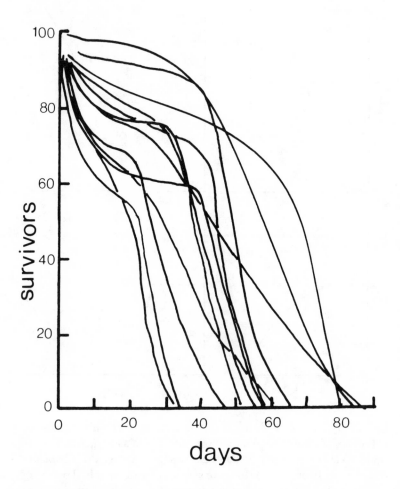

These trade-off curves represent the combinations of s_a and n that are feasible for organisms with a particular biology. Optimum combinations are most easily defined for the curve in Figure 1.4A&B,

for here there is only one point that will be tangential to a linear contour so the solution is unique. With this it is clear that the optimum will depend upon the slope of the contour and hence on s_j (see above). With a shallow slope (juvenile survival poor), low n, and good chances of post-reproductive survival, then repeated breeding (iteroparity) is optimum (see broken line). With a steep slope (juvenile survival good), high n, and poor chances of post-reproductive survival, a single breeding episode (semelparity) is favoured (see dotted line). Moreover, adult survival as influenced by factors other than investment in reproduction (extrinsic as opposed to intrinsic factors) will shift the position of the curve and hence the optimum. When the costs of reproduction are paid in terms of the post-reproductive survival of parents (cf. point (2) below), it can be shown that for a fixed s_j, reductions in s_a due to extrinsic causes favour enhanced investment in reproduction (Schaffer, 1979; Calow, 1983a,b).

Stearns (1983a,b) has tested these predictions by investigating the fecundity of different populations of mosquito fish (*Gambusia affinis*) in reservoirs on Hawaii. Fish from stable reservoirs were smaller, thinner and had lower fecundities and reproductive allocations than fish from fluctuating reservoirs. On this basis, it was presumed that adult mortality rates should be higher and/or more variable in fluctuating than stable reservoirs and vice-versa for juveniles. However, mortality was not measured directly and there were a number of complicating factors; in particular, there was evidence for considerable phenotypic plasticity in life-cycle traits responsive to short-term fluctuations in water level, and the possibility of founder effects could not be excluded.

For fishes there are no other good data that rigorously test between predictions using information on age-specific mortality. Moreover, there are a number of complications:

(1) Reproduction often entails behavioural patterns that render parents more prone to accident, disease and predators. The way this is related to investment in reproduction might be quite different from the physiological effects described above. For example, salmonid fishes probably incur considerable risks in getting to the spawning grounds, before any investment is made. Here there is a better correlation between extent of anadromy and post-reproductive life-span than between investment in gametes and post-reproductive life-spans (Bell, 1980; and see Calow, 1979). Thus *Oncorhyncus* is the most

anadromous genus, its members making extensive migration to the breeding grounds and often becoming sexually mature at sea. All are semelparous. *Salvelinus* is the least anadromous and its members are iteroparous. *Salmo* is intermediate in both respects. Thus in the extreme anadromous forms s_a reduces dramatically until the first egg is laid after the migration, thus yielding the concave-up curve as shown in Figure 1.6A (Bell, 1980), with the possibility of a physiological effect after the initial phase (denoted by ? in the figure). As the effect of migration becomes reduced, so the physiological effect becomes more pronounced (Figure 1.6B). Intermediate sigmoid effects are also possible (Figure 1.6C): these are a reverse of those proposed by Schaffer (1979).

Figure 1.6: Possible Trade-off Curves in Salmonidae; described more fully in the text.

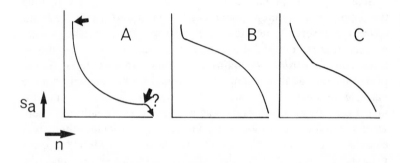

With any fitness isocline the only two solutions in case A in Figure 1.6 are 'reproduce maximally (semelparity) or minimally (iteroparity)' — see arrows in the figure. Depending on the fitness isoclines, the other curves in Figure 1.6 can lead to either semelparity or iteroparity, or non-unique optima since fitness isoclines can make contact at more than one point. This is the case for any non-monotonic curve with 'multiple bumps' (Schaffer & Rosenzweig, 1977).

(2) Figure 1.6 still assumes that though most of the costs are incurred before the release of the first gamete, they are nevertheless paid *after* that point ($n > 0$). But with this kind of cost it is also likely that the reproductive effort influences the survival chances of the adult *before* the release of the first gamete. This complicates the optimization analysis (Sibly & Calow, 1984), but the main effect is that

increased extrinsic adult mortality should lead to reduced rather than increased investment in reproduction. The reason for this is that the addition of extra intrinsic mortality costs to a high extrinsic level might mean that the parent fails to leave offspring at all. The situation when reproductive effort influences pre-reproductive survival is referred to as *direct costing*, and the situation when post-reproductive survival is influenced as *absorption costing* (Sibly & Calow, 1984). The extent to which direct or absorption costing apply to fishes has yet to be clearly defined.

(3) Environmental factors can also influence the positions and shapes of both the fitness isoclines and the trade-off curves. For example, Hirshfield (1980) found for Japanese medaka (*Oryzias latipes*) that, with increasing temperature, adult survivorship per unit effort invested in reproduction decreased but juvenile survivorship increased. Hence, fecundity should increase with temperature (e.g. Figure 1.7), which appears to fit the facts. Ration might also influence these parameters. In freshwater triclads, for example, the young of some species are less sensitive to low ration than others and this predicts that the fecundity of the former should itself be less sensitive to ration than the fecundity of the latter; this again fits the facts (Calow & Woollhead, 1977). Perhaps the distinction between fishes like haddock, in which fecundity is sensitive to ration, and sticklebacks in which fecundity is solely an increasing function of body size, can be explained in a similar way (see also Ware, 1984, and Chapter 9).

(4) The models described above are essentially deterministic whereas in the real world, life-cycle variables and trade-offs will change from time to time and place to place (as a result of the kinds of environmental variation noted in (3)) and this is likely to be particularly true of poikilothermic animals such as fishes. When these stochastic variations are small, it is sufficient to replace variables by their mean values. In the case of larger environmental variations, however, the solution is not so straightforward and depends crucially on the form of the stochasticity. There may be seasonally fluctuating resource availability and hence carrying capacity (Boyce & Daly, 1980), fluctuating age-specific mortality rates (Schaffer, 1974) and/or spatial patchiness when different patches show temporal variability in conditions (Andersson, 1971). The extent to which stochasticity influences the predictions from the deterministic models will depend upon the precise nature of this variability. Classifying the kinds of environmental variation that fish populations experience is therefore a pressing need.

Figure 1.7: Possible Effects of Increasing Temperature on Adult Survival (hence trade-off of s_a vs n) and Juvenile Survival (hence on position of F-isocline). For the latter, broken line represents reduced and solid line increased juvenile survival. Lower graph shows predicted relationships between n and temperature on the basis of these relationships. This is an interpretation of Hirshfield (1980).

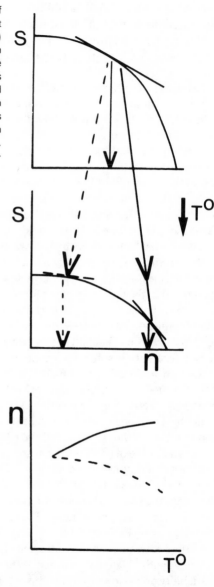

(5) Some fishes indulge in the care of progeny after release from parents (e.g. male stickleback; Chapter 9). The benefits here can involve juvenile survival and enhanced developmental rate (e.g. due to ventilation). The costs are in terms of parent survival and the time

needed to recover between breeding bouts. These trade-offs can be dealt with analytically and are considered in Sibly & Calow (1983) and Calow (1984b).

1.4 Interaction Between Growth and Reproduction

Though the mechanisms may be different (Calow, 1983b), growth and reproduction are both production processes and in a sense always compete for the same limited resources (Figure 1.1). The question of how much should be directed from somatic to gametic processes involves a consideration of when the switch should occur (how big should fishes grow?) and how much should be devoted to reproductive as opposed to somatic production at one time (all or a part?). These are fundamental questions for organismic biology, and yet they have proved difficult to answer. An attempt is made here only to illustrate that, in finding an answer, physiological information will be as important as ecological.

1.4.1 How Big (or for How Long) Should Fishes Grow?

An increase in size potentially has two opposing effects on fitness. It usually requires longer development (negative effect) but might yield increased fecundity or decreased mortality (positive effect). An important trade-off, therefore, is between fecundity and developmental time, with size acting as a hidden intermediary. This, therefore, depends on the relationships between size and fecundity and the time it takes to reach a particular size. Both of these are physiological relationships. The former is discussed in Chapter 9.

A related question is: what should happen if poor environmental conditions, e.g. of temperature and ration, reduce growth rate? This has been considered by Stearns & Crandall (1984). If fishes wait to achieve a specific size before beginning to reproduce, they lose valuable time and expose themselves to increased chances of mortality. On the other hand, if they start reproducing at a fixed time, they suffer a low fecundity associated with the smaller adult size. Intuitively, a compromise should be favoured and one factor which is likely to determine if this errs towards early breeding for the sake of saving time, or late breeding for the sake of high fecundity, is mortality risks. With high adult risks the former will be favoured whereas with low risks the latter will be favoured. If juvenile risks increase as growth rates are reduced, e.g. because the stress which constrains growth also kills juveniles, then maturity should be delayed, but the analysis of Stearns & Crandall indicates that whether this results in a smaller or larger size at maturity depends upon the nature of the sensitivity of juvenile mortality to the stressor. Particular solutions are also sensitive to the levels of fecundity and mortality that are specified. Stearns & Crandall (1984) cite some mainly indirect evidence in support of their model. The main point, though, is that metabolic plasticity can be considered just as much a trait that is subject to natural selection as any other physiological and morphological characters.

1.4.2 How Much Resource Should Be Invested in Reproduction?

This question has been considered rigorously by Leon (1976) and Alexander (1982). Their models are very general and assume only that fecundity is an increasing function of size, and mortality rate is a decreasing function of size. From this a strong prediction is that once reproduction is initiated, *all* resources should be devoted to it; the so-called bang-bang strategy. Yet many fishes continue to grow after (and sometimes even during) reproduction. Specific constraints that can lead to this result are summarized in Myers & Doyle (1983) and are: (a) if parental mortality risks increase, increasingly as a greater fraction of production is devoted to reproduction, which on the basis of physiological evidence seems plausible (see page 19) but still needs to be tested further, and (b) if the number of eggs per unit investment in reproduction reduces progressively as investment is increased. The physiological mechanisms behind (b) are reviewed by Myers & Doyle (1983) and include: (i) limitations of body space for carrying gametes; (ii) nutrient limitations; (iii) storage constraints — particularly in species which exclusively use stores for forming eggs; (iv) indirect energetic constraints.

1.5   Conclusions and Prospects

The main themes of this introductory chapter are that fish energetics (a) is sensitive to natural selection, and (b) influences the way that selection can work. Fundamental to (b) are the trade-offs attendant on the allocation of resources between different, often conflicting, demands. As well as attempting to define the adaptive significance of particular energy allocation patterns, it becomes increasingly important to define the forms of the trade-offs.

Unfortunately, this last aim is likely to prove experimentally taxing. The variation in allocation patterns observed within populations and genetically related groups is more likely to reflect phenotypic than genotypic plasticity and though the latter is interesting and important for fishes (Stearns & Crandall, 1984), it might bear little relationship to the all-important genotypic variability which is the stuff of selection. This point is made forcibly by Reznick (1983) on the basis of observations on guppies. Variation between populations might be more instructive (Reznick, 1983) but there is no certainty that differences due to the evolution of different optima, under different selection pressures, reflect the underlying nature of the genetic variability upon which the selection worked (Calow & Sibly, 1983).

There are two ways forward. First, it might be possible to construct plausible trade-offs on the basis of detailed information of the physiological mechanisms behind them, as was attempted on page 21 for the trade-off of s_a vs. n. Secondly, it should be possible to shift individuals 'outside' normal physiological limits by endocrine and other physiological manipulations. Both of these approaches deserve more attention and are dependent on the collection of sound physiological information on the allocation of energy and other resources in animals.

References

Alexander, R. McN. (1967), *Functional Design in Fishes*. London, Hutchinson

Alexander, R. McN. (1982), *Optima for Animals*. London, Edward Arnold

Andersson, W.W. (1971), Genetic equilibrium and population growth under density-dependent selection. *Amer. Natur. 105*, 489

Bell, G. (1980), The costs of reproduction and their consequences. *Amer. Natur. 116*, 45

Bell, G. (1984a), Measuring the cost of production. I. The correlation structure of the life table of a plankton rotifer. *Evolution 38*, 300

Bell, G. (1984b), Measuring the cost of reproduction. II. The correlation structure of the life tables of five freshwater invertebrates. *Evolution 38*, 314

Boyce, M.S. and Daly, D.J. (1980), Population tracking of fluctuating environments and natural selection for tracking ability. *Amer. Natur. 115*, 480

Brett, J.R. and Groves, T.D.D. (1979), Physiological energetics. In: Hoar, W.S., Randall, D.J. and Brett, J.R. (eds). *Fish Physiology*, vol. VIII, pp. 279–352. New York, Academic Press

Calow, P. (1978), *Life Cycles*. London, Chapman and Hall

Calow, P. (1979), The cost of reproduction — a physiological approach. *Biol. Rev. 54*, 23

Calow, P. (1982), Homeostasis and fitness. *Amer. Natur. 120*, 416

Calow, P. (1983a), Pattern and paradox in parasite reproduction. *Parasitology* **86**, 197

Calow, P. (1983b), Energetics of reproduction and its evolutionary implications. *Biol. J. Linn. Soc.* **20**, 153

Calow, P. (1984a), Economics of ontogeny — adaptational aspects. In: Shorrocks, B. (ed.). *Evolutionary Ecology*: pp. 81–104. British Ecological Society Symp. Oxford, Blackwell Scientific Publications

Calow, P. (1984b), Exploring the adaptive landscapes of invertebrate life cycles. *Ad. Invert. Repr.* **3**, 329

Calow, P. and Sibly, R.N. (1983), Physiological trade-offs and the evolution of life cycles. *Sci. Progr. Oxf.* **68**, 177

Calow, P. and Woollhead, A.S. (1977), The relationship between ration, reproductive effort and the evolution of life-history strategies — some observations on freshwater triclads. *J. Anim. Ecol.* **46**, 765

Charlesworth, B. (1980), *Evolution in Age-structured Populations*. Cambridge, Cambridge University Press

Fausch, K.O. (1984), Profitable stream positions for salmonids: relating specific growth rate to net energy gain. *Can. J. Zool.* **62**, 441

Hirshfield, M.J. (1980), An experimental analysis of reproductive effort and cost in the Japanese Medaka, *Oryzias latipes*. *Ecology* **61**, 282

Iles, T.D. (1974), The tactics and strategy of growth in fishes. In: Harden Jones, F.R. (ed.), *Sea Fisheries Research*, pp. 331–45. London, Elek

Ivlev, V.S. (1961), *Experimental Ecology of the Feeding of Fishes*. New Haven, Yale University Press

Koch, F. and Wieser, W. (1983), Partitioning of energy in fish: can reduction of swimming activity compensate for the cost of production? *J. exp. Biol.* **107**, 141

Larsen, L.O. (1973), Development in adult, freshwater river lampreys and its hormonal control. Ph.D. thesis, University of Copenhagen

Leon, J.A. (1976), Life histories as adaptive strategies. *J. theoret. Biol.* **60**, 301

Miller, P.J. (1979), A concept of fish phenology. *Symp. zool. Soc. Lond.* No. 44, 1

Miller, P.G. (1984), A tokology of gobioid fishes. In: Potts, G.W. and Wootton, R.J. (eds) *Fish Reproduction: Strategies and Tactics*, pp. 119–53. London, Academic Press

Myers, R.A. and Doyle, R.W. (1983), Predicting natural mortality rates and reproduction–mortality trade-offs from fish life-history data. *Can. J. Fish. Aquat. Sci.* **40**, 612

Parry, G.D. (1983), The influence of the cost of growth on ectotherm metabolism. *J. theoret. Biol.* **101**, 453

Reznick, D. (1983), The structure of guppy life histories: the trade-off between growth and reproduction. *Ecology* **64**, 862

Roff, D.A. (1983), An allocation model of growth and reproduction in fish. *Can. J. Fish. Aquat. Sci.* **40**, 1395

Schaffer, W.M. (1974), Optimal reproductive effort in fluctuating environments. *Amer. Natur.* **108**, 783

Schaffer, W.M. (1979), The theory of life-history evolution and its application to Atlantic Salmon. *Symp. zool. Soc. Lond. No. 44*, 307

Schaffer, W.M. and Rosenzweig, M.L. (1977), Selection for optimal life history II: multiple equilibria and the evolution of alternative reproductive strategies. *Ecology* **58**, 60

Shevchenko, V.V. (1972), The dynamics of protein and fat content in the organs and tissues of the North Sea haddock *Melanogrammus aeglefinus* in the process of seasonal growth and gonad maturation. *J. Ichthyol.* **12**, 830

Sibly, R. and Calow, P. (1983), An integrated approach to life-cycle evolution using selective landscapes. *J. theoret. Biol. 102*, 527

Sibly, R. and Calow, P. (1984), Direct and absorption costing in the evolution of life cycles. *J. theoret. Biol.* in press

Sibly, R. and Calow, P. (1985), The classification of habitats by selection pressures: A synthesis of life-cycle and r-K theory. BES Symp., Reading 1984

Stearns, S.C. (1983a), A natural experiment in life-history evolution: field data on the introduction of Mosquitofish (*Gambusia affinis*) to Hawaii. *Evolution 37*, 601

Stearns, S.C. (1983b), The genetic basis of differences in life-history traits among six populations of Mosquitofish (*Gambusia affinis*) that showed common ancestry in 1905. *Evolution 37*, 61

Stearns, S.C. and Crandall, R.E. (1984), Plasticity for age and size at sexual maturity: a life-history response to unavoidable stress. In: Potts, G.W. and Wootton, R.J. (eds) *Fish Reproduction: Strategies and Tactics*, pp. 13-33. London, Academic Press

Vahl, O. (1984), The relationship between specific dynamic action (SDA) and growth in the common starfish. *Oecologia 61*, 122

Ware, D.M. (1975), Growth, metabolism and optimal swimming speed of a pelagic fish. *J. Fish. Res. Bd Can. 32*, 33

Ware, D.M. (1981), Power and evolutionary fitness of teleosts. *Can. J. Fish. Aquat. Sci. 39*, 3

Ware, D.M. (1984), Fitness of different reproductive strategies in teleost fishes. In: Potts, G.W. and Wootton, R.J. (eds). *Fish Reproduction: Strategies and Tactics*, pp. 349–66. London, Academic Press

Waterlow, J.C. (1980), Protein turnover in the whole animal. *Invest. Cell. Pathol. 3*, 107

2 METABOLIC SCOPE IN FISHES

Imants G. Priede

2.1 Introduction

It is axiomatic in modern zoology to suppose that natural selection has shaped animals so as to perform their functions as efficiently as possible. Any adaptation, physiological or behavioural, which can be shown to save energy, is assumed to be adaptive. A fish, for example, that uses less energy for ventilating its gills than another of the same species will be able to use the energy saved to grow faster and produce more eggs. The efficient genotype is therefore selected (Alexander, 1967). It is now thought that much of animal foraging behaviour can be explained by so-called 'optimal foraging theory', whereby the animal behaves in such a manner as to maximize the ratio of energy income over energy expenditure.

During evolution there should therefore be a progressive reduction in energy expenditure by animals together with a general increase in food intake. This is clearly absurd since, taken to its logical conclusion, it would suggest an evolutionary progression towards sessile animals with minimal locomotor energy expenditure. Evolution has in fact proceeded from sessile forms towards more active forms of animal life. Homeothermy can be seen as a step towards some of the most inefficient life forms ever seen; dissipating vast amounts of energy in maintaining body temperature simply in order to remain active (Hammel, 1983).

It is evident that energetic efficiency is poorly correlated with evolutionary success. Energetic efficiency has no necessary link with Darwinian fitness since any efficiency ratio is acceptable as long as in the long term some net positive gain is made sufficient to permit reproduction.

In this chapter it is argued that in terms of energy the most important problem facing an animal trying to survive in the natural environment is simply to attain the power output necessary to live in its selected niche. For example, a certain minimum specific power output is necessary in order to fly. How efficiently that is achieved is secondary to the question of whether the animal is able to enter the flying niche at all. Competition will, within the flying species, perhaps favour the efficient phenotypes, but an inefficient super flyer may be highly competitive by virtue of speed and manoeuvrability.

For an aquatic animal the metabolic rate is primarily limited by the ability of the gills to extract oxygen from water. In comparison with air-breathing animals, availability of oxygen is low, there being approximately only 10 parts per million by weight of dissolved oxygen in air-saturated water. The ventilatory mechanism therefore has to pump vast amounts of water over the gills to sustain even modest levels of aerobic metabolism. The standard metabolic rate is the minimal maintenance or resting metabolic rate of unfed fish below which physiological function is in some way impaired. The maximum aerobic metabolic rate (R_{MAX}) is usually referred to as the *active metabolic rate* and the difference between the two is the *metabolic scope* within which the animal must function (Fry, 1947).

During the course of normal existence, the metabolic rate will fluctuate between these two extremes. If metabolism is depressed below the standard level, life cannot be sustained for long. Similarly, if the fish is forced to work at above the active or maximal metabolic rate, this cannot be sustained indefinitely and death may ensue. It is therefore critical for the animal to manage its metabolic processes so as to remain within its power capacities.

The metabolic scope of fish varies according to species and stage of development as well as being influenced by environmental variables, notably temperature. It is therefore convenient to normalize data with respect to standard and maximum metabolic rates:

$$S = (R - R_S)/(R_{MAX} - R_S)$$

where S = power or metabolic rate normalized with respect to metabolic scope; R = metabolic rate; R_S = standard metabolic rate; R_{MAX} = maximum aerobic metabolic rate (active). At standard metabolic rate a fish is then operating at zero S and at maximum power at an S value of 1 (Figure 2.1).

It is evident that, in the long term, any animal (or indeed any machine) must maintain positive energy balance but the rate of energy

Figure 2.1: Diagrammatic Representation of How Metabolic Rate Fluctuates in Time Between the Limits Set by Standard and Maximum (active) Metabolic Rates. Integration gives a frequency distribution of time spent at different metabolic rates. The *S* scale is a standardized power scale such that standard metabolic rate or power is 0 and maximum is 1. The graph on the left shows how, as the limits of metabolic scope are approached, probability of mortality increases.

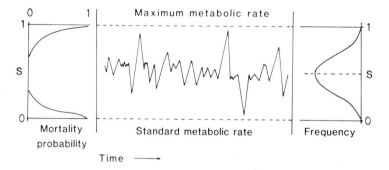

flow must be within the limits specified by the metabolic scope. The rate of energy flow per unit time is defined as power. In the SI system of units energy is measured in joules and power in joules per second which is equal to watts. Energy budgets strictly do not contain any time dimension but integration is usually carried out over a specified time span, perhaps one day or a year depending on the application.

A clear distinction must be drawn between energy budgets and power budgets. The problem with measuring power is the choice of time scale. Growth, for example, is difficult to measure accurately over a time scale of less than 10 days. Although it might be expressed in growth per day or even per hour, it will be an average taken over an extended time period and will not reflect minute-by-minute fluctuations in rates of protein synthesis. Rate of respiration in laboratory studies is usually expressed in terms of oxygen consumption per hour. This is commensurate with the temporal resolution of most respirometer systems. In ecological studies respiration rates are usually averaged over 24 h.

If it were possible to measure metabolic power over millisecond time intervals, fluctuations in energy flow during individual muscle contractions and individual heart beats would be detected. This would reveal something about the functions of individual organs but would not reflect the respiratory metabolism of the whole animal. A time base of somewhere between a second and a minute would reflect fluctuations in rate of oxygen uptake depending on ventilation rate and circulation time in the given species. The time constant of the whole respiratory transport chain from gas exchange at the gills through to production of ATP at mitochondria needs to be taken into consideration. During short bursts of muscle activity, ATP and creatine phosphate reserves will be depleted. Oxygen transport may lag a little behind this. If the metabolic scope is taken to be the limiting power specifications of the fish, it seems appropriate to choose a time base reflecting the integrated aerobic metabolism of the whole animal. Brett (1964) pointed out that fish can vastly exceed their active aerobic metabolic rate during short bursts of glycolysis, depleting carbohydrate reserves and building up levels of lactate. This constitutes an oxygen debt which must be repaid and is reflected in elevated rates of oxygen consumption following the bout of exercise. The capacity for anaerobic metabolism therefore represents a buffer permitting the upper limit of metabolic scope to be temporarily exceeded (Wokoma & Johnston, 1981). Similarly the standard metabolic rate is not an absolute lower limit on the aerobic scope. In conditions of low oxygen, for example, the metabolic rate may be depressed below normal. Tissues may function anaerobically for a time but again an oxygen debt is built up and must be repaid.

The definitions of maximum and minimum power outputs of a fish depend on the time frame of measurement. On the millisecond time scale, vast power outputs may be possible, reflecting maximum rates of ATP splitting during maximal contraction of the locomotor muscles. On the other hand, there may be very little metabolic activity for short instants of time between cycles of muscle contraction. The concept of metabolic scope is usually taken to refer to function in an aerobic steady state with no tendency to build up oxygen debt. Heart rate (Priede, 1974) and ventilation rate (Stevens & Randall, 1967) increase within a minute of the onset of heavy aerobic exercise. Integration over a time span of about a minute would seem appropriate for measurement of aerobic metabolic power. The buffering effect of oxygen storage and slight excursions into anaerobiosis may, however, introduce lags in oxygen uptake during fluctuations in metabolic rate. Similarly, extrusion of carbon dioxide may be delayed or episodic depending on the blood acid–base buffering status. This would seem to indicate a time base of about a half to one hour as appropriate. Hoar & Randall (1978) suggest that a time span of 200 min is necessary to ensure true steady-state measurements of oxygen uptake during sustained swimming in tunnel respirometers. This may be appropriate taking into account the limitations of the laboratory measurement technique. For fish in nature this would completely overlook the fact that they can swim considerable distances in a few minutes and show appropriate increases in heart rate and ventilation rate linked directly to this activity and then return to rest in a new part of the pond. The real fluctuations in oxygen uptake at the animal–environment interface are regulated on a time base of about a minute.

The immediate energetic objective of a fish is simply to regulate its metabolism on a minute-by-minute basis so as to remain within limits of metabolic scope. Temporary excursions in metabolic rate above or below these limits are permitted but at the cost of repayment of oxygen debt later on. If the fish can stay within the limits of scope and carry out all necessary functions such as the evasion of predators and the acquisition of food, it will survive.

In the engineering of reliable machines, a fundamental rule is to always operate components and systems well below their maximum power rating (Bazovsky, 1961). The active or maximum aerobic metabolic rate can be regarded as the continuous power rating of an animal. To ensure longevity, working well within the metabolic scope is desirable. It can be argued that there is a relationship between metabolic rate and mortality rate (Priede, 1977). Animals avoid working at high metabolic rates in order to minimize the probability of mortality. This power budgeting may be more important than any energy saved in the long-term energy budget.

2.2 Metabolic Scope and the Limits on Performance

The aerobic metabolic capacity (represented by R_{MAX}) of a fish is primarily limited by the ability of the gills to extract oxygen from the water. In a survey of gill areas of marine fishes Gray (1954) found a general correlation between gill area and activity of different species. The largest gill area was found in the pelagic menhaden *Brevoortia tyrannus*, and the smallest in a benthic sand flounder *Lophosetta maculata*; 1773 and 188 mm^2 per gram body weight, respectively. More recent studies on a wider range of species have confirmed this general pattern (Hughes, 1966; De Jager *et al.*, 1977).

Gill (fish–water interface) area may be the ultimate limitation, but the whole of the respiratory transport chain from diffusion across the gill epithelium through the circulatory system, cardiac performance and mechanisms of tissue gas exchange must be matched to this gas-exchange capacity. Significant metabolic costs may be attached to branchial pumps ventilating the gills, and the functioning of the cardiac pump.

The active oxygen consumption is defined as the maximum oxygen consumption observed during sustained cruising swimming. This maximum sustainable swimming speed is defined as the critical speed. In this state it is assumed that the oxygen uptake systems are working at maximum capacity. Generally the critical speed and active oxygen consumption increase with temperature to an optimum and then decrease again (Beamish, 1978). At low temperatures, tissue metabolism is depressed and it is unlikely that oxygen uptake capacity at the gills is a limiting factor. The gill area and oxygen transport system probably become limiting at higher temperatures where tissue metabolism will be greater and oxygen availability lower.

In the resting fish, not all the gill area is perfused by blood so when oxygen consumption increases, additional gill area must be perfused until, at the active metabolic rate, all the gill vasculature is recruited (Jones & Randall, 1978). Duthie & Hughes (1982) have carried out experiments in which the gill area of rainbow trout (*Salmo gairdneri*) was artificially manipulated. Gill filaments were destroyed by cautery and it was found that an approximately 30% reduction in gill area led to a 27% reduction in oxygen uptake and a 22% reduction in critical swimming speed (Figure 2.2). This clearly demonstrates the limitation imposed by oxygen uptake capacity of the gills. It is an important result since many toxins and pathogens have their main effect on the gill epithelium, reducing oxygen uptake capacity (Skidmore, 1970). The effect of the pathogen may not be acutely lethal; a reduction in performance may occur.

Most fish, especially in temperate climates, experience seasonal changes in temperature. The gill area will have been selected to permit appropriate levels of metabolic activity probably at the upper end of the temperature range of the species. Given a certain gill area it is unlikely that oxygen uptake capacity will be limiting at lower temperatures. Some species of fish can seasonally modify the characteristics of respiratory pigments and metabolic enzymes so as to compensate, to some extent, for temperature changes (Johnston *et al.*, 1975; Somero, 1975; Sidell, 1980).

The critical factor limiting performance and metabolic scope may vary considerably under different circumstances. Any component, from availability of oxygen in the environment to diffusion in tissues and mitochondrial metabolism, may prove limiting. It is inappropriate

Figure 2.2: Rainbow Trout (*Salmo gairdneri*): The Effect of Reduction of Effective Gill Area. Comparison of oxygen consumption and swimming speed in control and experimental fish. In the experimental fish the gill area was reduced by cauterizing the filaments on the second gill arch. This reduces maximum oxygen consumption and critical swimming speed in tests carried out in a tunnel respirometer. (Figure with permission from G.G. Duthie, unpublished observations.) $\dot{V}_{O_2}$ STAND = standard oxygen consumption (R_S). $\dot{V}_{O_2}$MAX = maximum (active) oxygen (R_{MAX}) consumption. U_{CRIT} = critical swimming speed.

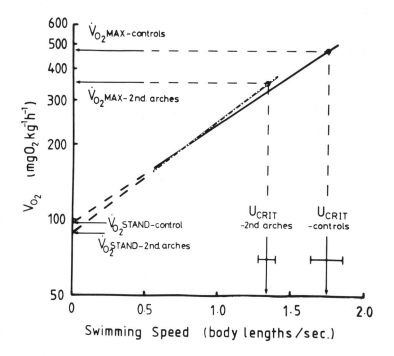

in the present context to examine all these processes in detail.

2.3 Scope and the Components of Respiratory Metabolism

Respiratory metabolism is usually divided into three separate components:

$$R = R_S + R_F + R_A$$

where R_S is the standard metabolic rate, R_F is specific dynamic action and R_A is metabolism due to locomotor activity (Chapter 1).

All these three activities must be accommodated within the metabolic scope if the animal is to survive.

In general, active fish with high active metabolic rates and large metabolic scopes for activity also have a high standard metabolic rate. There seems to be some metabolic cost involved in maintaining the additional metabolic machinery even if the animal is at rest. A large gill area, for example, will give rise to increased cost of osmotic and ionic regulation through increased diffusional interchange at the gills. The larger heart will require more metabolic activity simply to sustain it, and similarly enhanced maintenance activity is required throughout the body right down to the level of intracellular metabolism. Usually the increase in active metabolic rate is much greater than any corresponding increase in resting metabolism so the active animal shows a real increase in metabolic scope. The standard metabolism is mandatory and the individual animal cannot regulate the magnitude of this component. Exceptions may be certain cases of behavioural thermoregulation where fishes adjust their metabolic rates by moving between water bodies of different temperature. The other components of respiratory metabolism can, however, be adjusted by the animal.

Specific dynamic action is strictly defined as the increase in heat production observed following a meal (Kleiber, 1961) and is associated with animo acid mobilization and biochemical processes associated with digestion (Chapter 8). Since direct calorimetry is difficult in studies on fish, the increment in oxygen consumption following a meal is usually measured. This includes metabolism associated with gut motility and general post-feeding activity, and in consequence Beamish (1974) used the term apparent specific dynamic action (SDA) to refer to it. SDA is greater if the diet is rich in protein and estimates vary between 5 and 20% of ingested energy. SDA can represent a major component of the respiratory energy budget (Chapters 11 and 12).

The most obvious factor modulating respiratory metabolism is locomotor activity. Swimming activity is usually capable of using up the entire power capacity represented by the metabolic scope. In fish, therefore, there must be a continual conflict between the needs of locomotion and SDA. The two metabolic activities are mutually exclusive, and if the fish is swimming at its maximum speed, it presumably cannot proceed with any digestion. The nature of this metabolic conflict lies at the root of the power budgeting problem that fish face all the time.

Figure 2.3 shows estimates of the components of respiratory

Figure 2.3: Brown Trout (*Salmo trutta*): Estimates of Power Capacities and Actual Power Used in Summer (top) and Winter (bottom) in a Benthic, Lacustrine Environment. The open arrows indicate the power capacities; thickness proportional to power. The black arrows represent estimates of the actual rate of working (see text). Note that the total capacity of the subsystems exceeds that of the respiratory uptake and transport mechanisms to supply the power requirement. The figures on the left represent maximum possible oxygen uptake capacity.

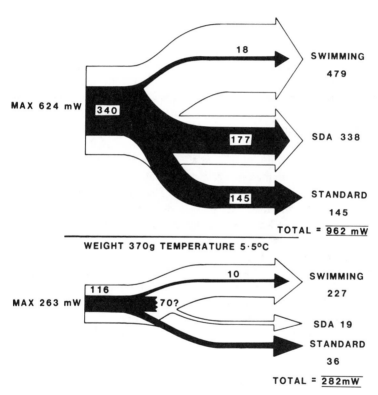

BROWN TROUT

metabolism in brown trout *Salmo trutta* L. The power is given in milliwatts and is calculated from oxygen consumption using the oxycalorific value of 13.6 J per milligram of oxygen consumed (Elliott & Davidson, 1975). The data in Figure 2.3 are calculated for a benthic feeding trout living in a Scottish loch. From tracking and telemetry studies, much is known about the activity levels of these fish in the wild (Young *et al.*, 1972; Holliday *et al.*, 1974; Priede & Young, 1977). Morgan (1973, 1974) measured the oxygen consumption of brown trout in a tunnel respirometer. From his results we obtain the figure of 624 mW for R_{MAX}. This corresponds to the active or maximum oxygen consumption. The study of Elliott (1976) on feeding and growth of brown trout provides further information. R_S or standard metabolism runs at 145 mW, and this figure is in very close agreement with independent estimates derived from Morgan (1973, 1974). This leaves 479 mW available for digestion and locomotion. Swimming at maximum cruising speed can take up all of that scope. Elliott (1976) gives estimates of metabolic rate during feeding at different levels, and from this R_F is estimated as 338 mW during feeding to satiation. The power budgeting problem for the fish is that it can work at only 624 mW when, if all its subsystems were running at full power, it would need to work at 962 mW. Heart-rate telemetry indicated that the fish worked at R less than 479 mW. Tracking of swimming speed indicated a mean locomotor power consumption (R_A) of 18 mW, the fish swimming very slowly for most of the time. The actual value of R_F was therefore running at not more than 177 mW. This indicated that the fish was feeding at less than maximal rate and remaining well within the limits of the metabolic scope with an S value of less than 0.6 (Priede, 1983).

For the brown trout at 5.5° C, digestive metabolism (R_F) is suppressed by the low temperature, and the situation is less critical (Figure 2.3). Total power capacity (R_{MAX}) is 263 mW, of which (from the evidence of heart-rate telemetry) the fish actually used not more than 116 mW (R). Only 10 mW was used in swimming (R_A), permitting digestion to proceed at the maximum rate possible at this low temperature. Measurement of heart rate is the only method which gives an indication of minute-by-minute fluctuations in metabolic rate of a free-living fish in the natural environment (Priede & Tytler, 1977; Priede, 1983). In studies on these trout it was evident that they spent most time at power outputs corresponding to about $S = 0.5$ and maximum power only occurred for less than 1% of the time. These fish were clearly adopting a very conservative power-budgeting strategy.

Figure 2.4 shows similar information for juvenile cod (*Gadus morhua* L.). This is based on data from Soofiani & Hawkins (1982) and Soofiani & Priede (1984). For the cod as well as the trout, the total power consumption of the subsystems (R), maintenance (R_S), locomotion (R_A) and digestion (R_F) exceeds the capacity of the oxygen uptake and transport systems both at 10° C and 15° C. The cod is interesting because the peak SDA (R_F) exceeds R_A at the maximum aerobic cruising speed. An unfed fish at maximum cruising speed does not display maximum oxygen consumption (Soofiani & Priede, 1984). The 'active' oxygen consumption is therefore not the maximum oxygen consumption (R_{MAX}). Thus, when a cod is swimming at its critical speed, S can be less than 1.

Figure 2.4: Atlantic Cod (*Gadus morhua*). Estimates of Power Capacities. Details are the same as for Figure 2.3 but no estimates of the actual field power budget are given.

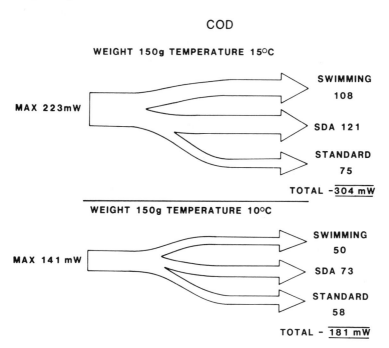

The cod and the trout therefore represent two metabolically different types of fish. For the cod, SDA is greater than the peak power of the aerobic swimming muscles. The metabolic scope is sufficient to accommodate peaks in digestion metabolism and locomotion has to be fitted in when possible. Many fish with small amounts of aerobic red muscle must be of this visceral metabolism type. The trout conforms to the accepted view that active metabolism corresponds to maximum aerobic metabolism and there is sufficient red muscle present to utilize the whole of the available metabolic scope. At the lower temperature SDA cannot be considered a major metabolic power load.

It is probably unusual for aerobic power capacity to exceed the total peak power of the subsystems (SDA and locomotion). All fish therefore have a power budgeting problem. Vahl & Davenport (1979) have pointed out that for the blenny (*Blennius pholis*) the demands of digestion or locomotion can each take up most of the metabolic scope, and that the metabolic scope is also affected by fluctuations in salinity. They show that the blenny must phase its feeding activity so as to avoid conflict of demands of locomotion, SDA and osmo-regulation.

It is well known in fish-culture practice that excessive feeding at high temperatures can so increase metabolism that mortality may ensue. Most feed manufacturers recommend cessation of feeding above a certain temperature. Feeding is also normally stopped before subjecting fish to any handling: the additive effects of SDA, stress and activity during handling can lead to metabolic power overload and significant mortality.

Although only three components are recognized in the respiration budget given above, it is evident that a number of factors such as stress, salinity, dissolved oxygen level, pH, etc. may affect the metabolic scope so as to modify the basic relationships that are outlined. Nevertheless, fish do have a power budgeting problem. There is strong evidence that a number of species regulate their feeding activity simply to keep their metabolic rate within the bounds of the metabolic scope. Any optimal foraging behaviour can only be a fine tuning superimposed upon this basic strategy.

At this point it should be mentioned that there may be instances where the metabolic scope is sufficient to accommodate all metabolic activities running simultaneously. This is particularly likely in fish living at below their normal ambient temperature. The respiratory mechanisms will be adapted for adverse conditions at higher temperatures, but in cool water with high oxygen solubilities and low metabolic rates, there may well be spare capacity available.

2.4 Natural Selection for Energetic Efficiency

It is readily observed that most fish have a streamlined body form which reduces the power required for swimming. This is a feature for which there is powerful natural selection since there are numerous examples of parallel evolution of the streamlined body form in widely divergent taxa.

There are two fundamentally different ways in which reduction in swimming power requirements can increase fitness:

(1) Energy saving. A reduction in power output over a period of time will permit an animal to accumulate a surplus of energy when compared with less efficient phenotypes in similar circumstances. The surplus energy may enhance fitness. One way in which this can occur is through an increase in reproductive output. If more eggs are laid, assuming equal probability of survival of the offspring, a coefficient of selection in favour of an efficient genotype can be calculated. There need not be a simple relationship between energy surplus and fitness, but this *type 1* selection assumes that the surplus does have some positive effect.

(2) Power budgeting. In this type of selection it is immaterial whether a surplus of energy is accumulated. The reduction in power requirement means that the fish can carry out its normal functions with a reduced probability of exceeding the limits of the metabolic scope. It is argued that this leads to a reduction in probability of mortality through an increased safety factor in metabolic functions. Fitness is therefore increased since there is an increased likelihood of survival to reproductive age and indeed survival through more breeding seasons. This is referred to as *type 2* selection.

Priede (1977) suggested that type 2 selection could be one- to ten-thousand times more powerful than type 1 selection. In slow swimming benthic-feeding brown trout in still water, the locomotor activity only accounts for about 4% of the energy budget. This means that small reductions in power requirements for swimming lead to insignificant savings in energy. It is difficult to understand how stabilizing selection for a streamlined shape could continue in such fish in the absence of a type 2 mechanism. Savings in the locomotor energy budget may be insignificant in slow-swimming fish, but it can be argued that in a fast-swimming pelagic fish the energy savings may be more important and type 1 selection might be correspondingly more powerful. These aspects of selection for swimming efficiency can be examined further with simple models.

Figure 2.5: The Relationship Between Metabolic Power Requirement and Swimming Speed in Some Hypothetical Fish: normal, slow and fast fish and more efficient phenotypes with a 10% reduction in power required for swimming. Note how the efficient phenotypes can swim faster using the same power as the normal types. These basic relationships are used in the model developed in the text.

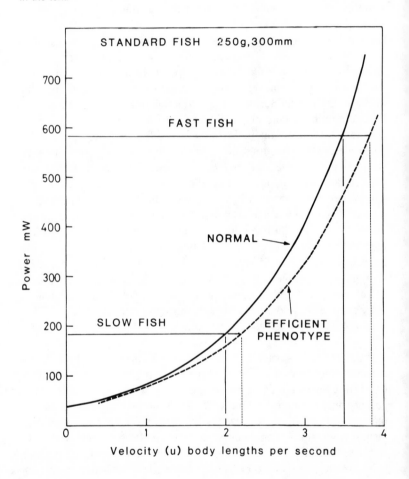

Figure 2.5 shows the power requirement for swimming in a typical fish weighing 250 g and 300 mm in length. The curve is based on the

data from Brett (1964) for sockeye salmon:

$$\log Y = 1.61 + 0.34u$$

where Y is the oxygen consumption in milligrams per kilogram per hour and u is the swimming speed in body lengths per second. Two hypothetical fish are considered: a slow fish with a critical speed of 2 body lengths per second, and a fast fish with a critical speed of 3.5 body lengths per second. The slow fish would correspond to a typical lake-dwelling salmonid and the fast fish to a pelagic marine species such as a mackerel. The fact that the fast fish would almost certainly have a higher standard metabolic rate is ignored for the sake of simplicity. The second curve in Figure 2.5 shows the power for swimming in 'efficient' fish with a power requirement reduced by 10%. The model will consider the process of selection for such an efficient fish.

Most fishes seem to spend most of the time swimming at moderate speeds. A log-normal distribution of swimming speeds is assumed for both fish with the same shape relative to the swimming performance capacity (Figures 2.6 and 2.7). This can be transformed using the velocity–power relationship into a frequency distribution of power consumption. Thus the mean power of the slow fish is 68.3 mW and of the fast fish 123.7 mW. There is a very narrow tail to the power distribution extending up towards the active metabolic rate. The logarithmic nature of the power curve makes this tail longer in the fast fish despite starting off with the same swimming speed distributions in both cases.

The metabolic cost of digestion has to be included. It is assumed that both fish grow at approximately the same rate but that the fast fish is consuming more food in order to compensate for the faster swimming speed. Apparent SDA is taken to be 20% and loss in faeces and urine is assumed to be 30% of food intake. These assumptions lead to the power budget shown in Table 2.1. In Figure 2.8, therefore, the initial frequency distribution of power outputs is shifted upwards by an amount equivalent to the SDA. It can be seen that this is highly significant in relation to scope in the slow fish but not so important in the fast fish.

The next step in developing the model is to insert the mortality probability curve. The curve used is arbitrary and is exactly the same as used by Priede (1977). The original curve was sketched under the premise that both ends must approach a probability of 1 at standard

Figure 2.6: Slow Fish. Frequency distribution of swimming speeds (f(u)), A, and corresponding metabolic rates (f (p)), B. p = power, u = specific swimming speed (body lengths per second). The p-u, C, curve is taken from Figure 2.5. The frequency distribution of swimming speeds is hypothetical but corresponds to the log-normal distribution closely resembling data from field studies.

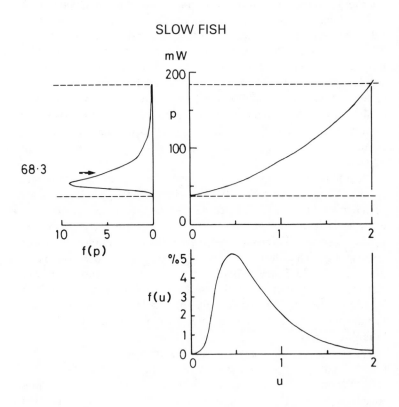

SLOW FISH

and maximum metabolic rate and combined with a typical power frequency distribution must give a realistic natural mortality rate.

$$\text{Natural mortality} = \sum_{p=R_s}^{p=R_{MAX}} fp \cdot Mp$$

Natural mortality rate is the product of time spent at different power levels (fp) and the corresponding mortality probabilities (Mp).

Figure 2.7: Fast Fish. Frequency distribution of swimming speeds and corresponding distribution of metabolic rates. Notation as for Figure 2.6.

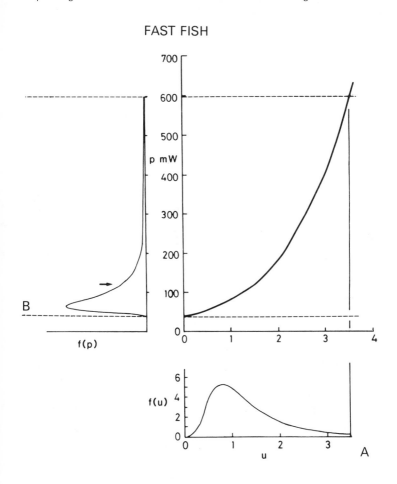

The mortality curves shown in Figure 2.8 are both the same when power (*p*) is expressed in standardized *S* units.

The model is completed by calculating the frequency distributions of metabolic rates of the efficient forms of both the fast and slow fish. As can be seen in Figure 2.8, this shifts the power distribution curves slightly to the left, representing a reduction in mean *S* value and a reduction in energy by expenditure. The result, in terms of energy saving and change in mortality rate, is given in Table 2.1. The energy

Table 2.1: Power Budgets for the Hypothetical Slow and Fast Fish (as in Figure 2.5). All values are in milliwatts unless otherwise stated; weights of fish are 250 g. The 'efficient' phenotypes of the slow and fast fish are assumed to have a locomotor power requirement reduced by 10%. A reduction of the costs of locomotion in the efficient fish leads to an increase in surplus power. Note the difference is greater in the fast fish than in the slow fish. The percentage reduction in mortality rate, however, is greater in the slow fish. (See text for explanation.)

	Slow Fish		Fast Fish	
	Normal	Efficient	Normal	Efficient
Standard metabolic rate (R_S)	38.5	38.5	38.5	38.5
SDA (R_F)	57.4	57.4	78.1	78.1
Locomotor activity (R_A)	29.9	26.9	85.3	76.8
Total respiration (R)	125.8	122.8	201.9	193.4
Active metabolic rate (R_{MAX})	184.3	184.3	596.3	596.3
Food intake	287.0	287.0	390.8	390.8
Faeces and urine	86.0	86.0	117.1	117.1
Surplus (growth)	75.2	78.2	71.8	80.3
Increase in surplus in the efficient form		3.0 mW or 3.0%		8.5 mW or 11%
Daily mortality	0.2093%	0.1588%	0.0550%	0.0386%
Increase in annual survivorship in the efficient form		20.25%		6.16%

saving for the slow fish represents about 1% of total intake, and growth is increased by 3.8%. This might well mean 3.8% more eggs. Daily mortality rate, however, falls by 24.1% and over 1 year this means 20.25% more survivors. The energy saving in the fast fish is rather more significant, growth being increased by 11%. Because of some of the initial assumptions in the model, the mortality rates are much lower than in the slow fish, but in the efficient form of fast fish an almost 30% reduction in mortality is predicted.

This model shows some of the characteristics of the interaction between type 1 and type 2 selection. Type 1 is clearly more important in active fish with a larger proportion of the energy budget devoted to swimming activity. There are, however, problems with the concepts of both type 1 and type 2 selection. In type 1 selection it is easy to show that under certain circumstances surplus energy can be accumulated. How this surplus is transformed into enhanced fitness in the evolutionary sense may be obscure. More energy can be diverted into reproduction but over-reproduction can be disadvantageous (Ware,

Figure 2.8: Frequency Distributions of Metabolic Rates in the Fast and Slow Fish. Comparison of the normal and efficient phenotypes. Thin line (no SDA) = basic curves transferred from Figures 2.6 and 2.7. Dashed line (normal fish) = the estimated frequency distribution allowing for metabolism associated with feeding in accordance with the power budget in Table 2.1. Thick solid line (efficient phenotype) = estimated frequency distribution for the efficient phenotype. This curve is shifted slightly to the left of the normal curve, since less power is required for swimming. The mortality curve is the same in both fast and slow fish, being related to *S* rather than absolute metabolic power. The part of the mortality curve rising to 1 at standard metabolic rate is omitted since it does not enter into the final calculations relating to type 2 selection.

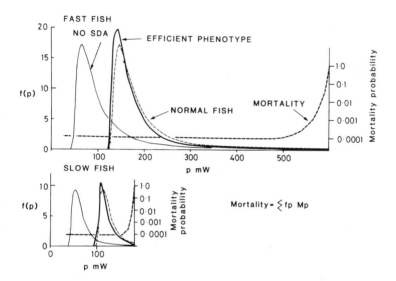

1982; Chapter 1).

In type 2 selection the shapes of the mortality–power curves are at present highly speculative. The model is very sensitive to changes in the mortality curve at its extremities, particularly approaching the maximum metabolic rate. The model is only intended to apply to the middle part of the life-span of the animal, after juvenile mortality has ceased and before senescence sets in. The juvenile stages of fishes are generally quite distinct, and with the notable exception of 'big bang' reproducers such as Pacific salmon, senility is not apparent in natural populations (Beverton & Holt, 1959). The type 2 model should therefore be applicable to most of the life-span of natural fish populations.

The precise nature of metabolic-rate-dependent mortality is obscure and involves a range of interrelated phenomena. It is well known that fish can suffer mortality as a result of severe exercise (Black, 1958; Wood *et al.*, 1983) and this has been the subject of numerous studies. Whether such mortality occurs in nature is difficult to discover. Physiological failure as discussed by Wood *et al.* (1983) may not be the proximate cause of death. If, for example, a fish is being chased by a predator, it might be captured before any physiological symptoms are manifest. Death through predation results from an inability to swim fast enough to escape. In that sense mortality through predation is metabolic-rate-dependent mortality. The predator can be regarded as the agent of mortality but not the ultimate cause.

The precise outcome of the type 2 model is critically dependent on the slope of the mortality–power curve. The buffering effect of anaerobic metabolic capacity may have important modifying effects on this relationship and at present there is insufficient evidence to advance much beyond the assumptions already made.

2.5 The Magnitude of the Metabolic Scope

The type 2 selection model assumes that there must be a considerable safety margin between the power requirements of living in a certain niche (R) and the actual power capacity of the fish (R_{MAX}). This suggests a selection pressure to reduce the power requirements of different physiological systems and/or to increase the metabolic scope. There must be costs attached to an increase in metabolic scope in that standard metabolic rate will tend to increase as a result of the acquisition of additional metabolic 'machinery'. There must therefore be some sort of dynamic equilibrium between selection pressures for increasing metabolic scope and selection pressures for minimizing metabolic costs. Where the equilibrium settles will depend on the lifestyle of the fish.

If it is assumed that the fishes originated from sessile ancestors, it is likely that the swimming capacity of the earliest fishes was limited. During the course of evolutionary history, the performance capacity of fish has been progressively elaborated. Primitive fishes were probably opportunistic scavengers. Such animals could obtain food at minimal metabolic cost but the major cost would be associated with digestion. These animals must have had a metabolic scope sufficient to accommodate SDA. Even the most inactive of fish must have sufficient metabolic scope to permit digestion of food.

If feeding is episodic, consisting of discrete large meals, then the metabolic scope must be sufficient to accommodate the peaks in metabolism following a meal. Once digestion of a meal has proceeded to completion, this would leave some metabolic scope available for locomotor activity, perhaps foraging for the next meal. The pattern of metabolism would evolve into alternating phases of activity and digestion but with the original size of the metabolic scope having been determined by the requirements for digestion rather than activity. This is precisely the pattern of aerobic metabolism suggested for the blenny by Vahl & Davenport (1979). Soofiani & Hawkins (1982) show that, for the juvenile cod, SDA alone can take up almost all the metabolic scope. Soofiani & Priede (1984) have found that the maximum aerobic swimming metabolism is significantly less than the peak values of SDA following a meal in these fishes.

Beamish (1974), in a study on largemouth bass, *Micropterus salmoides*, showed that peak oxygen consumption following a meal could rise to 80–100% of the maximum oxygen consumption. He attributed a significant proportion of this to excitement, but this evades the issue of what constitutes excitement metabolism. Soofiani & Hawkins (1982) took great care to avoid elevated levels of locomotor activity in their study. Different studies give a range of metabolic rates during feeding of between 1.5 and 5.8 times the standard metabolic rate (Brett & Groves, 1979).

Many species of fish are relatively inactive, only a small part of their energy budget being attributable to swimming activity. Studies on fish metabolism have been dominated by work on carefully chosen species that swim well in tunnel respirometers. Most species are incapable of sustaining continuous high-speed swimming. Anatomical examination shows that volumes of red aerobic muscle are usually small. Probably more than 50% of fish species conform to the viscerally dominated metabolic type where metabolic scope can just accommodate peaks in digestion metabolism. Only 1.3% of fish species are active marine epipelagic animals and only 0.6% are actively migrating diadromous species, but these have dominated studies on fish metabolism (Cohen, 1970). In the less active fish species, any locomotor activity is fitted in between the peaks in digestion activity. Hence the metabolic pattern of the hypothetical primitive fish is probably still the norm rather than the exception to the rule. If more fish, such as the cod, are found, in which sustained swimming metabolism does not utilize the whole metabolic scope, perhaps the value of the term 'active metabolic rate' will have to be reconsidered.

Figure 2.9: Cost of Transport for Fish with Different Efficiencies. Note how reduction in power requirement shifts the optimum (minimum cost) swimming speed upwards.

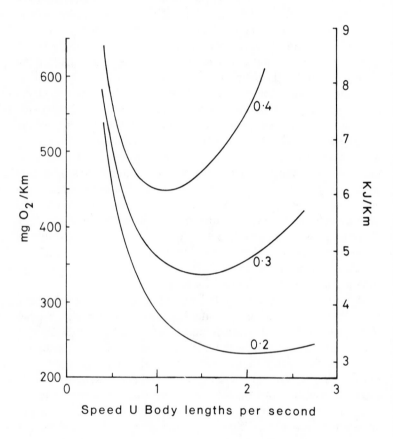

Figure 2.9 shows the cost of transport per kilometre at different swimming speeds. The basis of the curves is the relationship between swimming speed and oxygen consumption:

$$\log Y = C + bu$$

where $Y =$ oxygen consumption in milligrams per kilogram per hour; $C =$ log of standard metabolic rate; $u =$ swimming speed in body lengths per second; and b is a constant which varies between 0.2 and 0.5 according to species. A low value of b suggests an efficient swimmer, and a high value an inefficient swimmer requiring a greater increase in power as velocity increases. The curves in Figure 2.9 show the relationship between cost of swimming (Y/u) and velocity for different values of b. It can be seen that as velocity increases, so the cost of transport decreases until a certain optimum velocity is reached (Weihs, 1973). Also in Figure 2.9, it can be seen that the optimum swimming speed decreases with an increase in b value. Inefficient swimmers have lower optimum speed than do efficient swimmers. It would seem reasonable to assume that fish should be able to swim at least at their optimum speed if not any faster.

Priede & Holliday (1980) found that for the flatfish, plaice (*Pleuronectes platessa*), the metabolic scope at all temperatures was just sufficient to permit swimming at the optimum speed but no faster. Duthie (1982) later confirmed this observation for three other species of flatfish. There is, therefore, a complex interaction between type 1 and type 2 selection. Type 1 considerations indicate that the fish should increase its mean swimming speed to attain the hydro-dynamically defined optimum. Type 2 considerations indicate that if this is to be achieved, then there should be plenty of spare capacity in terms of metabolic scope.

Movements of plaice in the open sea have been tracked by Greer-Walker *et al.* (1978) using ultrasonic transmitters. They found that the swimming activity of these fish is characterized by periods of active swimming when the tidal flow is favourable, followed by rest on the bottom during unfavourable phases of the tidal cycle. During the swimming phase the fish apparently swims continually at close to its critical speed, possibly building up some oxygen debt. This is remarkable in view of the basic premise in the type 2 selection theory that fish should avoid functioning at close to their active metabolic rate. The critical speed in these fish, however, is also the optimum speed, and in flatfish there is good reason to suppose that an all-or-nothing swimming strategy is advantageous. When resting on the bottom, the fish has an opportunity to repay any oxygen debt and is also well camouflaged and hidden from any predators.

For most fish, extrapolation of the swimming speed–oxygen consumption relationship to zero speed gives the standard oxygen consumption (R_S). In flatfish this is not so: the resting oxygen consumption (R_S) is found to be about 60% of the intercept value at zero swimming speed (Priede & Holliday, 1980; Duthie, 1982). In the transition from the resting state to active swimming there is a significant cost of take-off or 'posture effect'. This is because flatfish have no swim bladder, are negatively buoyant and have to work actively to lift off the bottom. The starting cost means that swimming at low speeds is very inefficient and the best strategy for these fish is to swim at high speeds and take the opportunity to rest at low metabolic rates if time permits (Priede & Holliday, 1980). This seems to be a unique feature of the energetics of negatively buoyant flatfish, since no other fish-tracking studies have shown extended periods of swimming at close to critical swimming speeds in the field.

An important feature of the locomotor apparatus of most fish is a highly developed anaerobically functioning mass of white muscle (Bone, 1978). This can be used during bursts of speed at above the critical swimming speed (Brett, 1964). Soofiani & Priede (1984) found that following exhausting exercise, juvenile cod lying still in a respirometer would show maximum oxygen consumption; higher than peak SDA and higher than the peak oxygen consumption during maximum sustained swimming. It is interesting to note that there was a delay of an hour or two following exercise before the peak oxygen consumptions were observed. Initially, following a burst activity, lactate is probably stored in the muscles and is only released slowly into the circulation to be oxidized or converted to glucose in other tissues (Driedzic & Hochachka, 1978). These processes require energy, and repayment of oxygen debt should perhaps be considered as another component of the power budget additional to SDA and aerobic activity. All these three functions seem to be mutually exclusive. Development of an aerobic capacity therefore brings with it a necessity for an enhanced aerobic metabolic scope to accommodate peaks in repayment of oxygen debt.

Somero & Childress (1980) have shown that the activities of glycolytic enzymes necessary for anaerobic muscle function increase with fish size in contrast to a corresponding decrease in activity of citric acide cycle enzymes. This divergence in apparent aerobic and anaerobic capacity is difficult to explain but fits observations that fishes of different sizes are all able to attain the same relative speeds (body lengths per second) during peak bursts. In larger fishes the glycolytic enzyme activities in the white muscle increase substantially over levels found in smaller fish. This suggests that bigger fish will have more difficulty in repaying the oxygen debt. In larger fish, however, growth rates are slower, specific metabolic rates are lower and this may permit more of the scope to be used for oxygen debt repayment. Towards the end of a life-span a fish can also take more risks and perhaps run its metabolism at closer to the power limits.

For active pelagic fish, sustaining continual swimming in the face of conflicting demands of SDA and perhaps oxygen debt repayment must be a problem. Randall & Daxboeck (1982) have shown that in rainbow trout during exercise, blood flow is diverted away from the systemic circulation towards the working muscles. This suggests that digestion of meals and other non-muscular metabolic activity must be reduced or cease during exercise. In a continually swimming and feeding fish, the metabolic scope must be expanded to accommodate these requirements. How fast should such a fish be able to swim?

Consider a fish feeding on a uniformly distributed food resource. A good example would be a filter-feeding planktivore such as a basking shark (*Cetorhinus maximus*). The faster it swims the more food it captures per unit time. The fish has to attain a certain minimum speed in order to get enough food simply to sustain its metabolism. The higher the density of food, the lower is this minimum foraging speed. The metabolic scope must permit swimming at above the normal minimum foraging speed, taking into account typical plankton densities. Type 2 selection would ensure that there is performance capacity to meet this underlying energetic requirement. As food density decreases, theory suggests that velocities should increase, but there are limits to this relationship. In winter, when plankton densities are low, basking sharks cease their feeding activity and probably go to the bottom and remain in a state akin to torpor. In fact they lose all ability to feed since their gill rakers are shed in winter. The gill rakers regrow in time to recommence feeding in the spring (Parker & Boeseman, 1954). Most of the time during the summer, basking sharks forage at low speeds well below their probable aerobic swimming capacity (Harden-Jones, 1973; Priede, 1984).

If the fish increases its speed above the minimum foraging speed, it continues to capture more and more food per unit time. Food intake increases in a more or less linear fashion but power required for swimming increases logarithmically so eventually the two curves start to converge and the fish is swimming at above the optimum foraging speed. Ware (1975, 1978) has developed a set of equations relating theoretical swimming speed to body size and food density. The theoretical optimum speed is that which maximizes difference between rate of energy intake and power output. The fish maximizes surplus power (Ware, 1982). The optimum speed increases with increase in food density. This suggests that metabolic scope may well increase substantially to accommodate highly active foraging strategies. It should be noted that what is being maximized is surplus power, and that the surplus may represent a small percentage of total power consumption. There is a tendency towards higher and higher turnover in energy in return for relatively small net gains. Some earlier authors suggested that the ratio between energy income and expenditure is maximized (Weihs, 1975), but this seems to be less realistic than Ware's proposals.

Referring back to Figure 2.5, there are therefore a number of selective pressures in terms of foraging requirements, which are likely to lead to an increase in maximum metabolic rate from the 'slow fish' condition towards that of the fast fish capable of sustained swimming speeds in excess of 3 body lengths per second.

The most highly developed fish for high-performance swimming are the tunas in the family Scombridae. Of all the known fish species they are the ones which most closely approach an 'ideal' condition of adequate metabolic scope permitting unlimited simultaneous function of all metabolic activities. The gills of the tunas are very large in area and highly modified to withstand high ventilation rates (Muir & Kendall, 1968; Muir & Hughes, 1969). The heart size relative to body weight is equivalent to that of a mammal (Poupa *et al.*, 1981; Poupa & Lindstrom, 1983). High specific power output of the muscles is achieved through maintaining a high temperature in the aerobic locomotor muscles. Special heat-exchanger retes in the blood circulation to the muscles ensure that heat generated in the muscles is not dissipated to the surrounding water (Sharp & Dizon, 1978). The energetic costs of sustained high-speed swimming are great and therefore the rates of food intake and digestion must be high. It is interesting to note that there are retes in the blood supply to the liver and viscera which would help maintain SDA at a high rate in order to keep the locomotor system supplied with energy.

Tunas have the highest metabolic rates observed in any fishes. Limited laboratory data indicate that oxygen consumption is about 3–4 times that observed in well-studied active fish such as sockeye salmon. The metabolic rate approaches that of mammals (Graham & Laurs, 1982). The maximum oxygen consumption of skipjack tuna *Katsuwonus pelamis* is between 1000 and 2000 mg kg^{-1}h^{-1} (Gooding *et al.*, 1981). Extrapolating from the limited available data, this suggests a critical swimming speed of about 3.5 body lengths per second. A number of reliable observations in the field, however, indicate that skipjack can sustain a swimming speed of near 10 body lengths per second. Gooding *et al.* (1981) suggest that tuna may achieve this by using a more efficient swimming mode than most fish are able to achieve. Wardle & Videler (1980) have suggested kinematic means by which high efficiency might be achieved, but at maximum cruising special means of control of the onset of turbulence in the flow around the body are likely to be important (Wardle, 1977). Quantifying the maximum aerobic swimming performance of tunas presents a formidable technical challenge and at present we must accept the limitations of available data.

All the indications are that tunas have a most formidable swimming performance capacity. Evidence from tracking studies in the wild indicates that they spend most of their time swimming at well below the maximum sustainable speed. Carey & Olson (1981) report mean swimming speeds of yellowfin tuna in the Eastern tropical Pacific of 0.6–1.5 body lengths per second. Dizon *et al.* (1978) report mean swimming speeds of skipjack tuna of 1.03 to 2.3 body lengths per second. However, occasional periods with speeds in excess of 4.4 body lengths per second were observed. Observations in large laboratory tanks also show that tunas spend much of their time swimming at little above the minimum swimming speed (Magnuson, 1978). Tunas have a minimum swimming speed below which they cannot sustain themselves. They rely to a large extent on ram jet ventilation of the gills, are negatively buoyant and have to keep swimming in order to avoid sinking. Magnuson (1970, 1973) has developed equations to predict the minimum hydrostatic speed from hydrodynamic theory.

In the fish with low metabolic scope, it is clear that a choice has to be made between swimming activity and digestion of meals. If reasonable growth rates are to be sustained, swimming activity must be minimal to allow for the demands of SDA. In the fish with high metabolic scope, low mean swimming speeds (in relation to the critical speed) are also commonly observed. This is probably due to energy saving (type 1 selection) considerations rather than any power budgeting problem. Continuous swimming at high speeds would represent a tremendous drain on energy reserves and would require very high rates of feeding if this were to be sustained. The speed of about 1 body length per second represents a substantial energy cost when compared with the mean speeds commonly observed in benthic fish of about one-tenth of this. Kitchell *et al.* (1978) conclude that small tunas are growth-limited by food availability but that larger fish are limited by their ability to process available food. The latter may represent a limit on SDA.

It seems, therefore, that although the tunas have perhaps sufficient metabolic scope to feed and swim actively all the time, full metabolic scope is rarely utilized. However, they may, under certain circumstances, suffer power budgeting problems. In tropical seas they frequently encounter an oxygen minimum layer below the thermocline. The low levels of dissolved oxygen may impose limits on metabolic power.

2.6   Conclusions

There is a need for much more information on activity levels and metabolic rates of fish in the field in relation to their performance capacities. A major problem is the measurement of metabolic rate in the field. It is possible to calculate energy budgets for fish but usually power estimates are averaged over a time period of at least a few hours. Inevitably much information is lost with such poor resolution on the time scale. Fluctuations in locomotor power have been estimated from tracking fish using acoustic tags. Priede & Young (1977) first attempted to use heart rate as a measure of metabolic rate but there are objections to this since there may not be a simple relationship between heart rate and metabolic rate. However, if the limitations are acknowledged, much can be achieved. Alternative approaches that have been tried are acoustic telemetry of gill ventilation rates (Oswald, 1978) and radiotelemetry of electromyograms (Weatherley *et al.*, 1982).

Hitherto ecologists have tended to study the energetics of fish populations almost totally independently of physiologists investigating energy metabolism of fish tissues. The study of population dynamics tends to reduce fish to entities that have certain mathematical properties of reproduction and mortality without reference to real problems faced by individual animals trying to survive under different conditions. If a link can be established between metabolic rate and natural mortality, this will help unify population dynamics with ecological energetics. The power budget can also be understood in terms of measurable performance of different physiological systems such as respiration and circulation. Type 2 selection directly relates these perspectives to the underlying concepts of natural selection.

2.7 Acknowledgements

I thank Dr G.G. Duthie for critically reading drafts of this chapter. I also thank numerous colleagues for very helpful discussions of these ideas, and Professor G.N. Somero for providing me with study facilities at the Scripps Institution of Oceanography. This work was supported in part by a travel grant from the Royal Society.

References

Alexander, R. McN. (1967), *Functional Design in Fishes*, London, Hutchinson

Bazovsky, I. (1961), *Reliability Theory and Practice*. New Jersey, Prentice-Hall

Beamish, F.W.H. (1974), Apparent specific dynamic action of largemouth bass, *Micropterus salmoides. J. Fish. Res. Bd Can. 31*, 1763

Beamish, F.W.H. (1978), Swimming capacity. In Hoar, W.S. and Randall, D.J. (eds). *Fish Physiology*, vol. VII, pp. 101–87. New York, Academic Press

Beverton, R.J.H. and Holt, S.J. (1959), A review of the lifespan and mortality rates of fish in nature, and their relation to growth and other physiological characteristics. In *The Lifespan of Animals*, pp. 142–80. CIBA Foundation Symposium

Black, E.C. (1958), Hyperactivity as a lethal factor in fish. *J. Fish. Res. Bd Can. 15*, 573

Bone, Q. (1978), Locomotor muscle. In: Hoar, W.S. and Randall, D.J. (eds). *Fish Physiology*, vol. VII, pp. 361–424. New York, Academic Press

Brett, J.R. (1964), The Respiratory metabolism and swimming performance of young sockeye salmon. *J. Fish. Res. Bd Can. 21*, 1183

Brett, J.R. and Groves, T.D.D. (1979), Physiological energetics. In Hoar, W.S., Randall, D.J. and Brett, J.R. (eds). *Fish Physiology*, vol. VIII, pp. 279–352. New York, Academic Press

Carey, F.G. and Olson, R.J. (1981), Sonic tracking experiments with tunas. *Coll. Vol. Sci. Papers, Int. Comm. Cons. Atlantic Tuna 17* (2), 458

Cohen, D.M. (1970), How many recent fishes are there? *Proc. Calif. Acad. Sci. Ser. 4 38* (17), 341

De Jager, S., Smit-Onel, M.E., Videler, J.J., Van Gils, B.J.M. and Uffink, E.M. (1977), The respiratory area of the gills of some teleost fishes in relation to their mode of life. *Bjid. Dierk. 46*, 199

Dizon, A.E., Brill, R.W. and Yuen, H.S.H. (1978), Correlations between environment, physiology, and activity and the effects of thermoregulation in skipjack tuna. In: Sharp, G.D. and Dizon, A.E. (eds). *The Physiological Ecology of Tunas*, pp. 233–59. New York, Academic Press

Driedzic, W.R. and Hochachka, P.W. (1978), Metabolism of fish during exercise. In: Hoar, W.S. and Randall, D.J. (eds). *Fish Physiology*, vol. VII, pp. 503–43. New York, Academic Press

Duthie, G.G. (1982), The respiratory metabolism of temperature-adapted flatfish at rest and during swimming activity and the use of anaerobic metabolism at moderate swimming speeds. *J. exp. Biol. 97*, 359–73

Duthie, G.G. & Hughes, G.M. (1982), Some effects of gill damage on the swimming performance of rainbow trout (*Salmo gairdneri*). *J. Physiol. 327*, 21 p.

Elliott, J.M. (1976), The energetics of feeding, metabolism and growth of brown trout (*Salmo trutta* L.) in relation to body weight, water temperature and ration size. *J. Anim. Ecol. 45*, 923

Elliott, J.M. and Davidson, W. (1975), Energy equivalents of oxygen consumption in animal energetics. *Oecologia 19*, 195

Fry, F.E.J. (1947), Effects of environment on animal activity. *Publs Ont. Fish. Res. Lab.* 55, 1

Gooding, R.M., Neill, W.H. and Dizon, A.E. (1981), Respiration rates and low oxygen tolerance limits in skip-jack tuna *Katsuwonus pelamis*. *Fish. Bull.* 79, 31

Graham, J.B. and Laurs, R.M. (1982), Metabolic rate of the Albacore Tuna, *Thunnus alalunga*. *Mar. Biol.* 72, 1

Gray, I.E. (1954), Comparative study of the gill area of marine fishes. *Biol. Bull.* 107, 219

Greer-Walker, M., Harden-Jones, F.R. and Arnold, G.P. (1978), The movements of plaice (*Pleuronectes platessa* L.) tracked in the open sea. *J. Cons. int. Explor. Mer.* 38, 58

Hammel, H.T. (1983), Phylogeny of regulatory mechanisms in temperature regulation. *J. therm. Biol.* 8, 37

Harden-Jones, F.R. (1973), Tail beat frequency, amplitude, and swimming speeds of a shark tracked by sector scanning sonar. *J. Cons. int. Explor. Mer.* 35, 95

Hoar, W.S. and Randall, D.J. (1978), *Fish Physiology*, vol. VII. New York, Academic Press

Holliday, F.G.T., Tytler, P. and Young, A.H. (1974), Activity levels of trout (*Salmo trutta*) in Airthrey Loch and Loch Leven, Kinross. *Proc. R. Soc. Edinb.* 74B, 315

Hughes, G.M. (1966), The dimensions of fish gills in relation to their function. *J. exp. Biol.* 45, 177

Johnston, I.A., Davison, W. and Goldspink, G. (1975), Adaptations in Mg^{++}-activated myofibrillar ATPase activity induced by temperature acclimation. *FEBS Lett.* 50, 293

Jones, D.R. and Randall, D.J. (1978), The respiratory and circulatory systems during exercise. In: Hoar, W.S. and Randall, D.J. (eds). *Fish Physiology*, vol. VII, pp. 425–501. New York, Academic Press

Kitchell, J.F., Neill W.H., Dizon, A.E. and Magnuson, J.J. (1978), Bioenergetic spectra of skipjack and yellowfin tunas. In: Sharp, G.D. and Dizon, A.E. (eds). *The Physiological Ecology of Tunas*, pp. 357–68. New York, Academic Press

Kleiber, M. (1961), *The Fire of Life: An Introduction to Animal Energetics*. New York, Wiley

Magnuson, J.J. (1970), Hydrostatic equilibrium of *Euthynnus affinis*, a pelagic teleost without a gas bladder. *Copeia* 1970, 56

Magnuson, J.J. (1973), Comparative study of adaptations for continuous swimming and hydrostatic equilibrium of scombrid and xiphoid fishes. *Fish. Bull.* 71, 337

Magnuson, J.J. (1978), Locomotion by scombrid fishes: hydromechanics, morphology, and behaviour. In: Hoar, W.S. and Randall, D.J. (eds). *Fish Physiology*, vol. VII, pp. 239–313. New York, Academic Press

Morgan, R.I.G. (1973), Some aspects of the active metabolism of brown trout and perch. Ph.D. thesis, University of Stirling, Scotland

Morgan, R.I.G. (1974), The energy requirements of trout and perch in Loch Leven, Kinross. *Proc. R. Soc. Edinb.* 74B, 333

Muir, B.S. and Hughes, G.M. (1969), Gill dimensions for three species of tunny. *J. exp. Biol.* 51, 271

Muir, B.S. and Kendall, J.I. (1968), Structural modifications of gills of tunas and some other oceanic fishes. *Copsia* 1968, 388

Oswald, R.L. (1978), The use of telemetry to study light synchronization with feeding and gill ventilation rates in *Salmo trutta*. *J. Fish. Biol.* 13, 729

Parker, H.W. and Boeseman, M. (1954), The basking shark in winter. *Proc. Zool.*

Soc. Lond. 124, 185

Poupa, O. and Lindstrom, L. (1983), Comparative and scaling aspects of heart and body weights with reference to blood supply of cardiac fibres. *Comp. Biochem. Physiol. 76A*, 413

Poupa, D. O., Lindstrom, L., Maresca, A. and Tota, B. (1981), Cardiac growth, myoglobin, proteins, and DNA in developing tuna (*Thunnus thynnus thynnus* L.). *Comp. Biochem. Physiol. 70A*, 217

Priede, I.G. (1974), The effect of swimming activity and section of the vagus nerves on heart rate in rainbow trout. *J. exp. Biol. 60*, 305

Priede, I.G. (1977), Natural selection for energetic efficiency and relationship between activity level and mortality. *Nature (Lond..) 267*, 610

Priede, I.G. (1983), Heart rate telemetry from fish in the natural environment. *Comp. Biochem. Physiol. 76A*, 515

Priede, I.G. (1984), Satellite tracking of a basking shark (*Cetorhinus maximus*) together with simultaneous remote sensing. *Fish. Res. 2*, 201

Priede, I.G. and Holliday, F.G.T. (1980), The use of a new tilting tunnel respirometer to investigate some aspects of metabolism and swimming activity of the plaice (*Pleuronectes platessa* L.). *J. exp. Biol. 85*, 295

Priede, I.G. and Tytler, P. (1977), Heart rate as a measure of metabolic rate in teleost fishes: *Salmo gairdneri*, *Salmo trutta*, and *Gadus morhua*. *J. Fish Biol. 10*, 299

Priede, I.G. and Young, A.H. (1977), The ultrasonic telemetry of cardiac rhythms of wild brown trout (*Salmo trutta* L.) as an indicator of bio-energetics and behaviour. *J. Fish Biol. 10*, 231

Randall, D.J. and Daxboeck, C. (1982), Cardiovascular changes in the rainbow trout (*Salmo gairdneri* Richardson) during exercise. *Can. J. Zool. 60*, 1135

Sharp, G.D. and Dizon, A.E. (1978), *The Physiological Ecology of Tunas*. New York, Academic Press

Sidell, B.D. (1980), Responses of goldfish (*Carassius auratus* L.) muscle to temperature acclimation: alterations in biochemistry and proportions of different fiber types. *Physiol. Zool. 53*, 98

Skidmore, J.F. (1970), Respiration and osmoregulation in rainbow trout with gills damaged by zinc sulphate. *J. exp. Biol. 52*, 481

Somero, G.N. (1975), The roles of isozymes in adaptations to varying temperatures. In: C.L. Mackert (ed). *Isozymes II. Physiological Function*. pp. 221–34. New York, Academic Press

Somero, G.N. and Childress, J.J. (1980), A violation of the metabolism–size paradigm: activities of glycolytic enzymes in muscle increase in larger size fish. *Physiol. Zool. 53*, 322

Soofiani, N.M. and Hawkins, A.D. (1982), Energetic costs at different levels of feeding in juvenile cod, *Gadus morhua* L. *J. Fish Biol. 21*, 577

Soofiani, N.M. and Priede, I.G. (1985), Aerobic metabolic scope and swimming performance in juvenile cod, *Gadus morhua* L. *J. Fish Biol.* (in press)

Stevens, E.D. and Randall, D.J. (1967), Changes in blood pressure, heart rate, and breathing rate during moderate swimming activity in rainbow trout. *J. exp. Biol. 46*, 307

Vahl, O. and Davenport, J. (1979), Apparent specific dynamic action of food in the fish *Blennius pholis*. *Mar. Ecol. Prog. Ser. 1*, 699

Wardle, C.S. (1977), Effects of size on swimming speed in fish. In: Pedley, T.J. (ed.). *Scale Effects in Animal Locomotion*, pp. 299–313. New York, Academic Press

Wardle, C.S. and Videler, J.J. (1980), How fish break the speed limit. *Nature. (Lond.) 284*, 445

Ware, D.M. (1975), Growth, metabolism, and optimal swimming speed of a pelagic fish. *J. Fish. Res. Bd Can. 32*, 33

Ware, D.M. (1978), Bioenergetics of pelagic fish: theoretical change in swimming speed and ration with body size. *J. Fish. Res. Bd Can. 35*, 220

Ware, D.M. (1982), Power and evolutionary fitness of teleosts. *Can. J. Fish. Aquat. Sci. 39*, 3–13

Weatherley, A.H., Rogers, S.C., Pincock, D.G. and Patch, J.R. (1982), Oxygen consumption of active trout, *Salmo gairdneri* Richardson, derived from electromyograms obtained by radiotelemetry. *J. Fish Biol. 20*, 479

Weihs, D. (1973), An optimum swimming speed of fish based on feeding efficiency. *Israel. J. Technol. 13*, 163

Weihs, D. (1975), Optimal fish cruising speed. *Nature (Lond.) 245*, 48

Wokoma, A. and Johnston, I.A. (1981), Lactate production at high sustainable cruising speeds in rainbow trout (*Salmo gairdneri* Richardson) *J. exp. Biol. 90*, 361

Wood, C.M., Turner, J.D. and Graham, M.S. (1983), Why do fish die after severe exercise? *J. Fish Biol. 22*, 189

Young, A.H., Tytler, P., Holliday, F.G.T. and MacFarlane, A. (1972), A small sonic tag for measurement of locomotor behaviour in fish. *J. Fish Biol. 4*, 57

PART TWO:
FOOD AND FEEDING

3 THE APPLICATION OF OPTIMAL FORAGING THEORY TO FEEDING BEHAVIOUR IN FISH

Colin R. Townsend and Ian J. Winfield

3.1 Introduction

Studies of the composition of fish diets were first reported before the end of the last century (Forbes, 1888), and observations of feeding behaviour shortly afterwards (Triplett, 1901), but it was not until the 1940s that quantitative investigation of prey selection was made possible by the development of the first numerical index of electivity, the forage ratio (Hess & Schwartz, 1941). Much effort has since been devoted to the development of methods for quantifying electivity, including the well known index of Ivlev (1961). Lechowiez (1982) provides a critical review.

The past three decades have seen a flourishing of studies of food selection in fishes, including both investigations of natural diets (e.g. Hrbacek *et al.*, 1961; Brooks & Dodson, 1965; Hall, 1970, Hall *et al.*, 1979) and, more recently, direct observations of feeding behaviour (e.g. Vinyard, 1980; Townsend & Risebrow, 1982; Vinyard, 1982;

Paszkowski & Tonn, 1983). Furthermore, recent work has shown the value of incorporating laboratory behavioural observations and analyses of natural diet into the same study (e.g. Mittelbach, 1981; Winfield *et al.*, 1983). These largely descriptive studies have revealed the principal characteristics of fish, prey and environment which are important *proximate* determinants of the observed patterns of prey selection.

In contrast, research into the strategic goals, or *ultimate* reasons for observed predation behaviour, is at an early stage. It is in this area that *optimal foraging theory* has made a fundamental contribution. Foraging has both profits (energy gained from prey) and costs (energy expended in searching for and handling prey). The theory of optimal foraging is based on the evolutionary premise that individuals within a population that forage most efficiently and maximize their net rate of energy income will possess greater fitness and contribute more genes to future generations (Calow & Townsend, 1981; Chapter 1). For any given array of prey types, differing in energy content, ease of capture and abundance, it is not difficult to determine, by means of a simple mathematical model, the optimal diet composition. Similarly, for any particular distribution of prey in a patchy environment, it is not hard to discover the optimal allocation of search effort: the one that would maximize the rate of energy income. Optimal foraging theory allows us to generate precisely quantified predictions of what would constitute the ideal behaviour, against which the actual observed performance of the predator can be compared. Agreement between prediction and observation indicates that we have some understanding of the most important aspects of the predator–prey interaction. Discrepancies between predicted and observed behaviour should be interpreted with caution but may serve to indicate other factors which need to be taken into consideration and thus provide further predictions to be tested.

In this chapter we will discuss both proximate and ultimate aspects of feeding behaviour in fish. An attempt will be made to complement the most important findings of the last 30 years with recent advances arising from the application of optimal foraging theory. Our examples are drawn entirely from fresh water because it is only here that optimization theory has been invoked.

3.2 The Decision What to Forage

3.2.1 The Strategic Goals: Application of Optimal Foraging Theory

Predators must make three types of decision: what, where and when to forage. Behavioural ecologists have paid considerable attention to the first decision, concerning prey choice, both in terms of the behavioural mechanisms involved and ultimate causes underlying them. The usefulness of the optimal foraging approach depends on how good the researcher's guesses are regarding the nature of costs, benefits and constraints in a given situation. Such factors are most apparent when the predator is deciding *what* to eat. Consequently, it is in this area that formal optimal foraging theory has been most extensively developed.

The value of a particular foraging strategy should ideally be measured directly in terms of its effect on the fitness of the predator: the contribution its genes makes to future generations. However, a shorter term objective such as maximizing the rate of energy intake is also important and is likely to be closely linked with eventual reproductive success (Chapter 1). It is with respect to this more easily measured goal that ecologists have employed optimal foraging theory. Fish are particularly suited to such investigations because diet influences growth rate and body size influences fecundity (Nikolskii, 1969).

Models of optimum prey choice begin with the basic trade-off between the benefit and costs associated with a particular prey type. Constraints on the predator's performance are generally ignored initially, a point to which we shall return later. Note also that the situation is typically further simplified by the assumption that maximization of the net rate of energy gain is the sole goal of the predator: this is the *energy maximization premise*. Such an assumption is not unreasonable for fish which feed predominantly on animal food because their diet will almost inevitably contain an adequate supply of nutrients. (In contrast, nutrients and not energy may be limiting in the diet of herbivores — Chapter 6.)

The benefit to a predator of a particular category of prey item may thus be measured in terms of net energy yield (gross energy extracted minus energetic costs of handling and digesting the food). If the predator is to maximize its net *rate* of energy gain, then the energy yielded by the prey must be set against the time needed to acquire the energy: the *handling time* for an item in the prey category in question. The *energy value* of a prey category can thus be conveniently calculated by dividing energy yield by handling time (Townsend & Hughes, 1981). The energy maximization premise dictates that a predator should consume those prey categories which, if included in the diet, will tend to maximize the energy intake per unit time, but reject those categories which would tend to lower this value.

Consider the simplest possible situation in which a predator encounters just two kinds of prey. Suppose the predator searches for T_s seconds and encounters the two prey categories at rates λ_1 and λ_2. The two categories of prey have energy yields of E_1 and E_2 and handling times of H_1 and H_2. They therefore have energy values (or profitabilities) of E_1/H_1 and E_2/H_2.

If the predator eats prey from both categories it will obtain the following energy in T_s seconds:

$$E = T_s (\lambda_1 E_1 + \lambda_2 E_2)$$

and this will take a total time of

$$T = T_s + T_s (\lambda_1 H_1 + \lambda_2 H_2)$$

(i.e. total time = search time + total handling time).

The predator's net *rate* of energy intake is:

$$E/T = \frac{\lambda_1 H_1 + \lambda_2 E_2}{1 + \lambda_1 H_1 + \lambda_2 H_2}$$

(note that T_s has cancelled out).

Suppose prey category 1 is the more profitable type. If the predator is to maximize E/T, it should specialize on type 1 if:

$$\frac{\lambda_1 E_1}{1 + \lambda_1 H_1} > \frac{\lambda_1 E_1 + \lambda_2 E_2}{1 + \lambda_1 H_1 + \lambda_2 H_2}$$

(i.e. energy gain from eating just prey category 1 > energy gain from eating both).

This can be rearranged to give:

$$E_2/H_2 < \lambda_1 \left[\frac{E_1 - E_2 H_1}{H_2} \right] \qquad (3.1)$$

(note that λ_2 has cancelled out).

Inequality (1) contains λ_1, the rate of encounter with prey category 1, but not λ_2, the rate of encounter with prey category 2. Whether or not it is worth while feeding purely on the first prey category is therefore seen to depend on the difference in energy value

between the two prey categories *and* on the encounter rate with
the first (more valuable) prey category, but to be independent of the
encounter rate with the second (less valuable) category.

These arguments can be applied to any number of prey categories if
they can be ranked in order of energy value. The general form of
inequality (3.1) for many prey categories is

$$E_m/H_m < \sum_{j=1}^{m-1} \lambda_j \left[E_j - \frac{E_m H_j}{H_m} \right]$$

where j and m represent the jth and mth prey categories, respectively.

The general predictions of the optimal foraging model are that the
most valuable prey category should always be accepted when
encountered, but as the rate of encounter with it falls, the diet should
expand to include the next most valuable prey category. As the
encounter rates with the more valuable food types continue to decline,
the diet should expand further to include sequentially the progressively
poorer prey categories. A further prediction is that the decision to
specialize should be an all-or-none response and increasing the
encounter rate with a poorer prey category should have no effect on
whether it is included in the optimal diet. This is because the term
λ_2 (encounter rate with the poorer prey type) does not appear in
inequality (3.1).

The power of optimal foraging theory is that given a knowledge of
the parameters these predictions can be made truly quantitative and
hence precisely testable.

3.2.2 Tests of Theories of Optimal Prey Choice in Fish

Despite the interest which optimal foraging theory has generated,
there have been surprisingly few careful tests of its predictions.
However, the theory has been useful in predicting prey choice by
animals as diverse as a crab (Elner & Hughes, 1978), a small perching
bird (Krebs *et al.*, 1977), a wading bird (Goss-Custard, 1970), moose
(Belkovsky, 1978) and two species of fish: bluegill sunfish (*Lepomis
macrochirus*) (Werner & Hall, 1974), and brown trout (*Salmo
trutta*) (Ringler, 1979).

Werner & Hall (1974) studied bluegill sunfish feeding on three size
classes of the cladoceran *Daphnia magna* under carefully controlled
laboratory conditions. The fish–cladoceran interaction is particularly
suited to studies of foraging because, given fish of sufficient size,
handling time is effectively constant (usually in the order of 1–2 s) for

a broad range of prey sizes. Furthermore, the nature of attacks towards cladocerans of different sizes is also effectively constant because their escape ability is uniformly low. Thus the *relative* energy values of the different prey categories will be closely related to their body sizes. However, because of the dependence of visibility on body size, encounter rates are dramatically influenced by prey size. Thus, Werner & Hall (1974) had to take into account different visibilities when estimating encounter rates. They performed three experiments, one at low, one at intermediate and one at high *Daphnia* density. The densities were selected carefully so that an 'optimal forager' would be predicted to include all three prey categories in its diet at the low density, but to exclude the smallest *Daphnia* at the intermediate density and to exclude both small and medium *Daphnia* at the high density. The results of their experiments are shown in Figure 3.1.

Figure 3.1: Observed and Predicted Diets of Bluegill Sunfish Preying on Three Size Classes of *Daphnia* Prey. (After Werner & Hall, 1974.)

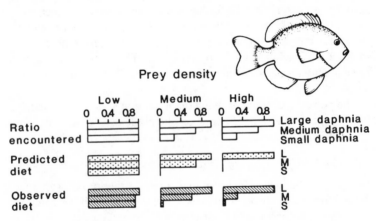

When total prey abundance was low, the fish exhibited no selection and hence the prey types were consumed in direct proportion to the rates at which they were encountered. At medium prey density, the fish ate relatively more of the two larger size classes of prey, thus selecting against the smallest prey. When prey were presented at an even higher density, the fish ate the largest size class almost exclusively. Thus, the theory predicted quite accurately the observed behaviour of the fish. Note, however, that the less profitable prey were never completely excluded from the diet as the theory suggests they should be. This discrepancy has been noted in all tests of optimal diet theory, whether performed on fish or on other animals.

Figure 3.2: The Observed Diet of Trout During a 6-day Feeding Experiment, Together with Theoretical Diets Assuming Random Feeding (items consumed according to their abundance in the drift) and Optimal Foraging (items taken in accordance with their ranked size). (After Ringler, 1979.)

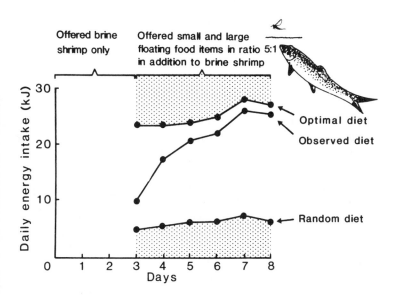

Ringler (1979) observed the performance of brown trout feeding on drifting food items in an experimental stream. Prey categories were freshly thawed brine shrimps, crickets and mealworms which had previously been frozen for storage. A broad spectrum of prey size categories, and hence energy values, was thus assembled. The inert nature of the prey, together with their small size relative to that of the trout, served to keep the interaction as simple as possible and resulted in almost constant handling times. Food items were introduced into the experimental stream for short periods repeated at daily intervals over a period of several days. The numbers and sizes of prey taken by the fish were determined by direct observation. Ringler then calculated the hypothetical daily energy intake for trout feeding randomly and for trout feeding optimally. These values, together with the actual performance of the fish, are shown in Figure 3.2. Initially the performance of the fish was only marginally better than that which would have been achieved from a totally random diet. With increasing experience of the prey categories, however, their performance quickly approached that considered to be optimal. Like sunfish (Werner & Hall, 1974), the trout achieved this performance by selecting for the larger items and against the smaller, less profitable prey. Note that again the optimal diet was not quite attained because the trout continued to take small numbers of the smaller prey items. It should also be noted that Ringler's results clearly demonstrate the potential importance of predator learning ability, a factor which has been largely ignored in optimal foraging theory. This phenomenon will be discussed in more detail later.

3.2.3 Appraisal of Theories of Optimal Prey Choice in Fish

One particularly important contribution of optimality theory has been its provision of a general framework within which many apparently discrete topics may be interrelated. The theory is at its most powerful when it is used as a tactical tool for organizing and thence directing new lines of research (Werner & Mittelbach, 1981).

Krebs & Davies (1981) consider another advantage of optimality models to be their requirement that underlying assumptions are made explicit — for example, prey are instantaneously recognizable, they only differ in energy value, the predator has perfect knowledge of prey availability, and so on. Similar assumptions have long been made implicitly by behavioural ecologists but have seldom been clearly stated. Making correct judgements about the validity of assumptions is critical, and it is worth while looking again at some of the basic assumptions of optimal foraging theory to assess their likely validity in interactions involving fish.

The first assumption is that fish can actually rank a variety of food categories according to their energy values. Such a capability requires in turn that fish are able to associate the energy value of the prey with some recognition stimuli. Evidently, in the experiments described in the previous section, the bluegill sunfish and brown trout could do this through a straightforward link between prey body size and energy value. In addition, sunfish (*Lepomis* spp.) have been shown to be able to correlate the appearance of prey with another important feature which influences energy value — their escape ability (Vinyard, 1980). The ranking assumption seems likely to be valid, although fish may not be infallible. Misidentification probably occurs occasionally and will result in a sub-optimal diet.

Figure 3.3: Influence of Learning on an Index of Energy Value (expressed here as biomass consumed ÷ handling time) of the Evasive Copepod *Cyclops vicinus* for Three European Species of Cyprinid Fish, Bream (triangles), Roach (circles) and Rudd (squares) (Winfield, 1983).

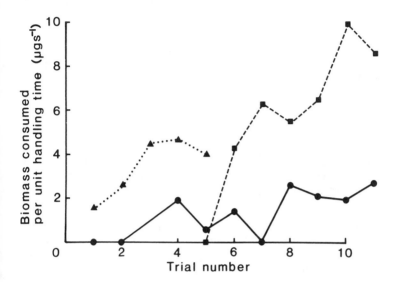

A further assumption of the model is that the predator's performance does not vary with experience, and hence that energy values and encounter rates remain constant. However, recent research has shown that this is often not the case and that such parameters may be influenced dramatically by learning. Foraging efficiency has been found to increase sharply with experience when bluegill sunfish are feeding on *Daphnia* or *Chironomus* (Werner *et al.*, 1981). Similarly, energy value of the evasive copepod *Cyclops vicinus* to three species of European zooplanktivorous fish has been demonstrated to increase significantly with increasing experience as a result of reductions in handling time (Winfield, 1983) (Figure 3.3). Reaction distances, and consequently encounter rates, have also been found to increase with learning in a variety of fish (Beukema, 1968; Ware, 1971; Winfield, 1983). The potential importance of learning to optimal foraging models is considerable. When the handling time for a given prey type was reduced with experience of that prey in one such model, the predicted optimum diet of a hypothetical predator was found to change completely, depending on encounter rates and learning efficiency (Hughes, 1979). Despite the enormity of the problem, future models of foraging behaviour must begin to incorporate the effects of learning if further progress is to be achieved. Werner & Mittelbach (1981) suggest that, although broad generalizations from simple models may prove impossible, an appropriate approach to this problem may be to incorporate specific learning effects for specific situations.

In addition to being able to rank prey values, it is also assumed that predators can memorize this ranking for at least the duration of a foraging bout. If knowledge of such ranks is learnt and is not merely innate, then the fish may require periodical reinforcement of all of the prey values in order not to forget them. Such sampling will tend to lower the value of the diet with respect to the short-term optimum. This phenomenon could explain the earlier observation that a variety of predators consistently take inferior prey in small numbers rather than avoiding them entirely as the theory predicts.

Simple models of optimum prey choice are really only applicable to the equilibrium state when the fish has a perfect knowledge of prey energy values and encounter rates. It is doubtful whether such a situation ever truly exists in nature where prey populations are constantly changing and the environment is extremely complex. However, it has been suggested (Krebs & Davies, 1981; Townsend & Hughes, 1981) that a predator may in fact not require particularly detailed information. The predator may be able to reach a near-optimum performance simply by following a set of relatively simple 'rules-of-thumb' such as select larger prey, select prey that are easy to catch and handle, relax selectivity when hungry. A distinction may be drawn between how an animal actually arrives at a diet (the behavioural mechanisms or proximate aspects of the interaction) and its strategic 'goals' (the ultimate aspects of the interaction). Note that while optimal foraging theory only relates directly to the latter, it may raise interesting questions about the former. The value of simple 'rules-of-thumb' will be particularly great in the complexity of the natural habitat where a diversity of subtle factors act together to influence prey energy values and encounter rates. Such factors must also be appreciated by the researcher before he or she can successfully apply foraging models to fish feeding in their natural habitat on their natural diets.

3.2.4 *Extension of Optimal Foraging Theory from the Laboratory to the Field*

The ultimate objective of the ecologist is to gain an understanding of the relationship between an animal and its natural environment. An obvious progression, therefore, is for optimal foraging theory to move on from simple laboratory conditions to a consideration of the interaction between fish and their prey as it occurs in nature. Such a development must accommodate a considerable increase in the complexity of the interactions.

The laboratory tests described previously were performed using simple arrays of simple prey types. In contrast, the natural diets of fish are usually composed of numerous categories of quite complex prey. For example, bluegills (Werner & Hall, 1974) were tested in the laboratory with individuals of just one species of *Daphnia* (which hence differed only in terms of body size), but their natural diet can include macroinvertebrates such as damselfly and chironomid larvae in addition to many other species of Microcrustacea. Daphnids are uniformly easy to catch, and hence suited to laboratory studies, because defence mechanisms are virtually absent: they do not hide, escape, or possess physical or chemical defences. However, many prey species in the natural habitat do possess such features. Consequent effects on energy values and encounter rates must be determined before optimal foraging theory can be successfully, and legitimately, applied to field conditions. Recent research has shown a movement in this direction (Mittelbach, 1981).

Empirical studies have revealed numerous features of prey which can influence their selection by fish. Such factors are as diverse as shape (Jacobs, 1965), visibility (Mellors, 1975; Zaret & Kerfoot, 1975), normal locomotory activity (Zaret, 1980a,b; Wright & O'Brien, 1982), escape ability (Szlauer, 1964; Vinyard, 1980; Winfield *et al.*, 1983), prey reproductive state (Sandstrom, 1980; Vuorinen *et al.*, 1983; Winfield & Townsend, 1983); taste (Kerfoot, 1982) and, of course, body size (Hrbacek *et al.*, 1961; Brooks & Dodson, 1965; Wong & Ward, 1972). A comprehensive review of these factors in the fish–zooplankton interaction may be found in O'Brien (1979). The mathematics of optimal foraging theory requires that a knowledge of all of the above factors be distilled into their quantitative effect on just two parameters, the prey's energy value (E/H) and its encounter rate with the predator (λ).

The Influence of Prey Characteristics. Considerable problems may be met when attempting to measure the energy value, or more particularly the net energy yield, of the variety of prey types encountered in the natural habitat. When attacks on the different prey types are similar in character (e.g. when sunfish attack cladocerans, or trout attack dead macroinvertebrates), and hence energy costs of the attacks are also similar, the energy content of the prey will provide a reliable index of the net energy yield. Thus the actual energy costs of the attacks need not be measured. However, this short cut would not be justified when different prey are attacked using different tactics entailing different energy costs. One particularly widespread example of this phenomenon is that of a fish feeding on a mixture of easily captured cladocerans and more evasive copepods. Kettle & O'Brien (1978) observed that small lake trout use a much more energetic strike when attacking the more evasive prey type. The only solution to this problem is to measure the actual amount of energy expended during an attack. Such an approach was used to study the effect of prey evasion capability on the energy costs of foraging by young Sacramento perch (*Archoplites interruptus*) (Vinyard, 1982). Vinyard estimated the energy costs of attacks towards both easily captured *Daphnia* and more evasive *Diaphanosoma* of identical size. He did this by video-recording the attack patterns of the fish and computing the power required for the accelerations observed. His estimates indicate that the fish expended *four* times as much energy in the attacks on *Diaphanosoma* which evaded capture nearly *half* of the time, suggesting that the energy cost of capture for these evasive prey is approximately 8 times that of *Daphnia*. Even though the costs of attacks on both prey types were just a small fraction of their total energy content, and hence reduced net energy yield only slightly, the fish were found to select for the less costly prey.

The visibility to fish of even the most complex prey is relatively easily measured in the laboratory, but transformation of this factor to natural encounter rates has been a major stumbling block in the extension of foraging models to the field (Werner & Mittelbach, 1981). The ecologist can relatively easily measure what prey are in the habitat, but actual encounter rates depend on the physical appearance of prey and also the prey's behaviour in terms of both its movement (Wright & O'Brien, 1982) and cryptic habits.

The Influence of Predator Characteristics. The effects of subtle differences in prey characteristics on the optimal foraging equation are equalled, if not exceeded, by similarly subtle differences in the features of different predators. Consequently, one of the areas in which optimal foraging theory can make a major contribution is in studies of resource partitioning and niche relationships in both inter- and intraspecific situations. Ecologists have long assumed that the morphology of an animal's feeding apparatus has a direct relationship with its foraging ability. One well-known example of this approach is the study of bill size and resource partitioning in Darwin's finches (Lack, 1947). The value of optimal foraging theory in this context is that it can provide direct evidence for the importance of quantitative aspects of an animal's feeding apparatus.

The quantitative relationships between mouth morphology and feeding efficiency have been particularly well studied in zooplank-tivorous fish. More particularly, considerable attention has been paid to the effects of prey body size and escape ability, two factors which have been found to be of paramount importance in this interaction (see O'Brien, 1979).

As one would expect intuitively, there is a clear relationship between the largest size of prey a fish can efficiently handle and the maximum dimensions of its mouth aperture (see, for example, Wong & Ward, 1972). Mouth size varies both interspecifically and, due to the existence of different sized individuals within the same species, intraspecifically. Consequently, different predator individuals may have quite different optimum prey sizes. Such an effect has been demonstrated interspecifically for bluegill (*Lepomis macrochirus*), green sunfish (*L. cyanellus*) and largemouth bass (*Micropterus salmoides*) (Werner, 1977), and intraspecifically for bluegill (Werner, 1974) and 15-spined stickleback (Kislalioglu & Gibson, 1976).

The relationship between mouth morphology and proficiency at capturing evasive prey is considerably more complex and has not yet been subjected to thorough investigation. Nevertheless, factors such as mouth protrusibility (McComas & Drenner, 1982), size and shape of aperture, volume of buccal cavity and its speed of expansion (Drenner *et al.*, 1978) have all been demonstrated to have some effect on the outcome of an attack. Such differences in the mouth morphologies of young bream (*Abramis brama*), roach (*Rutilus rutilus*) and rudd (*Scardinius erythropthalmus*) resulted in the evasive copepod *Cyclops vicinus* having a unique energy value for each predator (Winfield, 1983).

It should be noted that differences in the visual apparatus of different predators may have implications for optimal· diet considerations through their effects on encounter rates. No research has yet been undertaken on this possibility.

The Influence of Environmental Characteristics. Finally, the extension of theories on optimal prey choice from the laboratory to the natural habitat must take into account a number of factors not directly related to the predators and prey themselves. Such factors can influence both prey energy values and encounter rates.

Prey energy values will be influenced by any environmental factors which affect the predator's attack efficiency. Factors which adversely affect the predator's preying ability will, for example, reduce the prey's energy value by increasing the energetic cost of the attack. Such an effect will be particularly important if it has a differential impact on the profitability of different prey species. Poor optical conditions (low light or high turbidity levels, for example), may impair attack efficiency on evasive prey (see Vinyard & O'Brien, 1976) while having a negligible effect on the performance of more easily captured prey types. Similarly, low temperature may disproportionately improve the predator's performance on evasive prey by impairing the latter's locomotory abilities (Schmitt & O'Brien, 1982). The degree of physical structure in the environment may also have an appreciable effect on energetic aspects of the predator–prey interaction (Glass, 1971; Crowder & Cooper, 1982).

Environmental factors may also influence encounter rates with prey. For example, poor optical conditions may effectively provide a refuge for relatively larger prey because attenuation of contrast will make them relatively less visible (Lythgoe, 1966). Temperature could also alter the optimal diet through an effect on encounter rates. Werner & Hall (1974) found that at the same prey density and distribution, bluegill included more prey categories in their diet at 14°C than at 23°C. This could be attributed to longer search times resulting from the fish swimming more slowly at the lower temperature.

3.3 The Decision Where to Forage

3.3.1 Two Levels of Decision: Foraging within and between Habitats

Mobile predators such as fish will encounter areas of their environment which differ in the abundance and/or quality of food they contain. Considerable attention has been given to how the optimal forager should behave under such conditions.

When deciding where to forage, the predator can be regarded as making decisions on two levels: first, it must decide within which *habitat* it will forage, and, secondly, which areas or *patches* to exploit within that habitat. Although it is convenient to approach the study of foraging location through these two levels, it should be remembered that in reality they are merely two points on a gradient of environmental complexity. Consequently, many of the problems faced by the predator, and the tactics and strategies it employs to overcome them, will be common to both intra- and interhabitat foraging behaviour.

With the notable exception of recent studies on sunfish (see below), most studies to date (both theoretical and empirical) have been primarily concerned with decisions about the *intra*habitat location of foraging. Such studies of so-called 'patch' use (all carried out on predators other than fish — Krebs *et al.*, 1974; Zach & Falls, 1976; Cowie, 1977; Hubbard & Cook, 1978; Townsend & Hildrew, 1980) only consider variations in the abundance of prey: other factors such as prey type, prey size and degree of environmental structure are held constant. Hence these factors are not included as components of the foraging model. In contrast, all of these factors, and more besides, may vary *between* habitats and hence will influence the energetic consequences of decisions regarding habitat use.

The development of theory relating to interhabitat foraging location is thus complex, and empirical investigations of it are difficult, but such studies should be pursued for two reasons. First, resources in fish communities are commonly partitioned on the basis of habitat use (Werner, 1977). Thus foraging theory as it applies to interhabitat feeding patterns is relevant to studies of such fundamental ecological problems as niche relationships and the factors determining community structure. The second reason is particularly relevant to the present chapter. Such studies are essential if we are successfully to extend the theory concerning prey choice to the fish in its natural environment.

3.3.2 *The Intrahabitat Location of Foraging — Optimal Patch Use*

Assuming that an animal can perceive the area within which it is foraging as a patchwork of different food densities, then the energy maximization premise requires that the animal should apportion its foraging time between patches so as to maximize its net rate of energy gain (Townsend & Hughes, 1981). The profitability of a patch will fall as the predator removes prey from it, and each predator faces the crucial decision about how long to forage in a given patch before moving on in search of another. Consideration of this problem has led to the formulation of the *marginal value theorem* (Charnov, 1976), which states that the forager should abandon a patch when the instantaneous net rate of energy gain falls to a value equal to the average net rate of energy gain in the habitat as a whole.

Experimental tests of optimal patch use theory have been performed on animals as diverse as great tits, parasitic wasps and net-spinning caddis larvae (Krebs *et al.*, 1974; Hubbard & Cook, 1978; Townsend & Hildrew, 1980; respectively), and in each case the forager approached the predicted optimum behaviour. However, as was also the case in tests of optimal diet, the foragers never quite achieved the optimal performance and perhaps we should not be surprised, given the sophisticated gathering of information (perfect knowledge of the environment is assumed), computation of parameters in the optimization equation and calculation of the optimal solution which the theory apparently demands.

Several behavioural mechanisms or 'rules-of-thumb' have been proposed which would enable a predator to perform using less sophisticated information and still exploit an array of food patches with near-optimum efficiency. Numerous authors (Hassell & May, 1974; Krebs *et al.*, 1974; Murdoch & Oaten, 1975; Townsend & Hildrew, 1980) have argued that a predator may have a fixed 'giving-up time' which acts like a clock that is reset after each prey capture: if the giving-up time runs out without a prey being caught, the predator leaves the patch and searches for another. An appropriate value for giving-up time can produce foraging so near the optimum that the experimenter could not tell the difference. In similar vein, Ollason (1980) has proposed a simple memory model in which the predator leaves the patch if it is not finding prey as quickly as it remembers doing. Finally, Pyke *et al.* (1977) and Waage (1979) have suggested that near optimal patch exploitation may be achieved by an increased rate of turning after a successful attack and hence a slower rate of movement away from high-density food patches, a phenomenon they have termed 'area-restricted search'.

There have been fewer experimental tests of optimal patch use than of optimal diet, probably because it is quite difficult to set up a truly patchy prey distribution and to find a predator that actually perceives its environment as patchy. In the case of fish, no tests of optimal patch use have been performed at all. The assumptions of the optimal patch-use model are that prey are distributed in more or less fixed and discrete patches, that patches differ in density, that the predator perceives the patchiness, and that patches are depleted during a foraging bout. It is likely that in many fish–prey interactions, and particularly those involving zooplankton, all these assumptions do not necessarily hold. The pattern of patchiness may change too rapidly for the predator to be able to monitor it and, in addition, the periods of foraging (at dawn and dusk for many fish) may be too short for patches to be seriously depleted or for fish to respond to depletion.

Nevertheless, a number of studies have yielded information relevant to intrahabitat foraging behaviour in fish, and particularly about the value of foraging in groups. Recall that the theory of optimal patch use requires that predators have perfect knowledge of their environment (in order to determine when their current net rate of energy intake is equal to the average for the habitat as a whole). The dynamic nature of the natural environment obviously means that without periodic sampling of the habitat the fish's perception would quickly become out of date. In this context it is interesting to note that when attempting to locate patchily distributed food, fish in larger shoals found food more quickly than those in smaller shoals (Figure 3.4) (Pitcher *et al.*, 1982) and that fish in larger shoals carried out more sampling forays and were thus better able to respond quickly to changes in the relative profitabilities of patches (Pitcher & Magurran, 1983). This greater sampling effort is carried out at the expense of a reduction in short-term energy returns: goldfish (*Carassius auratus*) do not forage in complete accordance with the energy maximization premise because they also invest time and energy in sampling inferior food patches. The need to sample is one constraint that must be introduced to the next generation of models of optimum patch use.

3.3.3 The Interhabitat Location of Foraging — Optimal Habitat Use

Strategies of habitat use will be influenced by a variety of factors in addition to those impinging on optimal patch use. The optimal solution to the energy maximization problem when several habitats are available is influenced not only by the predator's own features and those of its prey but also by the physical nature of the environment. Important habitat features are those which influence prey energy values (e.g. dense vegetation may decrease predator attack efficiency and hence increase the energetic costs of capturing prey) or encounter rates (e.g. clear open water will render prey more visible and hence

Figure 3.4: Relationship Between the Number of Seconds Spent Foraging Before Food was Found by a Single Test Individual Foraging in a Shoal of Increasing Size for Both Minnows (*Phoxinus phoxinus*) and Goldfish (*Carassius auratus*). The median and quartiles of 16 replicates are shown for each shoal size. (From Pitcher *et al.*, 1982.)

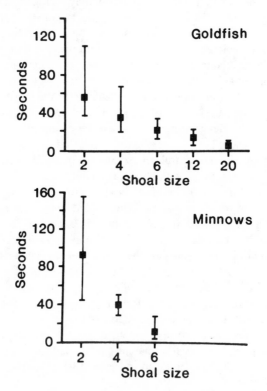

increase encounter rates).

Earl Werner and his associates have extended their laboratory work on bluegill sunfish by seeking to apply the energy maximization premise to the problem of habitat use in the field. Their principal concern has been to find whether the fish allocate their search effort to the habitat which provides the greatest net rate of energy income, given three contrasting habitats in which to forage (open water, macrophyte stands and bare sediment).

In the laboratory, Werner *et al.* (1983a) estimated the foraging efficiencies of bluegills in the three different habitats which they are known to use in nature. Three size classes of bluegills were observed foraging in laboratory replicas of the open water, vegetation and bottom sediment habitats. The habitat types were modelled in aquaria containing natural substrates and prey types (*Daphnia* for the open water, *Chironomus* larvae for the bottom sediment, and Coenagrionidae nymphs for the vegetation). Multiple regression techniques were then used to obtain equations predicting prey encounter rates and handling times in each habitat as a function of prey size, prey density and fish size. Thus, given data on the size–frequency distribution and abundance of prey in the field, optimal diet composition and habitat use could now be predicted by substituting the values in a suitable foraging model.

Seasonal changes in prey populations of the open water, vegetation and bare sediment habitats were exploited by Werner *et al.* (1983a) in an elegant test of predictions concerning habitat use on the basis of the energy maximization premise. If fish make habitat-use decisions so as to maximize their rate of net energy gain, one would expect them to shift foraging locations in response to changes in relative habitat profitabilities. Werner *et al.* (1983a) performed the test by monitoring prey populations in three habitats of a circular pond at the Kellogg Biological Station in Michigan during the summer months (July to September). Initially, profitabilities of the open-water habitat were high due to the presence of a *Daphnia pulex*; energy return rates were predicted to be 7 to 27 times greater than those for the sediments or vegetation. The profitability of the open water declined precipitously over the following weeks and remained low into September. Profitability of the vegetation was low throughout: predicted feeding rates were only 8–23% of those in open water or sediment.

The actual seasonal trends in habitat use shown by all three size classes of bluegills were in agreement with prediction (Figure 3.5). Between 21 and 25 July, the fish switched from a diet dominated by *D. pulex* to one containing more than 80% *Chironomus*. Note, however, that this result was only obtained in water from which piscivorous largemouth bass had been excluded. This point will be returned to later.

The study of Werner *et al.* (1983a) (and also an earlier investigation of Mittelbach (1981), which produced comparable results) has provided additional information on what has so far been a largely neglected aspect of optimal habitat use — the fish's sampling and learning capacities. Numerous studies have now demonstrated the potential contribution learning can make to foraging capability (Ware, 1971; Godin, 1978; Winfield & Townsend, 1983; Winfield *et*

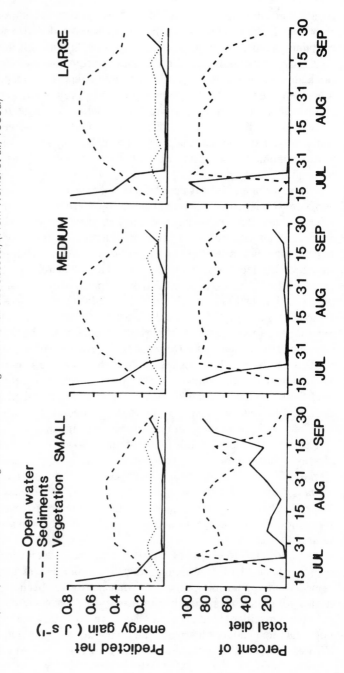

Figure 3.5: Seasonal Patterns in (a) Predicted Habitat Profitabilities (Predicted Net Rate of Energy Gain), and (b) Actual Percentage of the Diet Originating in Each Habitat for Three Size Classes of Bluegill Sunfish. Piscivores were absent. (The vegetation habitat is omitted from (b) for clarity: only 8–13% of the diet originated from the vegetation for all size classes.) (From Werner *et al.*, 1983a.)

al., 1983). Werner *et al.* (1981) pointed out that learning may be of particular importance during the fish's sampling operations. When sampling a habitat, and hence new prey categories of which it has had little recent experience, a fish is likely to operate at sub-optimal efficiency. The fish will tend to underestimate the true profitability of the new habitat. The effect of this could be to delay switching until the value of the current foraging habitat became considerably lower than the potential value of the new habitat. Thus the actual performance of the fish may lag considerably behind that predicted by optimal foraging theory. It may be noted that, in accordance with this contention, the behaviour of large bluegills in Mittelbach's (1981) study of Lawrence Lake did in fact switch approximately 2 weeks later than predicted. However, once the switch has been made, the fish's rate of net energy gain should quickly increase with its increasing foraging efficiency. Such an effect was apparent in bluegills switching from foraging in vegetation to foraging in open water or bare sediment habitats in another study (Werner *et al.*, 1981).

3.3.4 An Additional Goal: Minimizing Predation Risk

In the previous section we noted that bluegill sunfish showed seasonal switches in habitat use which conformed with behaviour predicted to maximize the net rate of energy income. This experiment had been performed in the absence of largemouth bass (*Micropterus salmoides*), an important natural predator of bluegills. Werner *et al.* (1983b) carried out the same procedure but in the presence of the piscivore, and found a dramatic difference in the bluegill foraging patterns as a consequence. Largemouth bass were chosen to be of such a size that the smallest of three classes of sunfish was very vulnerable to bass predation whereas the larger classes were more or less invulnerable. During the experiment only the small size class suffered significant mortality from bass (59% dead as compared with 28% in the absence of the predators).

In the presence of predators, how did the three size classes of bluegill distribute their foraging effort in the three habitats? The presence of largemouth bass had no significant effect on the foraging pattern of the more or less invulnerable medium-sized and large bluegills. They still chose habitats to maximize energy return rates. In contrast, the small bluegills obtained a greater fraction of their diet from the vegetation habitat (Figure 3.6), despite the fact that net rates of energy income were only one-third of those in open water.

Evidently the fish were balancing feeding efficiency against the risk

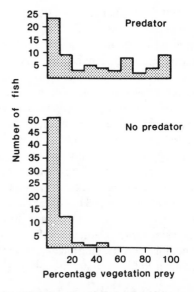

Figure 3.6: The Influence of the Presence of Largemouth Bass, which Predate Bluegill Sunfish, on the Importance of Prey of Small Bluegill Derived from Vegetation Habitat. (From Werner *et al.*, 1983b.)

of predation by piscivores. The small bluegills were faced essentially with the choice of two contrasting habitats: open water with high feeding profitability but also high risk of predation, *or* vegetation with lower feeding profitability but also lower predation risk. The relative advantages and disadvantages were such that the small bluegills elected to forage in the vegetation.

It is important to understand that introduction of the concept of predation risk does not question the appropriateness of the optimality approach; rather it questions the validity of the energy maximization premise in this particular case.

Werner *et al.*'s (1983b) experimental test of the prediction that predation risk would influence the balance of benefits and costs associated with particular foraging locations provided some further intriguing results. As expected, the small bluegills, forced to forage in sub-optimal habitat by the presence of predators, grew significantly more slowly as a consequence. Because of the reduced utilization of open water by small fish, the resources in this habitat were released to the larger size classes, which showed greater growth as a result. It is clear that the differential impact of a predator can have far-reaching consequences on the relative success of different prey size classes and on their partitioning of food and habitat resources.

Overt threat of predation has also been found to cause sticklebacks (*Gasterosteus aculeatus*) to forage sub-optimally with respect to rate of net energy gain: hungry sticklebacks normally prefer to attack high-density areas of *Daphnia*, but when a model piscivorous kingfisher (*Alecedo atthis*) was flown over their tank, they preferred to attack low-density areas (Milinski & Heller, 1978 (Figure 3.7)). Milinski & Heller (1978) hypothesized that when a stickleback feeds on a high density of *Daphnia*, it has to concentrate hard to overcome the confusing effect of the numerous prey in its field of vision, and hence it is less able to keep watch for predators. When a predator is in the vicinity, a hungry fish chooses to be more vigilant and puts a greater premium on survival than on feeding.

Figure 3.7: Hungry Sticklebacks Prefer to Forage in High-density *Daphnia* Areas but After a Model Kingfisher has been Flown Over the Aquarium They Opt for Low Prey Density Areas. (From Milinski & Heller, 1978.)

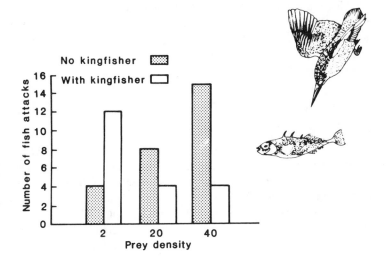

Despite such empirical evidence that fish do trade off foraging efficiency with risk of predation, foraging models have yet to be extended to include the effects of predation pressure. Hence, unlike many areas of ecology, this is one field in which theory lags considerably behind empirical observation. The reason for this state of affairs is clear: feeding efficiency and predation risk are measured in quite different and incompatible units (energy gain and likelihood of death, respectively), and incorporation of both factors into a single equation is thus difficult. There are two distinct ways in which this problem might be overcome (Dill, 1983). First, by a series of experiments, one could measure how much energy reward a fish is willing to sacrifice in order to enjoy a certain reduction in predation risk (or vice versa). Both factors, now expressed in a common currency, could then be incorporated into a relatively simple model. Alternatively, and perhaps more ambitiously, the effects of both energy intake and predation risk on some direct estimates of fitness could be estimated. An appropriate measure, according to Dill (1983), would be *expected lifetime reproductive output*, which integrates age-specific fecundity (itself strongly influenced by feeding history) and age-specific mortality (see also Chapter 1). This latter approach is being adopted by Werner and his associates in future studies of habitat choice in sunfish (Dill, 1983).

Small bluegills reduce their risk of predation by foraging in a sub-optimal feeding habitat in which their predators are relatively rare, and which in addition offers a greater chance of escape from attack. However, as might be expected, this strategy is not without its own costs, and results in a significantly reduced growth rate. An alternative, and perhaps less costly, method of reducing predation risk will be considered in the next section.

3.4 The Decision When to Forage

3.4.1 Diel Changes in Foraging Efficiency

The majority of fish are primarily visual feeders and hence one would expect their foraging performance to be influenced by ambient light levels. Numerous studies support this contention: O'Brien (1979) collates data from bluegill sunfish, white crappie (*Pomoxis annularis*), lake trout (*Salvelinus namaycush*) and grayling (*Thymalus* sp.) which all show a relationship between ambient light intensity and the distance at which prey are detected. It is of interest to note that bluegill sunfish display a particularly sharp drop in reaction distance when light levels fall below 10 lux, a level associated with twilight. Given this general dependence on light, it is not surprising that many species show peaks of feeding activity during the daylight hours (see Keast & Welsh, 1968; Mathur & Robbins, 1971; Noble, 1972; Elliott, 1976; Cordes & Page, 1980).

Many species of Microcrustacea commonly exhibit diel vertical migrations, moving down towards the bottom of the water body during the day (Hutchinson, 1967; Zaret, 1980a,b). The relevance of this behaviour to the present chapter is that it may cause the prey of zooplanktivores to vary in accessibility over the 24-hour period, being least available during the day. Thus, due to the restrictions of prey availability and a need for sufficient light, some fish are restricted to foraging at dawn and/or dusk (see Narver, 1970; Doble & Eggers, 1978; Eggers, 1978; Hall et al., 1979). Werner & Hall (1974) consider that the bluegill is among these species, and hence in its natural habitat may only be exposed to zooplanktonic prey for brief periods at dawn and dusk.

The relationship between light intensity and foraging efficiency has also been investigated in two European species of cyprinids, with somewhat surprising results. Minnows (*Phoxinus phoxinus*) and young common bream (*Abramis brama*) were both found to be able to feed on *Daphnia* at very low light levels (Harden Jones, 1956; Townsend & Risebrow, 1982; respectively). The observations for minnows were more exhaustive than those for bream (Figure 3.8) and showed that the foraging ability of this predator is not appreciably impaired until the ambient light level falls below that of a starlit sky (approximately 10^{-3} lux). Although no data are available for the 10^{-4} to 10^{-1} lux range of light intensities, the pattern for bream is similar. The consequences of this excellent low light level vision for natural foraging periodicities will be returned to later.

3.4.2 Diel Changes in Predation Risk

In addition to influencing feeding efficiency of the fish, diel changes in light levels will also have an effect on the foraging performance of piscivores. It is unlikely, for example, that piscivorous diving birds such as terns and grebes will be able to feed under the poor light conditions that exist during the night. Eggers (1978) and Bohl (1980) contend that the same is true of piscivorous fish: certainly Carlander & Cleary (1949) found yellow perch (*Perca flavescens*) and pike (*Esox lucius*) to be more active during the day, although they found the converse to be true for walleyes (*Stizostedion vitreum*). Among European authors, Linfield & Rickards (1979) consider the pike to be a largely diurnal predator.

Despite the paucity of data, it seems reasonable to conclude that predation pressure exerted by piscivores will be reduced during the hours of darkness. This will be particularly marked in the relatively simple fish community of Europe, where pike and perch (*Perca fluviatilis*) are often the only piscivorous fish species.

3.4.3 A Temporal Strategy to Reduce Predation Risk

Small bluegill sunfish reduce the risk of their predation by the

Figure 3.8: The Relationship Between Light Intensity and Foraging Abilities of Under-yearling Bream (open circles — data from Townsend & Risebrow, 1982) and Minnows (closed circles — data from Harden Jones, 1956). (Means ± standard errors are shown.)

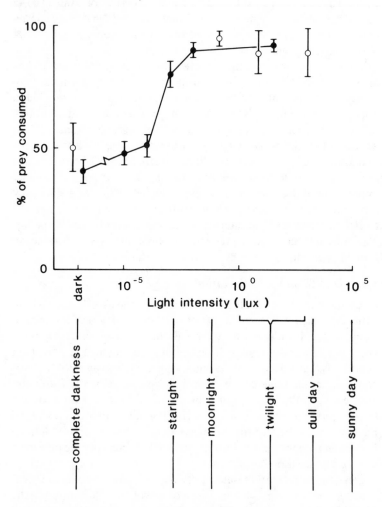

energetically costly activity of foraging in a sub-optimal feeding habitat where they are less likely to encounter and be captured by predators (Werner *et al.*, 1983a,b). There exists an alternative mechanism by which susceptible fish could reduce their risk of predation. Predation pressure in the food-rich open-water habitat is likely to be considerably reduced during the night. Thus fish with sufficiently good low light level vision could solve their conflicting problems by foraging in the open water at night, using the vegetation zone purely as a daytime refuge. Whereas bluegills do not have the optical physiology to do this, minnows and bream apparently do possess this ability. Do they employ such a foraging strategy?

Although no relevant fieldwork has been performed on minnows, two independent studies have been made on zooplanktivorous bream (Bohl, 1980; Winfield, 1983). In each case, young bream (and the young of several other cyprinid species) were found to spend the daylight hours in close proximity to refugia afforded by the marginal vegetation of the water body. It was only at night that the small fish ventured out into the open water to exploit its zooplankton. In each case maximum feeding activity was shown to coincide with the night-time migration out into the open water. Both field sites contained populations of optically orientated piscivorous fish and birds.

These field data are purely observational. Further support for the predator-avoidance hypothesis can only be gained from experimental manipulations similar to those recently performed by Werner *et al.* (1983a) on bluegills. It would also be of interest to determine if the foraging efficiency of young bream would be greater if they foraged during the day, or if their vision is so efficient that night-time levels are sufficient to achieve maximum foraging efficiency.

3.5 Concluding Remarks

Optimal foraging theory has been influential in shaping ideas about the organization and strategic goals of foraging. In the initial stages of its development, the theory has been highly simplistic and in particular has failed to take into account the constraints under which real animals operate. Whereas the hypothetical model optimal forager can be visualized as possessing perfect knowledge of all essential parameters, this is clearly not the case for real animals. However, even using this simplistic approach, optimal foraging theory was useful in providing an index of perfect adaptation, a yardstick against which actual performance could be judged.

In many ways it is surprising that real predators have often conformed quite closely with predicted optimal behaviour. Neverthe-less, it seems that in every case the predator is making use of behavioural tricks or 'rules-of-thumb' that allow it to approach the optimal behaviour but that rely on a smaller amount of less sophisticated information than the model optimal forager is imagined to possess. In this respect the optimal foraging approach has worked by showing up small discrepancies between observed and predicted performance, and this, in turn, has led to studies of the actual mechanism by which foragers approach optimal behaviour. Thus, the search for ultimate answers about *why* organisms behave as they do has fed back to a renewed look at the proximate mechanisms involved in *how* they do it.

In which direction should optimal foraging theory now move? We believe that one way forward is for the theorists to produce refined models, specific to particular organisms in particular environments, which explicitly include constraints on what real predators can actually do. In addition, it is becoming increasingly clear that a predator's learning capacity and experience greatly modify its performance. Theory needs to take this into account. In essence, such an exhaustive model that is relevant to a particular situation becomes a specific statement of proximate mechanism rather than a general statement of ultimate goal.

Another extension of the optimal foraging approach that is to be encouraged is the progression from studies of simple interactions in laboratory aquaria to what is actually happening in the natural environment. In this context, fish studies are pioneering the way, particularly those of Earl Werner and his associates. Our understanding of foraging decisions by fish has advanced considerably as a result of the advent of optimal foraging theory, and in the future the approach promises to elucidate patterns of resource partitioning and fish community organization.

Finally, the fish studies described in this chapter have clearly shown that energy maximization is of fundamental importance but, nevertheless, is sometimes overridden by the risk of being predated while foraging. The incorporation of both energy maximization and predation risk into the same optimal foraging models, although beset with difficulties (not least in coming up with a common currency for the equations), would be a significant step forward.

References

Belkovsky, G.E. (1978), Diet optimisation in a generalist herbivore: the Moose. *Theor. Pop. Biol. 14*, 105

Beukema, J.J. (1968), Predation by the three-spined stickleback (*Gasterosteus aculeatus* L.): the influence of hunger and experience. *Behaviour 31*, 1

Bohl, E. (1980), Diel pattern of pelagic distribution and feeding in planktivorous fish. *Oecologia 44*, 368

Brooks, J.L. and Dodson, S.I. (1965), Predation, body size and composition of plankton. *Science 150*, 28

Calow, P. and Townsend, C.R. (1981), Ecology, energetics and evolution. In: Townsend, C.R. and Calow, P. (eds) *Physiological Ecology: An Evolutionary Approach to Resource Use*, pp. 3–19. Oxford, Blackwell

Carlander, K.D. and Cleary, R.E. (1949), The daily activity patterns of some freshwater fishes. *Amer. Midl. Nat. 41*, 447

Charnov, E.L. (1976), Optimal foraging: the marginal value theorem. *Theor. Pop. Biol. 9*, 129

Cordes, L.E. and Page, L.M. (1980), Feeding chronology and diet composition of two darters (*Percidae*) in the Iroquois River System, Illinois. *Amer. Midl. Nat. 104*, 202

Cowie, R.J. (1977), Optimal foraging in great tits (*Parus major*). *Nature* (Lond.) *268*, 137

Crowder, L.B. and Cooper, W.E. (1982), Habitat structural complexity and the interaction between bluegills and their prey. *Ecology 63*, 1802

Dill, L.M. (1983), Adaptive flexibility in the foraging behaviour of fishes. *Can. J. Fish. Aquat. Sci. 40*, 398

Doble, B.D. and Eggers, D.M. (1978), Diel feeding chronology, rate of gastric evacuation, daily ration, and prey selectivity in Lake Washington juvenile sockeye salmon (*Oncorhynchus nerka*). *Trans. Am. Fish. Soc. 107*, 36

Drenner, R.W., Stickler, R. and O'Brien, W.J. (1978), Capture probability: the role of zooplankton escape in selective feeding of planktivorous fish. *J. Fish. Res. Bd Can. 35*, 1370

Eggers, D.M. (1977), The nature of prey selection by planktivorous fish. *Ecology 58*, 46

Eggers, D.M. (1978), Limnetic feeding behaviour of juvenile sockeye salmon in Lake Washington and predator avoidance. *Limnol. Oceanog. 23*, 1114

Elliott, G.V. (1976), Diel activity and feeding of schooled largemouth bass fry. *Trans. Am. Fish. Soc. 105*, 624

Elner, R.W. and Hughes, R.N. (1978), Energy maximization in the diet of the shore crab, *Carcinus maenas. J. Anim. Ecol. 47*, 103

Emlen, J.M. (1966), The role of time and energy in food preference. *Amer. Natur. 100*, 611

Forbes, S.A. (1888), On the food relation of freshwater fishes: a summary and discussion. *Bull. Ill. State Lab. Nat. Hist. 2*(8), 475

Glass, N.R. (1971), Computer analysis of predation energetics in the largemouth bass. In: Patten, B.C. (ed.). *Systems Analysis and Simulation in Ecology*, pp. 325–63. New York, Academic Press

Godin, J.J. (1978), Behaviour of juvenile pink salmon (*Oncorhynchus gorbuscha* Walbaum) toward Novel Prey: influence of ontogeny and experience. *Env. Biol. Fish. 3*, 261

Goss-Custard, J.D. (1970), The responses of redshank (*Tringa totanus* L.) to spatial variations in the density of their prey. *J. Anim. Ecol. 39*, 91

Hall, D.J. (1970), Predator–prey relationships between yellow perch and *Daphnia* in a large temperate lake. *Trans. Am. Microsc. Soc. 90*, 106

Hall, D.J., Werner, E.E., Gilliam, J.F., Mittelbach, G.G., Howard, D., Doner, C.G., Dickerman, J.A. and Steward, A.J. (1979), Diel foraging behaviour and prey selection in the golden shiner (*Notemigonus crysoleucas*). *J. Fish. Res. Bd Can. 36*, 1029

Harden Jones, F.R. (1956), The behaviour of minnows in relation to light intensity.

J. exp. Biol. 33, 271

Hassell, M.P. and May, R.M. (1974), Aggregation of predators and insect parasites and its effect on stability. *J. Anim. Ecol. 43*, 567

Hess, A.D. and Schwartz, A. (1941), The forage ratio and its use in determining the food grade of streams. *Trans. Fifth North American Wildlife Conference*, pp. 162–4

Hrbacek, J., Dvorakova, M., Korinek, V. and Prochazkova, L. (1961), Demonstration of the effect of the fish stock on the species composition of zooplankton and the intensity of metabolism of the whole plankton association. *Verh. Int. Verein. Theor. Angew. Limnol. 14*, 192

Hubbard, S.F. and Cook, R.M. (1978), Optimal foraging by parasitoid wasps. *J. Anim. Ecol. 47*, 593

Hughes, R.N. (1979), Optimal diets under the energy maximization premise: the effects of recognition time and learning. *Amer. Natur. 113*, 209

Hutchinson, G.E. (1967), *A Treatise on Limnology. Volume 2. Introduction to Lake Biology and the Limnoplankton.* New York, Wiley

Ivlev, V.S. (1961), *Experimental Ecology of the Feeding of Fishes.* (Transl. by D. Scott.) New Haven, Yale University Press

Jacobs, J. (1965), Significance of morphology and physiology of *Daphnia* for its survival in predator-prey experiments. *Naturwissenschaften 52*, 141

Keast, A. and Welsh, L. (1968), Daily feeding periodicities, food uptake rates, and dietary changes with hour of day in some lake fishes. *J. Fish. Res. Bd Can. 25*, 1133

Kerfoot, W.C. (1982), A question of taste: crypsis and warning colouration in freshwater zooplankton communities. *Ecology 63*, 538

Kettle, D. and O'Brien, W.J. (1978), Vulnerability of arctic zooplankton species to predation by small lake trout. *J. Fish. Res. Bd Can. 35*, 1495

Kislalioglu, M. and Gibson, R.N. (1976), Prey 'handling-time' and its importance in food selection by the 15-spined stickleback, *Spinachia spinachia* (L). *J. exp. mar. Biol. Ecol. 25*, 151

Krebs, J.R. and Davies, N.B. (1981), *An Introduction to Behavioural Ecology.* Oxford, Blackwell

Krebs, J.R., Ryan, J.C. and Charnov, E.L. (1974), Hunting by expectation or optimal foraging? A study of patch use by chickadees. *Anim. Behav. 22*, 953

Krebs, J.R., Erichsen, J.T., Webber, M.I. and Charnov, E.L. (1977), Optimal prey selection in the great tit (*Parus major*). *Anim. Behav. 25*, 30

Lack, D. (1947), *Darwin's Finches.* Cambridge, Cambridge University Press

Lechowiez, M.J. (1982), The sampling characteristics of electivity indices. *Oecologia 52*, 22

Linfield, R.S.J. and Rickards, R.E. (1979), The zander in perspective. *Fish. Man. 10*, 1

Lythgoe, J.N. (1966), Visual pigments and underwater vision. In: Bainbridge, R., Evans, G.C. and Rackham, O. (eds). *Light as an Ecological Factor*, pp. 375–91. Oxford, Blackwell

McComas, S.R. and Drenner, R.W. (1982), Species replacement in a reservoir fish community: silverside feeding mechanisms and competition. *Can. J. Fish. Aquat. Sci. 39*, 815

Mathur, D. and Robbins, T.W. (1971), Food habits and feeding chronology of young white crappie, *Pomoxis annularis*, Rafinesque, in Conowingo Reservoir. *Trans. Am. Fish. Soc. 100*, 307

Mellors, W.K. (1975), Selective predation of ephippial *Daphnia* and the resistance of ephippial eggs to digestion. *Ecology 56*, 974

Milinski, M. and Heller, R. (1978), Influence of a predator on the optimal foraging behaviour of sticklebacks (*Gasterosteus aculeatus* L.) *Nature 275*, 642

Mittelbach, G.G. (1981), Foraging efficiency and body size: a study of optimal diet and habitat use by bluegills. *Ecology 62*, 1370

Murdoch, W.W. and Oaten, A. (1975), Predation and population stability. *Adv. Ecol. Res. 9*, 2

Narver, D.W. (1970), Diel vertical movements and feeding of under-yearling sockeye salmon and the limnetic zooplankton in Babine Lake, British Columbia. *J. Fish. Res. Bd Can. 27*, 281

Nikolskii, G.V. (1969), *Theory of Fish Population Dynamics.* (Translated by J.E.S. Bradley.) Edinburgh, Oliver and Boyd

Noble, L.N. (1972), A method of direct estimation of food consumption with application to young yellow perch. *Prog. Fish. Culture 34*, 191

O'Brien, W.J. (1979), The predator–prey interaction of planktivorous fish and zooplankton. *Amer. Sci. 67*, 572

Ollason, J.G. (1980), Learning to forage — optimally? *Theor. Pop. Biol. 18*, 44

Paszkowski, C.A. and Tonn, W.M. (1983), An experimental study of foraging site selection in young-of-the-year yellow perch, *Perca flavescens. Env. Biol. Fish. 8*, 283

Pitcher, T.J. and Magurran, A.E. (1983), Shoal size, patch profitability and information exchange in foraging goldfish. *Anim. Behav. 31*, 546

Pitcher, T.J., Magurran, A.E. and Winfield, I.J. (1982), Fish in larger shoals find food faster. *Behav. Ecol. Sociobiol. 10*, 149

Pyke, G.H., Pulliam, H.R. and Charnov, E.L. (1977), Optimal foraging: a selective review of theory and tests. *Quart. Rev. Biol. 52*, 137

Ringler, N.H. (1979), Selective predation by drift-feeding brown trout (*Salmo trutta*). *J. Fish. Res. Bd Can. 36*, 392

Sandstrom, O. (1980), Selective feeding by baltic herring. *Hydrobiol. 69*, 199

Schmidt, D. and O'Brien, W.J. (1982), Planktivorous feeding ecology of arctic grayling (*Thymallus arcticus*). *Can. J. Fish. Aquat. Sci. 39*, 475

Szlauer, L. (1964), The refuge ability of plankton-animals before models of plankton-eating animals. *Polish Arch. Hydrobiol. 13*, 89

Townsend, C.R. and Hildrew, A.G. (1980), Foraging in a patchy environment by a predatory net-spinning caddis larva: a test of optimal foraging theory. *Oecologia 47*, 219

Townsend, C.R. and Hughes, R.N. (1981), Maximising net energy returns from foraging. In: Townsend, C.R. and Calow, P. (eds). *Physiological Ecology: An Evolutionary Approach to Resource Use* pp. 86–108. Oxford, Blackwell

Townsend, C.R. and Risebrow, A.J. (1982), The influence of light level on the functional response of a zooplanktonivorous fish. *Oecologia 53*, 293

Triplett, N.B. (1901), The educability of the perch. *Amer. J. Psychol. 12*, 354

Vinyard, G.L. (1980), Differential prey vulnerability and predator selectivity: effects of evasive prey on bluegill (*Lepomis macrochirus*) and pumpkinseed (*L. gibbosus*) predation. *Can. J. Fish. Aquat. Sci. 37*, 2294

Vinyard, G.L. (1982), Variable kinematics of Sacramento perch (*Archoplites interruptus*) capturing evasive and nonevasive prey. *Can. J. Fish. Aquat. Sci. 39*, 208

Vinyard, G.L. and O'Brien, W.J. (1976), Effect of light intensity and turbidity on the reactive distance of bluegill sunfish. *J. Fish. Res. Bd Can. 33*, 2845

Vuorinen, I., Rajasilta, M. and Salo, J. (1983), Selective predation and habitat shift in a copepod species — support for the predation hypothesis. *Oecologia 59*, 62

Waage, J.K. (1979), Foraging for patchily distributed hosts by the parasitoid, *Nemeritis canescens. J. Anim. Ecol. 48*, 353

Ware, D.M. (1971), Predation by rainbow trout (*Salmo gairdneri*). The effect of experience. *J. Fish. Res. Bd Can. 28*, 1847

Werner, E.E. (1974), The fish size, prey size, handling time relation in several sunfishes and some implications. *J. Fish. Res. Bd Can. 31*, 1531

Werner, E.E. (1977), Species packing and niche complementarity in three sunfishes. *Amer. Natur. 111*, 553

Werner, E.E. and Hall, D.J. (1974), Optimal foraging and size selection of prey by the bluegill sunfish (*Lepomis macrochirus*). *Ecology 55*, 1042

Werner, E.E. and Mittelbach, G.G. (1981), Optimal foraging: field tests of diet choice and habitat switching. *Amer. Zool. 21*, 813

Werner, E.E., Mittelbach, G.G. and Hall, D.J. (1981), The role of foraging profitability and experience in habitat use by the bluegill sunfish. *Ecology 62*, 118

Werner, E.E., Mittelbach, G.G., Hall, D.J. and Gilliam, J.F. (1983a), Experimental tests of optimal habitat use in fish: the role of relative habitat profitability. *Ecology 64*, 1525

Werner, E.E., Gilliam, J.F. Hall, D.J. and Mittelbach, G.G. (1983b), An experimental test of the effects of predation risk on habitat use in fish. *Ecology 64*, 1540

Winfield, I.J. (1983), Comparative studies of the foraging behaviour of young co-existing cyprinid fish. Ph.D. Thesis, University of East Anglia

Winfield, I.J. and Townsend, C.R. (1983), The cost of copepod reproduction: increased susceptibility to fish predation. *Oecologia 60*, 406

Winfield, I.J., Peirson, G., Cryer, M. and Townsend, C.R. (1983), The behavioural basis of prey selection by underyearling bream (*Abramis brama* (L.)) and roach (*Rutilus rutilus* (L.)). *Freshwat. Biol. 13*, 139

Wong, B. and Ward, F.J. (1972), Size selection of *Daphnia pulicaria* by yellow perch fry in West Blue Lake, Manitoba. *J. Fish. Res. Bd Can. 29*, 1761

Wright, D.I. and O'Brien, W.J. (1982), Differential location of *Chaoborus* larvae and *Daphnia* by fish: the importance of motion and visible size. *Amer. Midl. Nat. 108*, 68

Zach, R. and Falls, J.B. (1976), Ovenbird (Aves: Parulidae) hunting behaviour in a patchy environment: an experimental study. *Can. J. Zool. 54*, 1863

Zaret, T.M. and Kerfoot, W.C. (1975), Fish predation on *Bosmina longirostris*: body-size selection versus visibility selection. *Ecology 56*, 232

Zaret, T.M. (1980a), The effect of prey motion on planktivore choice. In: Kerfoot, W.C. (ed.) *Evolution and Ecology of Zooplankton Communities*, pp. 594–603. Hanover, University Press of New England

Zaret, T.M. (1980b), *Predation and Freshwater Communities*. New Haven, Yale University Press

4 ENERGETICS OF FEEDING AND DIGESTION

T.J. Pandian and E. Vivekanandan

4.1 Introduction

During the last 25 years, the literature on energetics of feeding and digestion in fish has grown rapidly and much has been summarized in several reviews (Ricker, 1946; Barrington, 1957; Winberg, 1961; Warren & Davies, 1967; Windell, 1967; Kapoor et al., 1975; Fänge & Grove, 1979). This chapter, however, concentrates on new, neglected and/or inadequately covered aspects of energetics of feeding and digestion.

4.2 Preference for Carnivorous Diet

4.2.1 Advantages of Carnivory

Of 20 000 fish species recognized (Nelson, 1976), Love (1980) has meticulously summarized the diets of about 600 species. A detailed analysis of Love's summary shows that about 85% of species are carnivores, 6% herbivores, 4% omnivores, 3% detritivores, and 2% scavengers and parasites. The food-capture mechanism of herbivores or filter feeders is more complex than that of the other groups. They have to bite, nibble or scrape algae from a hard substrate and crush the hard plant cells, or strain an enormous volume of water to retain a small quantity of plankton. Avoidance of the laborious process of foraging is therefore one possible advantage of carnivory in fishes.

4.2.2 *Protein Requirement*

Fishes also require a high protein level in their diet (Table 4.1), two to four times greater than that of terrestrial homeotherms (Love, 1980). Mammals suffer a slight retardation of growth when restricted to such a high protein diet (Just, 1980), because of the energy needed to eliminate the excess nitrogen. In contrast, fishes can rapidly and continuously excrete 60 to 80% of their total nitrogenous waste as ammonia through the gills (Ashley, 1972; Pandian, 1975).

The energy demands of fishes differ from those of terrestrial homeotherms, and these differences affect the protein requirements and transformation:

(1) in comparison with terrestrial birds and mammals, fishes expend less energy in maintaining the body at a different temperature from that of the environment;

(2) locomotion and maintenance of position in water require less energy than they do in air (Cowey, 1980);

(3) as protein acts both as a nutrient and as an energy source, the lowered energy requirement not only decreases energy intake but also increases the protein energy to total energy ratio.

Optimum protein energy to total energy ratio is around 1:2 for *Salmo gairdneri* (Lee & Putnam, 1973), whereas it is around 1:10 for ruminants (Williamson & Payne, 1980).

The mean protein requirement of omnivorous, herbivorous and detritivorous fishes is equal ($\approx$ 49% of dry food ingested) to that of the carnivores (Table 4.1). Many authors (Bowen, 1979; Pierce *et al.*, 1981) have indicated that nutritional constraint is imposed on herbivorous and detritivorous fishes in terms of low protein levels. Evidently the protein source for these fishes is the attached micro-organisms, which are high in protein content (Goldman & Kimmel, 1978). It is probably obligatory for herbivorous and detritivorous fishes to ingest these micro-organisms to satisfy their demand for high protein, unless they have a chance of ingesting high proteinaceous algae.

Recent studies on proteolytic activity in the digestive tract of fishes with different feeding habits reveal that herbivorous and omnivorous fishes are capable of digesting proteins as effectively as the carnivores. Hofer & Schiemer (1981) observed that the digestive tracts of most herbivorous and omnivorous fishes are distinctly longer than those of carnivores, and the specific proteolytic activity is negatively correlated with the relative length of the gut. But the duration of exposure of

Table 4.1: Protein Requirements of Some Fishes.

Species	Protein requirement (% dry weight of feeder)	Reference
Carnivores		
Chrysophrys aurata	40	Sabaut & Luquet, 1973
Pleuronectes platessa	50	Cowey *et al.*, 1972
Anguilla japonica	45	Nose & Arai, 1972
Oncorhynchus tshawytscha	48	De Long *et al.*, 1958
Salmo gairdneri	43	Satia, 1974; Zeitoun *et al.*, 1974
S. salar	40	Rumsey & Ketola, 1975
Omnivores		
Ictalurus punctatus	40	Dupree & Sneed, 1966
Cyprinus carpio	38	Ogino & Saito, 1970
Tilapia aurea	36	Davis & Stickney, 1978
Herbivores		
Ctenopharyngodon idella	42	Dabrowski, 1977
Detritivores		
Mugil auratus	70	Vallet *et al.*, 1970
M. capito	70	Vallet *et al.*, 1970

ingested food to proteases rises with the increasing gut length; thus the total proteolytic activity (micrograms per fish per day) produced by the carnivore *Puntius dorsalis*, the omnivore *P. filamentosus* and the herbivore *Sarotherodon mossambicus* was of the order 1117, 3476 and 6435. Hence the more intensive proteolytic digestion occurs in herbivorous and omnivorous fishes.

In summary, it is likely that the following have led the majority (85%) of fishes to resort to carnivory and also to considerably elevate their minimum protein requirement to > 40%:

(1) the relatively lower energy cost of maintaining body temperature and position in water;
(2) the ease with which ammonia is excreted;
(3) the capacity to digest a proteinaceous diet efficiently irrespective of feeding habit.

4.3 Feeding Rate of Fishes

Temperature has an important effect on the food intake of fishes belonging to different climatic zones. Data provided by a number of

Figure 4.1: Relationship Between Temperature and Feeding Rate of Fishes from Temperate and Tropical Zones. Dotted lines indicate general range of variability within each zone. Information is listed in the order of latitude zones in which feeding experiments were conducted. For instance, 1 represents Jobling (1983a), who conducted the experiments at the latitude of 70°N; likewise, 2 represents Edwards *et al.* (1969); 3: Healey (1972); 4: Johnson (1966); 5a,b: Elliott (1975); 6a,b: Wallace (1973); 7a: Fischer (1972); 7b: Haniffa & Venkatachalam (1980); 8a–c: Brett *et al.* (1969); 9: Jones & Green (1977); 10: Swenson & Smith (1973); 11: Beamish (1972); 12: Tyler (1970); 13a: Tyler & Dunn (1976); 13b: Frame (1973); 14: Carline & Hall (1973); 15a–c: Boehlert (1981); 16a–c: Pierce & Wissing (1974); 16d: Gerking (1955); 17: Cohen & Grosslein (1981); 18: Durbin *et al.* (1981); 19: Stilwell & Konler (1982); 20, 21: Popova & Sytina (1977); 22: Hatanaka *et al.* (1956a); 23: Hatanaka *et al.* (1956b); 24: Hatanaka *et al.* (1957); 25a,b: Hatanaka & Murakawa (1958); 26a,b: Yoklavich (1981); 27: Peters & Angelovic (1971); 28a,b: Peters & Boyd (1972); 29: Minton & McLean (1982); 30: Andrews & Stickney (1972); 31: Hopkins & Baird (1981); 32: Legand & Rivation (1969); 33: Magnuson (1969); 34: Singh & Bhanot (1970); 35: Sumitra Vijayaraghavan *et al.* (1982); 36: Tang & Hwang (1966); 37: Pandian (1967); 38: Belliyappa *et al.* (1983); 39a: Raghuraman (1973); 39b: Ponniah & Pandian (1977); 40: Vivekanandan (1977); 41: Arunachalam *et al.* (1976); 42: Mathavan *et al.* (1976); 43a–e: Vivekanandan *et al.* (1977); 44: De Silva & Perera (1976).

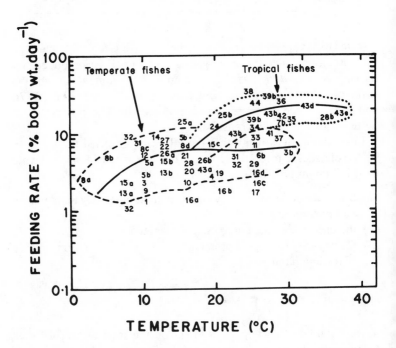

authors for temperate fishes (32 species inhabiting the area between 70°N and 27°N) and tropical fishes (12 species in areas between 21°N and 7°N) are plotted on a semi-logarithmic graph in Figure 4.1. The feeding rate of temperate fishes ranged from 1.8 to 17.3% body weight day^{-1} (mean, 5.9%) and that of tropical fishes from 4.1 to 36.0% (mean, 16.7%).

The feeding rate for the tropical fishes is therefore about 180% greater than the mean value for the temperate species. Brett & Groves (1979) plotted the logarithm of standard metabolism (milligrams oxygen per kilogram per hour) for tropical, temperate and polar fishes against temperature, and concluded that tropical fishes operate at a higher maintenance level in accordance with the higher environmental temperatures. The tropical fishes incur an energy expenditure of 2.1 KJ kg^{-1} h^{-1}, which is an elevation of 70% over the mean value (1.2 KJ kg^{-1} h^{-1}) for temperate species. The present finding suggests that temperature elevates not only the maintenance level but also the feeding rate in tropical fishes. However, the elevation in feeding rate (180%) is nearly 2½ times greater than that (70%) observed for the maintenance metabolism. Clearly, a considerable fraction of the 110% difference may contribute to faster, if not more efficient, growth, a characteristic of tropical fishes.

4.4 Predation and Feeding Rate

4.4.1 Prey Density

Data available for the larvae of *Oncorhynchus gorbuscha* (Parsons & LeBrasseur, 1973), *Clupea harengus* (Rosenthal & Hempel, 1973) and adults of *Salmo gairdneri* (Ware, 1972) and *Stizostedion vitreum vitreum* (Swenson & Smith, 1976) have been used (Figure 4.2) to investigate the relationship between prey density and feeding rate. For either of the two selected sizes of prey (0.001 and 0.01% of predator's body weight), the feeding rate increases with increasing prey density and reaches a steady state at a certain level.

4.4.2 Prey Size

It may be noted from Figure 4.2 that, besides prey density, prey size has an important effect on the feeding rate. Even at low density, a larger prey enhances the feeding rate, and probably satiates the fish more quickly than a smaller one in higher density. Given the opportunity to eat only the smaller prey, the fish may not be satiated at all even after 24 h of continuous feeding. For instance, the food requirements of *O. gorbuscha* (≈ 10 g) is 7% body weight day^{-1}, which it can obtain in 4 h by feeding on less concentrated (10 g m^{-3}) *Calanus plumchrus* (2.5 mg live weight per individual). Conversely, it may not procure the required daily ration even after 24 h of active grazing on *Pseudocalanus minutus* (0.1 mg per individual) in high density (90 g m^{-3}) (Parsons & LeBrasseur, 1973). From the data on *O. gorbuscha*, *C. harengus* and *S. gairdneri*, it can be concluded that to get satiated quickly the fish has to take prey whose size is equivalent to 0.05 to 0.1% of its own weight.

Figure 4.2: A Schematic Representation of the Effect of Prey Concentration on Feeding Rate of Predatory Fishes. The continuous lower line represents the small prey (= 0.001% of predator's body weight), and the upper one the larger prey (= 0.1% of predator's weight). The horizontal dotted lines indicate the maximum predation rates (and hence feeding rate). Mpr^1: maximum predation rate for the larger prey; Mpr^2: maximum predation rate for the smaller prey.

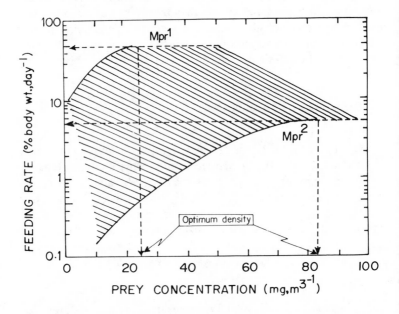

A preference for larger prey has been postulated for planktivorous fish (Werner & Hall, 1974; Mittelbach, 1981; Eggers, 1982). Though handling a large prey takes a longer time, the profitability in terms of feeding rate is greater for the fish, which feeds selectively on this size class. Fishes seem able to assess the energy value and size of prey and thereby a fish is able to rank and forage the most energy-rich food; these aspects are examined in Chapter 3. For instance, the profitability of *Lepomis macrochirus* ($\approx$ 0.4 g dry body weight) was 14 μg s^{-1} for a prey of 15 μg dry weight, which is handled in 1.06 s; a threefold increase of profitability (40 μg s^{-1}) is realized when an individual of equal body weight handles a larger prey (74 μg dry weight) in 1.84 s (Mittelbach, 1981). Similar data reported for a different body weight of *L. macrochirus* (Mittelbach, 1981), *Alosa pseudoharengus* and *Coregonus hoyi* (Crowder, 1981) are plotted in Figure 4.3. The profitability for the predator increases with weight of the prey and reaches an asymptote as the prey size approaches 0.05–0.1% of the predator's weight. Though predators can handle prey up to 1% of their body size (Blaxter, 1965) depending upon the mouth gape, prey which are more than 0.1% of the predator's body size may incur a heavy loss in handling time for the predator. In predator–prey interactions, the density and size of prey influence the satiation time and feeding rate of predators, thus altering the relationship between energy expenditure and gain. Whereas low prey density induces the predator to search a larger volume of water, the small (in relation to the predator's size) prey decreases the profitability of feeding.

4.5 Temporal Cost of Satiation

4.5.1 Carnivores and Herbivores

The strategy of non-random selection of prey, for example in terms of size and density, effectively minimizes the time cost of feeding in carnivores compared with other trophic levels. The carnivores spend 1–3 h every day for satiation, during which period they handle food equivalent to 0.4–12.% body weight h^{-1} (Table 4.2). Certain species such as the puffer (*Fugi vermiculatus*) and the filefish (*Stephanolepis cirrhifer*) (Ishiwata, 1968) become satiated in 15 m or even less. Compared with carnivores, the temporal cost of feeding is substantially higher for herbivores, which obtain food by scraping or rasping the substrate, and sucking the loosened materials into the mouth, or by biting pieces out of the substrate. Considerable quantities of coral, rock, shell fragments and sand may have to be ground and swallowed along with the algae. Species that feed regularly on leaves and stems of higher plants have to crush the hard plant tissues. Feeding in these herbivores takes up a major portion of the time during which the fish is

Figure 4.3: Profitability Curves for Predators Feeding on Different Prey Size. The predators are *Lepomis macrochirus* (Ia: 0.4 g dry weight; Ib: 1.2 g dry weight; Ic: 4 g dry weight; Mittelbach, 1981), *Alosa pseudoharengus* (II: 1 g dry weight) and *Coregonus hoyi* (III: 1 g dry weight) (Crowder, 1981). For comparison, weight of predators and prey are calculated on a dry weight basis considering water content as 80%. The shaded portions indicate the optimum prey size (0.05–0.1% of predator's size) for the maximum profitability.

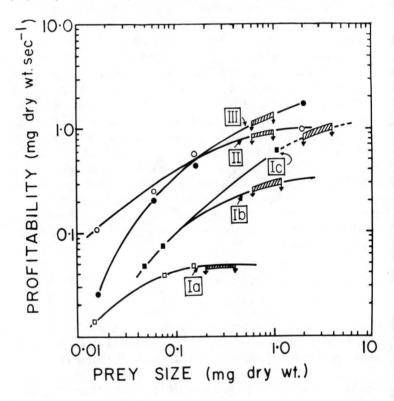

active, as the interbite intervals are largely involved with searching and handling (culling and chewing) food (Montgomery, 1980). The time cost of feeding for the two categories of herbivore (algal scrapers and plant crushers) ranges from 8 to 14 h, and the material that could be handled per hour during this period is less than for carnivores. The tuna *Katsuwonus pelamis* (1.6 kg live body weight) is an exception. This handles only 0.36% weight h^{-1} (Table 4.2). However, the relatively large size of *K. pelamis* used by Magnuson (1969) in the experiment may account for this discrepancy.

Table 4.2: Temporal Cost of Satiation for Fishes Belonging to Different Trophic Levels.

Species	Time cost of feeding (h day^{-1})	Material handled (% body weight h^{-1})	Reference
Carnivores			
Macropodus cupanus	1	1.13	Mathavan *et al.*, 1980
Oncorhynchus nerka	1	1.00	Brett, 1971
Heteropneustes fossilis	2	2.22	Marian *et al.*, 1982
Katsuwonus pelamis	3	0.36	Magnuson, 1969
Salmo trutta	3	0.68	Elliott & Persson, 1978
Ophiocephalus striatus	3	5.00	Vivekanandan, 1976
Petromyzon marinus	1	11.6	Farmer *et al.*, 1975
Herbivores			
Sparisoma aurofrenatum	8	—	Nursall, 1981
Microspathodon dorsalis	8	0.97	Montgomery, 1980
Eupomacentrus rectifraenum	12	0.57	Montgomery, 1980
Tilapia nilotica	14	0.02	Moriarty & Moriarty, 1973
Sarotherodon mossambicus	24	0.55	Hofer & Schiemer, 1981
Haplochromis nigripinnis	14	0.02	Moriarty & Moriarty, 1973
Detritivores			
Mugil cephalus	12	3.00	Moriarty, 1976
M. cephalus	24	3.00	Odum, 1973
Filter feeders			
Engraulis mordax	24	0.31	Leong & O'Connell, 1969
Brevoortia tyrannus	24	0.35	Durbin *et al.*, 1981

4.5.2 *Detritivores*

As detritivores feed on a nutritionally dilute substratum, the temporal cost of feeding to satiation is longer. The detritivores swim along near the bottom, suck in loose surface material, spit out larger particles and swallow the rest. Feeding is therefore almost continuous. By comparing the amount of plant detrital pigment in the contents of the cardiac stomach with that in the sediments upon which the fish feed, it is estimated that *Mugil cephalus* filters almost 100 g dry sediment to obtain 1 g dry nutrient in its digestive tract (Odum, 1973). Assuming more or less continuous feeding and a turnover rate of 5 times in 24 h, a 100 g mullet would filter 1500 g dry sediment day^{-1} or about 550 kg dry sediment year^{-1}.

4.5.3 *Filter Feeders*

Compared to predators, which consume their daily ration in a few large meals, the filter feeders randomly and continuously filter the water column, and consume food as a continuous stream of very small food particles. It is doubtful whether some filter feeders are satiated even after 24 h of continuous filtering. The anchovy *Engraulis mordax* (body weight, 4 g) requires 59 mg of dry food to satisfy its daily requirements; it is capable of filtering at the rate of 78 litres h^{-1} and even by filtering this enormous quantity of water, it could not acquire the required quantity of plankton under natural conditions. On the other hand, it is able to achieve satiation within an hour by particulate feeding. Durbin *et al.* (1981) also observed no evidence of satiation in the filter-feeding *Brevoortia tyrannus*, even with the largest ration. Evidence of satiation would include a decrease in swimming speed and/or a switch to intermittent particulate feeding. However, the fish fed continuously at a constant rate as long as food was available, and continued to search for food after the food supply was stopped. Furthermore, the metabolic cost per increment in swimming speed of filter-feeding *B. tyrannus* was about 2½ times higher than the cost of swimming in other types of fishes.

Since the energy cost associated with feeding is consistently high in filter feeders, adults of these species resort to other modes of feeding. For instance, *Engraulis mordax* turns to particulate feeding and gulping (Janssen, 1976), and wavyback tuna (*Euthynnus affina*) to particulate feeding (Walters, 1966). Except for the ammocete (lamprey larva), there is probably no fish larva that solely depends on filter feeding.

4.6 Feeding in Air-breathing Fishes

The energetics of air-breathing fishes has received much attention recently. Because of the need to visit the water surface to exchange atmospheric oxygen, the energetics of air-breathing fishes is influenced by various environmental factors, particularly the depth of water. By rearing the obligatory air-breathing fish *Ophiocephalus* (*Channa*) *striatus* (body weight, 0.75 g; total length, 4.5 cm) in aquaria containing different depths of water, Pandian & Vivekanandan (1976) forced the fish to swim longer or shorter distances per surfacing to exchange atmospheric oxygen. For instance, in aquaria containing the maximum depth of water (40 cm), the fish surfaced once in 46 s; the up-and-down swimming activity required a total period of 10 s to cover the distance of about 80 cm. In aquaria containing 5 cm water depth, the fish surfaced once in 66 s, and the up-and-down swimming activity required only 0.5 s to cover the distance of 10 cm. As a result, the surfacing frequency (1879 times day^{-1}) and thereby the distances travelled (1318 m day^{-1}) to exchange atmospheric oxygen by the fish in 40 cm depth were much higher than those by the fish in 5 cm depth (1317 times day^{-1}; 114 m day^{-1}). This 'exercise' imposes a considerable energy drain on the deeper water fish, and consequently under these conditions the fish increase their feeding rates (Figure 4.4). The feeding rate levels off with a further increase in water depth, though surfacing frequency tends to increase linearly.

In the event of food becoming a limiting factor at a given depth, the obligatory air-breathing fish *C. striatus* (Vivekanandan, 1976) and *Polyacanthus cupanus* (body weight, 0.5 g; Ponniah & Pandian 1977) adjust to the food shortage by reducing the surfacing frequency. Both these fishes reduced their surfacing frequency by about 60% when they were offered food equivalent to about 25% of maximum feeding rate. These experiments reveal that not only depth of water but also food availability regulate the feeding rate and surfacing frequency (Figure 4.5). Also these fishes establish a surfacing frequency (times per day) to feeding rate (gram-calories per gram per day) ratio of about 10:1 in different depths of water and different ration levels.

Starving or near-starving air-breathers reduced their surfacing frequency by about 70%; they presumably have lowered metabolic rate through reduction in specific dynamic action (Jobling, 1983b, and Chapter 8). Whereas starving gill-breathers are able to reduce

Figure 4.4: Effects of Depth of Water on Surfacing Frequency (●—●) and Feeding Rate (○—○) of the Obligatory Air-breathing Fish *Channa striatus* (0.75 g) (Pandian & Vivekanandan, 1976; Vivekanandan, 1977).

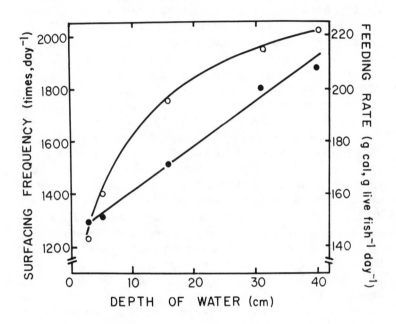

oxygen consumption to a very low level (0.11 ml O_2 g^{-1} h^{-1}: *Tilapia mossambica* starved for 20 days at 28°C; Mathavan *et al.*, 1976), obligatory air-breathers are in the disadvantageous position of not being able to reduce their metabolic rate below 0.25 ml O_2 g^{-1} h^{-1} (*C. striatus* starved for 21 days at 27°C; Vivekanandan, 1976), as they have to surface (e.g. *C. striatus*, 550 times day^{-1}) for survival.

The impact of depth of water on surfacing frequency and feeding rate diminishes as the fish grows. Vivekanandan (1977) observed that the surfacing frequency and feeding rate of *C. striatus* are depth dependent in the weight classes between 0.75 and 20 g (see also Arunachalam *et al.*, 1976). The larger weight classes (> 20 g) effectively minimize their surfacing frequency at a given depth of water (at 31 cm depth, the surfacing frequency of 0.75 and 41 g *C. striatus* were 1798 and 512 times day^{-1}, respectively). This is because (1) the larger fishes have relatively lower metabolic rate than smaller ones, and (2) they may be capable of 'gulping' more oxygen per surfacing than the smaller ones.

Figure **4.5**: (a) Effect of Imposed Ration Level on Surfacing Frequency of *C. striatus* (■, Vivekanandan, 1976) and *Polyacanthus cupanus* (●, Ponniah & Pandian, 1977). (b) Effects of Imposed Surfacing Frequency on Feeding Rate of *Channa striatus* Exposed to Different Depths of Water (Pandian & Vivekanandan, 1976).

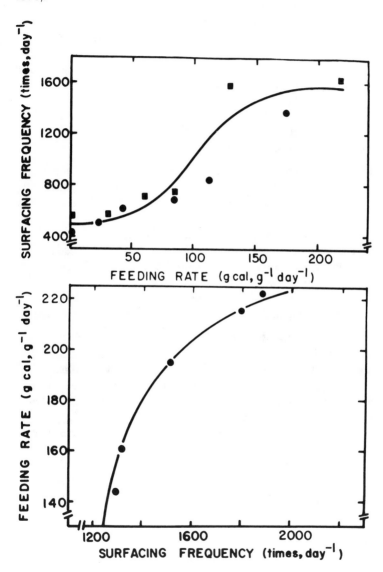

4.7 Digestion of Food

The laborious process of foraging in herbivores and detritivores (as discussed in section 4.5) is followed by an equally laborious digestive process, which necessitates morphological adaptations of the intestine. The intestine is longest in herbivores, being 7–20 times the body length (Bryan, 1975). In detritivores such as *Mugil cephalus*, it is 3.7 times the body length (Odum, 1973). The surface area of the intestine in carnivorous teleosts is usually increased by folds of the mucosa but the intestine length is less than the body length.

4.7.1 Structural and Secretory Adaptations

Some of the obstacles experienced by the herbivores and detritivores in digesting their food are:

(1) the fibres consisting of cellulose, hemicellulose, and lignin are not digestible and these prevent access of intestinal enzymes to the cytoplasmic contents of plant material (see, for example, Van Dyke & Sutton, 1977);
(2) inhibitors present in the plant substance nullify the enzyme activity; for instance, a powerful trypsin inhibitor present in the filamentous green alga *Chaetomorpha brachygona* inactivates the trypsin of *Chanos chanos* (Benitez & Tiro, 1982);
(3) excessive amounts of inorganic material in the detritus offer no energy value and reduce the efficiency of the digestive process.

In an attempt to overcome these obstacles, the herbivores and detritivores have developed structural and/or secretory adaptations (Table 4.3). The presence of spiral folds in the oesophagus, which increases the surface area, and numerous mucous glands enhance digestive activity in the milkfish *Chanos chanos*. Besides, *C. chanos* has strong protease activity in the intestine. *Ctenopharyngodon idella* and *Tilapia rendalli* have developed strong pharyngeal teeth to crush cell walls and release the cytoplasmic contents of plants. In certain species like *T. mossambica*, the pH of stomach fluid is as low as 1.0 when digestion is in progress (Moriarty, 1973; Payne, 1978). This is 50% less than the pH of fluids in the stomachs of actively digesting carnivorous fish (Barrington, 1957).

4.7.2 Micro-organisms and Absorption Efficiency

In addition to these structural and secretory adaptations, micro-organisms attached to the plants and bacteria adsorbed on the detritus

Table 4.3: Digestive Adaptations of Herbivorous and Detritivorous Fishes.

Species	Adaptation		Reference
	Structural	Functional	
Herbivores			
Carassius auratus	Spiral folds in oesophagus increase surface area	Digestive activity in oesophagus	Sarbahi, 1951
Chanos chanos		Protease activity in oesophagus; tryptic activity in intestine, when fed on unicellular alga	Benitez & Tiro, 1982
Ctenopharyngodon idella	Well developed pharyngeal teeth	Digestive activity in oesophagus; tryptic activity in intestine	Stronganov & Buzinova, 1969; Van Dyke & Sutton, 1977; Hsu & Wu, 1979
Haplochromis nigripinnis		Gastric acidification	Moriarty, 1973
Tilapia mossambica		Gastric acidification	Bowen, 1981
		Amylase activity in all parts of alimentary canal; tryptic activity in intestine	Nagase, 1964
T. nilotica		Gastric acidification	Moriarty, 1973
T. rendalli	Well developed pharyngeal teeth	Gastric acidification	Caulton, 1976
Detritivores			
Cyprinus carpio	Buccal sorting of inorganic material	Digestive activity in oesophagus	Kawai & Ikeda, 1971; Moriarty, 1976
Mugil cephalus	Gastric grinding		Payne, 1978

Table 4.4: Absorption Efficiency of Fishes Belonging to Different Trophic Levels.

Species	Food	Absorption efficiency of fishes (%)	Reference
Carnivores			
Petromyzon marinus	blood	97	Farmer *et al.*, 1975
Anabas scandens	goat liver	98	Vivekanandan *et al.*, 1977
Channa striatus	tilapia	98	Vivekanandan, 1977
Megalops cyprinoides	prawn	97	Pandian, 1967
Lepomis macrochirus	mealworm	97	Gerking, 1955
Pleuronectes platessa	whiteworm	97	Edwards *et al.* 1969
Paralichthys dentatus	whiteworm	97	Peters & Angelovic, 1971
Micropterus salmoides	emerald shiners	97	Beamish, 1972
Tilapia mossambica	goat liver	95	Mathavan *et al.*, 1976
Polyacanthus cupanus	mosquito larva	92	Ponniah & Pandian, 1977
Heteropneustes fossilis	goat liver	90	Arunachalam *et al.*, 1976
Salmo gairdneri	tubifex	85	Brocksen & Bugge, 1974
Herbivores			
Anabas scandens	alga	88	Vivekanandan *et al.*, 1977
Tilapia mossambica	alga	79	Mathavan *et al.*, 1976
Holacanthus bermudensis	alga	72	Menzel, 1959
Ctenopharyngodon idella	macrophyte	70	Haniffa & Venkatachalam, 1980
C. idella	duckweed	61	Van Dyke & Sutton, 1977
Tilapia rendalli	macrophyte	58–48	Caulton, 1978
T. zilli	macrophyte	45	Buddington, 1979
T. nilotica	alga	45	Moriarty & Moriarty, 1973
Cebidichthys violaceus	macrophyte	31–52	Edwards & Horn, 1982
Detritivores			
T. mossambica	detritus	42	Bowen, 1979
Dorosoma cepedianum	detritus	42	Pierce, 1977

not only supplement the nutrient value of the otherwise low-energy-yielding food but also facilitate digestion. Death of the micro-organisms and bacteria lead to the release of their cytoplasm in the gut of the herbivores and detritivores and alter their digestion rate (Pandian, 1975). Mathavan *et al.* (1976) also observed that supple-mentation of animal matter increased the digestive fraction of algal food and hence the absorption efficiency of *T. mossambica*. When fed exclusively on plants or detritus, the absorption efficiency of herbivores or detritivores was lower than that recorded for carnivores (Table 4.4). A herbivore like *T. mossambica*, when offered animal, plant or detrital feed, exhibited an absorption efficiency of 95, 79 or 42%, respectively. The higher absorption efficiency exhibited by a herbivore, when fed exclusively on an animal diet, and the high protein requirement (Table 4.1) suggest that animal matter is essential for herbivorous and detritivorous fishes, and that these fishes neither will nor can consume and absorb a sufficient quantity of plant/detrital material to meet their metabolic energy demands (see also Menzel, 1959; Kitchell & Windell, 1970).

4.8 Gastric Evacuation

4.8.1 Quality of Digestion

The fact that herbivores could absorb only about 62% (x, $n = 9$, Table 4.4) of plant material invokes the obvious doubt about the quality of digestion in these fishes. Moriarty & Moriarty (1973) studied the passage of food through the alimentary canal of *Tilapia nilotica* and *Haplochromis nigripinnis*, which feed continuously throughout the day. Only part of the ingested plant material is retained in the stomach, and much of the meal is passed straight into the intestine and is poorly absorbed despite the structural and/or secretory adaptations developed by these fishes. In contrast, predatory fishes, which consume large items of food, gradually erode the outer layers of the bolus and continuously pass food into the intestine for further digestion and absorption (Fänge & Grove, 1979). The difference between the digestive mechanism of these two groups of fishes is reflected in the gastric evacuation time. Fänge & Grove (1979) have reviewed the gastric evacuation rates of a number of fish species. From their tabulation, it is apparent that in a temperature range of 22–5°C, herbivores evacuate 100% of the maximum plant food consumed in about 6 h (range, 3–10 h) and the carnivores evacuate their stomach contents in about 22 h (range, 6–48 h). Any factor which reduced the food retention time in the gastrointestinal tract reduces the efficiency of absorption. Due to improper and/or partial digestion, the herbivores retain food for a shorter duration in the stomach, which in turn, reduces the absorption efficiency.

Similarly, the digestive tract of the detritivorous *Mugil cephalus* is only capable of partially triturating plant detritus and blue–green algae. About 20-30% of the food is passed through the digestive tract in an undigested condition (Odum, 1973). In addition to this, the detritivore passes an enormous quantity of sediments of low energy value through its stomach. These diets are normally evacuated from the stomach most rapidly, in an attempt to maintain rates of energy turnover at a relatively constant level (Jobling, 1981).

4.8.2 Rates of Feeding and Gastric Evacuation

Since herbivores feed for 8–14 h in a day (Table 4.2), the stomach has to continuously initiate or accelerate peristalsis, which in turn may cause a rapid evacuation rate (Fänge & Grove, 1979). If the arrival of food is sufficient to do this, it is natural to suspect the maximum peristaltic activity and most rapid evacuation rate when the rate of feeding is high. But the results obtained by different authors for the relationship between quantity of food consumed and evacuation rate are contradictory. As mentioned earlier, Fänge & Grove (1979) have summarized the time (in hours) required for 100% evacuation of stomach and the respective feeding rates (% body weight day^{-1}) of different size groups of fishes for 22 species belonging to different trophic levels and temperature regimes. From this, the gastric evacuation rate of individual species is recalculated in terms of milligrams of food per gram live fish per hour and plotted in a log–log graph against feeding rate of the respective species (Figure 4.6). The gastric evacuation rate increased in proportion to the feeding rate. In this straight-line relationship, the gastric evacuation rates were 0.1, 0.5 and 5.0 mg g^{-1} h^{-1} for the feeding rates of 0.2, 1.0 and 10.0% body weight day^{-1}, respectively. Therefore the available data suggest that an important determinant of gastric evacuation rate in fishes is the feeding rate.

Factors that influence gastric evacuation rate may in turn affect appetite. For instance, the herbivores and detritivores rapidly empty their stomachs and forage continuously. The continuous foraging and handling of the plants and hard substrates increase the time cost of feeding. The feeding rate is influenced by the quality and quantity of food handled, in addition to factors like temperature. In turn it

Figure 4.6: Effect of Feeding Rate on Gastric Evacuation Rate of 22 Species of Fishes Exposed to Different Temperature and Feeding Regimes; data from the compilation of Fänge & Grove (1979) recalculated and plotted.

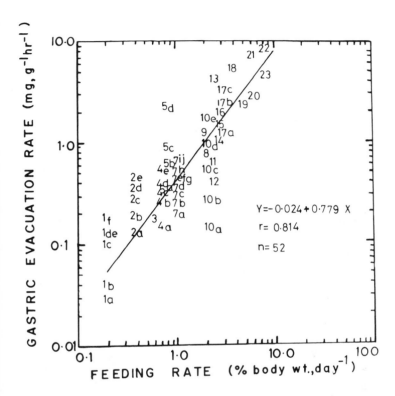

determines the gastric evacuation and completes a cyclic pattern of feeding and digestion (Figure 4.7).

As the digestion of food particles begins at its outer surface, the gastric evacuation rate should be higher for a meal consisting of small food particles than for one of similar size made up of a few large items (Jobling, 1980, 1981). The slow evacuation rate for fishes fed on larger food particles may nullify the higher profitability realized by the fishes selectively feeding on larger prey (Figure 4.3). Consequently, there is a narrow range of prey size which is optimum for realizing the highest profitability of feeding as well as the fastest gastric digestion rate.

Figure 4.7: Cyclic Pattern of Feeding and Digestion Energetics in Fishes.

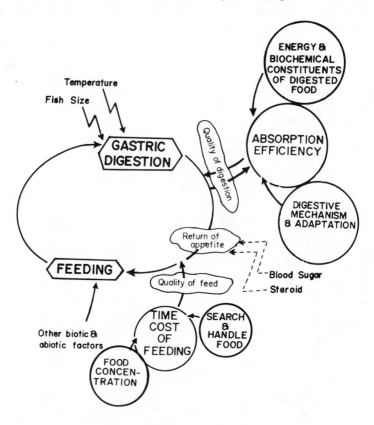

4.9 Concluding Remarks

High protein requirement and high efficiency for protein digestion are typical features of fish energetics. A few herbivorous fishes have developed special adaptations for feeding and digestion. From the point of aquaculture, a requirement for highly proteinaceous feed by fish may prove disadvantageous (Chapter 12). Yet aquaculture practice has been used for centuries; a survey indicates that most of the fishes successfully cultured are herbivores like carps and mullets. Does this mean that only a natural habitat can support carnivorous fishes? Before answering this question, much needs to be done to understand the theoretical and practical aspects of profitability in terms of capturing and handling feed, and the optimum ranges for the size and density of food organisms to minimize the energy cost of feeding and digestion, and to maximize energy channelled for growth.

References

Andrews, J.W. and Stickney, R.R. (1972), Interactions of feeding rates and environmental temperature on growth, food conversion and body composition of channel catfish. *Trans. Am. Fish. Soc. 101*, 94

Arunachalam, S., Vivekanandan, E. and Pandian, T.J. (1976), Food intake, conversion and swimming activity in the air-breathing catfish *Heteropneustes fossilis. Hydrobiologia 51*, 213

Ashley, L.M. (1972), Nutritional pathology. In: Halver, J.E. (ed.). *Fish Nutrition*, pp. 439–537. New York, Academic Press

Barrington, E.J.W. (1957), The alimentary canal and digestion. In: Brown, M.E. (ed.). *The Physiology of Fishes*. vol. I, pp. 109–61. New York, Academic Press

Beamish, F.W.H. (1972), Ration size and digestion in largemouth bass, *Micropterus salmoides* Lacepede. *Can. J. Zool. 50*, 153

Belliyappa, G., Krishnan, N. and Reddy, S.R. (1983), Effects of body size on the rate and pattern of ammonia excretion in an air-breathing fish. *Proc. Indian Acad. Sci. 92*, 31

Benitez, L.V. and Tiro, L.B. (1982), Studies on the digestive proteases of the milkfish *Chanos chanos. Mar. Biol. 71*, 309

Blaxter, J.H.S. (1965), The feeding of herring larvae and their ecology in relation to feeding. *Rep. Calif. Coop. Oceanic Fish. Invest. 10*, 79

Boehlert, G.W. (1981), Ontogenetic changes in growth and their relationship with temperature and habitat change. In: Cailliet, G.M. and Simenstad, C.A. (eds). *Gutshop '81*, pp. 115–23. Seattle, Washington Sea Grant Publication

Bowen, S.H. (1979), A nutritional constraint in detritivory by fishes. The stunted population of *Sarotherdon mossambicus* (Peters) in Lake Sibaya, South Africa. *Ecol. Monogr. 49*, 17

Bowen, S.H. (1981), Digestion and assimilation of periphytic detrital aggregate by *Tilapia mossambica. Trans. Am. Fish. Soc. 110*, 239

Brett, J.R. (1971), Satiation time, appetite, and maximum food intake of sockeye salmon, *Orcorhynchus nerka. J. Fish Res. Bd Can. 28*, 1635

Brett, J.R. and Groves, T.D.D. (1979), Physiological energetics. In: Hoar, W.S., Randall, D.J. and Brett, J.R. (eds). *Fish Physiology*, vol. VIII, pp. 279–352. New York, Academic Press

Brett, J.R., Shelbourn, J.E. and Shoop, C.T. (1969), Growth rate and body composition of fingerling sockeye salmon, *Oncorhynchus nerka*, in relation to temperature and ration size. *J. Fish. Res. Bd Can. 26*, 2363

Brocksen, R.W. and Bugge, J.P. (1974), Preliminary investigations on the influence of temperature on food assimilation by rainbow trout *Salmo gairdneri* Richardson. *J. Fish Biol. 6*, 93

Bryan, P.G. (1975), Food habits, functional digestive morphology, and assimilation efficiency of the rabbitfish *Siganus spinus* (Pisces, Siganidae) on Guam. *Pac. Sci. 29*, 269

Buddington, R.K. (1979), Digestion of an aquatic macrophyte by *Tilapia zilli. J. Fish Biol. 15*, 449

Carline, R.F. and Hall, J.D. (1973), Evaluation of method for estimating food consumption rates of fish. *J. Fish. Res. Bd Can. 30*, 623

Caulton, M.S. (1976), The importance of pre-digestive food preparation to *Tilapia rendalli* Boulenger when feeding on aquatic macrophytes. *Trans. Rhod. Scient. Ass.* 57, 22

Caulton, M.S. (1978), Tissue depletion and energy utilization during routine metabolism by sub-adult *Tilapia rendalli. J. Fish Biol. 13*, 1

Cohen, E.G. and Grosslein, M.D. (1981), Food consumption by silver hake (*Merluccius bilinearis*) on Georges Bank with implications for recruitment. In: Cailliet, G.M. and Simenstad, C.A. (eds). *Gutshop '81*, pp. 97–105. Seattle, Washington Sea Grant Publication

Cowey, C.B. (1980), Protein metabolism in fish. In: Buttery, P.J. and Lindsay, D.B. (eds). *Protein Deposition in Animals*, pp. 271–88. London, Butterworths

Cowey, C.B., Pope, J.A., Adron, J.W. and Blair, A. (1972), Studies on the nutrition of marine flatfish. Growth of plaice, *Pleuronectes platessa*, on diets containing proteins derived from plants and other sources. *Mar. Biol. 10*, 145

Crowder, L.B. (1981), Species interactions and community structure of fishes in Lake Michigan. In: Cailliet, G.M. and Simenstad, C.A. (eds). *Gutshop '81*, pp. 151–7, Seattle, Washington Sea Grant Publication

Dabrowski, H. (1977), Studies on the utilization by rainbow trout (*Salmo gairdneri*) of feed mixtures containing soya bean meal and an addition of amino acids. *Aquaculture 10*, 297

Davis, A.T. and Stickney, R.R. (1978), Protein requirements of *Tilapia aurea. Trans. Am. Fish. Soc. 107*, 479

De Long, D.C., Halver, J.E. and Mertz, E.T. (1958), Protein requirements of Chinook salmon, *Oncorhynchus tschawytscha. J. Nutr. 65*, 589

De Silva, S.S. and Perera, P.A.B. (1976), Studies on the young grey mullet, *Mugil cephalus* L. I. Effects of salinity on food intake, growth and food conversion. *Aquaculture 7*, 327

Dupree, H.K. and Sneed, K.E. (1966), Response of channel catfish fingerlings to different levels of major nutrients in purified diets. *Tech. Pap. U.S. Bur. Sport Fish. Wildl. 9*, 1

Durbin, A.D., Durbin, E.G., Verity, P.G. and Sinayda, T.J. (1981), Voluntary swimming speeds and respiration rates of a filter feeding planktivore, the Atlantic menhaden, *Brevoortia tyrannus* (Pisces: Clupeidae). *Fish. Bull. 78*, 877

Edwards, R.R.C., Finlayson, S.M. and Steele, J.H. (1969), The ecology of O-group plaice and common dabs in Loch Ewe. II. Experimental studies of metabolism. *J. exp. mar. Biol. Ecol. 3*, 1

Edwards, T.W. and Horn, M.H. (1982), Assimilation efficiency of a temperate zone intertidal fish (*Cebidichthys violaceus*) fed diets of macroalgae. *Mar. Biol. 67*, 247

Eggers, D.M. (1982), Planktivore preference by prey size. *Ecology 63*, 381

Elliott, J.M. (1975), The growth rate of brown trout, *Salmo trutta* L., fed on maximum rations. *J. Anim. Ecol. 44*, 805

Elliott, J.M. and Persson, L. (1978), The estimation of daily rates of food consumption for fish. *J. Anim. Ecol. 47*, 977

Fänge, R. and Grove, D. (1979), Digestion. In: Hoar, W.S., Randall, D.J. and Brett, J.R. (eds). *Fish Physiology*, vol. VIII, pp. 162–260. New York, Academic Press

Farmer, G.J., Beamish, F.W.H. and Robinson, G.H. (1975), Food consumption of the adult landlocked sea lamprey, *Petromyzon marinus* L. *Comp. Biochem. Physiol. 50*, 753

Fischer, Z. (1972), The elements of energy balance in grass carp (*Ctenopharyngodon idella* Val.). Part II. Fish fed with animal food. *Polish Arch. Hydrobiol. 19*, 65

Frame, D.W. (1973), Conversion efficiency and survival of young winter flounder *Pseudopleuronectes americanus* under experimental conditions. *Trans. Am. Fish. Soc. 102*, 614

Gerking, S.D. (1955), Influence of rate of feeding and body composition on protein metabolism of bluegill sunfish. *Physiol. Zool. 28*, 267

Goldman, C.R. and Kimmel, B.L. (1978), Biological processes associated with suspended sediment and detritus, in lakes and reservoirs. In: Cairns, J. Jr, Benfield, E.F. and Webster, J.R. (eds). *Current Perspectives on River–Reservoir Ecosystems*, pp. 19–44. Virginia, North American Benthological Society

Haniffa, M.A. and Venkatachalam, V. (1980), Effects of quality of food energy budget and chemical composition of the grass carp *Ctenopharyngodon idella. Proc. Int. Symp. Conservation Inputs from Life Sciences*, Univ. Jelangasaan, Malaysia, pp. 163–73

Hatanaka, M. and Murakawa, G. (1958), Growth and food consumption in young amberfish *Seriola quinqueradiata* (T et S). *Tohoku J. Agric. Res. 9*, 69

Hatanaka, M., Kosaka, M. and Sato, Y. (1956a), Growth and food consumption in young plaice. Part I. *Limanda yokohamae* (Gunther). *Tohoku J. Agric. Res. 7*, 151

Hatanaka, M., Kosaka, M. and Sato, Y. (1956b), Growth and food consumption in plaice. Part II. *Kareius bicoloratus* (Basilevsky). *Tohoku J. Agric. Res. 7*, 163

Hatanaka, M., Sekino, K., Takahashi, M. and Ichimura, T. (1957), Growth and food consumption in young mackerel, *Pneumatophorus japonicus* (Houttuyn). *Tohoku J. Agric. Res. 8*, 351

Healey, M.C. (1972), Bioenergetics of a sand goby (*Gobius minutus*) population. *J. Fish. Res. Bd Can. 29*, 187

Hofer, R. and Schiemer, F. (1981), Proteolytic activity in the digestive tract of several species of fish with different feeding habits. *Oecologia* (Berl.) *48*, 342

Hopkins, T.L. and Baird, R.C. (1981), Trophodynamics of the fish *Valenciennellus tripunctatus*. I. Vertical distribution, diet and feeding chronology. *Mar. Ecol. Prog. Ser. 5*, 1

Hsu, Y.L. and Wu, J.L. (1979), The relationship between feeding habit and digestive proteases of some freshwater fishes. *Bull. Inst. Zool. Acad. Sinica 18*, 45

Ishiwata, N. (1968), Ecological studies on the feeding of fishes. IV. Satiation curve. *Bull. Jpn Soc. Sci. Fish. 34*, 691

Janssen, J. (1976), Feeding modes and prey size selection in the alewife (*Alosa pseudoharengus). J. Fish. Res. Bd Can. 33*, 1972

Jobling, M. (1980), Gastric evacuation in plaice, *Pleuronectes platessa* L.: Effects of dietary energy level and food consumption. *J. Fish Biol. 17*, 187

Jobling, M. (1981), Mathematical models of gastric emptying and the estimation of daily rates of food consumption for fish. *J. Fish Biol. 19*, 245

Jobling, M. (1983a), Effect of feeding frequency on food intake and growth of arctic charr, *Salvelinus alpinus* L. *J. Fish Biol. 23*, 177

Jobling, M. (1983b), Towards an explanation of specific dynamic action (SDA). *J. Fish Biol. 23*, 549

Johnson, L. (1966), Experimental determination of food consumption of pike *Esox lucius* for growth and maintenance. *J. Fish Res. Bd Can. 23*, 1495

Jones, B.C. and Geen, G.H. (1977), Food and feeding of spiny dogfish (*Squalus acanthias*) in British Columbia Waters. *J. Fish. Res. Bd Can. 34*, 2067

Just, A. (9180), Influence of dietary composition on site of absorption and efficiency of utilization of metabolizable energy in growing pigs. In: Mount, L.C. (ed.) *Energy Metabolism*, pp. 27–30. London, Butterworths

Kapoor, B.G., Smit, H. and Verighina, I.A. (1975), The alimentary canal and digestion in teleosts. *Adv. Mar. Biol. 13*, 109

Kawai, S. and Ikeda, S. (1971), Studies on digestive enzymes of fishes. I. Carbo-hydrases in digestive organs of several fishes. *Bull. Jpn Soc. Sci. Fish. 37*, 333

Kitchell, J.F. and Windell, J.T. (1970), Nutritional value of algae to bluegill sunfish *Lepomis macrochirus. Copeia, 1970*, 186

Lee, D.J. and Putnam, G.B. (1973), The response of rainbow trout to varying protein, energy rations in a test diet. *J. Nutr. 103*, 916

Legand, M. and Rivation, J. (1969), Cycles biologiques des poissons mésopelagiques dans l'ouest de l'océan Indien. *Oceanographic 7*, 29

Leong, R.J.H. and O'Connell, C.P. (1969), A laboratory study of particulate and filter feeding of the northern anchovy (*Engraulis mordax*). *J. Fish. Res. Bd Can. 26*, 557

Love, R.M. (1980), *The Chemical Biology of Fishes*, p. 943. London, Academic Press

Magnuson, J.J. (1969), Digestion and food consumption by skipjack tuna *Katsuwonus pelamis. Trans. Am. Fish. Soc. 98*, 379

Marian, M.P., Ponniah, A.G., Pitchairaj, R. and Narayanan, M. (1982), Effect of feeding frequency on surfacing activity and growth in the air-breathing fish, *Heteropneustes fossilis. Aquaculture 26*, 237

Mathavan, S., Vivekanandan, E. and Pandian, T.J. (1976), Food utilization in the fish *Tilapia mossambica* fed on plant and animal foods. *Helgolander wiss. Meeresunters. 28*, 66

Mathavan, S., Muthukrishnan, J. and Hellenal, G.A. (1980), Studies on predation on mosquito larvae by the fish *Macropodus cupanus. Hydrobiologia 75*, 255

Menzel, D.W. (1959), Utilization of algae for growth by the angelfish *Holacanthus bermudensis. J. Cons. perm. int. Explor. Mer. 24*, 308

Minton, J.W. and McLean, R.B. (1982), Measurements of growth and consumption of sauger (*Stizostedion canadense*): implication for fish energetics studies. *Can. J. Fish. Aquat. Sci. 39*, 1396

Mittelbach, G.G. (1981), Foraging efficiency and body size. A study of optimal diet and habitat use by bluegills. *Ecology 62*, 1370

Montgomery, W. (1980), Comparative feeding ecology of two herbivorous damselfishes (Pomacentridae: Teleostei) from the Gulf of California. *J. exp. mar. Biol. Ecol. 47*, 9

Moriarty, D.J.W. (1973), The physiology of digestion of bluegreen algae in the cichlid fish *Tilapia nilotica. J. Zool. (Lond.) 171*, 25

Moriarty, D.J.W. (1976), Quantitative studies on bacteria and algae in the food of the mullet *Mugil cephalus* L. and the prawn *Metapenaeus bennettae* (Racek and Dall). *J. exp. mar. Biol. Ecol. 22*, 131

Moriarty, D.J.W. and Moriarty, C.M. (1973), The assimilation of carbon from phytoplankton by two herbivorous fishes: *Tilapia nilotica* and *Haplochromis nigripinnis. J. Zool. (Lond.) 171*, 41

Nagase, G. (1964), Contribution to the physiology of digestion in *Tilapia mossambica*: digestive enzymes and the effects of diets on their activity. *Zeit. Vergl. Physiol. 49*, 270

Nelson, J.S. (1976), *Fishes of the World*, p. 416. New York, Wiley

Nose, T. and Arai, S. (1972), Protein requirements of the eel *Anguilla japonica. Bull. Freshwater Fish. Res. Lab. Tokyo 22*, 145

Nursall, J.R. (1981), The activity budget and use of territory by a tropical blennid fish. *Zool. J. Linn. Soc. 72*, 69

Odum, W.E. (1973), Utilization of the direct grazing and plant detritus food chains by the striped mullet *Mugil cephalus*. In: Steele, J.H. (ed.). *Marine Food Chains*, pp. 222–40, Edinburgh, Oliver & Boyd

Ogino, C. and Saito, K. (1970), Protein requirements of the carp *Cyprinus carpio. Bull. Jpn Soc. Scient. Fish. 36*, 250

Pandian, T.J. (1967), Intake, digestion, absorption and conversion of food in the fishes *Megalops cyprinoides* and *Ophiocephalus striatus. Mar. Biol. 1*, 16

Pandian, T.J. (1975), Mechanism of heterotrophy. In: Kinne, O. (ed.) *Marine Ecology*, vol. II, *Physiological Mechanisms*, Part 1, pp. 61–249. London, Wiley

Pandian, T.J. and Vivekanandan, E. (1976), Effect of feeding and starvation on growth and swimming activity in an obligatory air-breathing fish. *Hydrobiologia* *49*, 33

Parsons, T.R. and LeBrasseur, R.J. (1973), The availability of food to different trophic levels in the marine food chain. In: Steele, J.H. (ed.) *Marine Food Chains*, pp. 325–43. Edinburgh, Oliver & Boyd

Payne, A.I. (1978), Gut pH and digestive strategies in estuarine grey mullet (Mugilidae) and *Tilapia* (Cichlidae). *J. Fish Biol. 13*, 627

Peters, D.S. and Angelovic, J.W. (1971), Effect of temperature, salinity and food availability on growth and energy utilization of juvenile summer flounder, *Paralichthys dentatus. Proc. III. Nat. Symp. Radioecology*, p. 545

Peters, D.S. and Boyd, M.T. (1972), The effect of temperature, salinity and availability of food on the feeding and growth of the hogchoker, *Trinectes maculatus* (Bloch & Schneider). *J. exp. mar. Biol. Ecol. 7*, 201

Pierce, R.J. (1977), Life history and ecological energetics of the gizzard shad. (*Dorosoma cepedianum*) in Acton Lake, Ohio, *Diss. Abs. Int.* 38

Pierce, R.J. and Wissing, T.E. (1974), Energy cost of food utilization in the bluegill (*Lepomis macrochirus*). *Trans. Am. Fish. Soc. 103*, 38

Pierce, R.J., Wissing, T.E. and Megrey, B.A. (1981), Aspects of the feeding ecology of gizzard shad in Acton Lake, Ohio. *Trans. Am. Fish. Soc. 110*, 391

Ponniah, A.G. and Pandian, T.J. (1977), Surfacing activity and food utilization in the air-breathing fish *Polyacanthus cupanus* exposed to constant pO_2. *Hydrobiologia 53*, 221

Popova, O.A. and Sytina, L.A. (1977), Food and feeding relations of eurasian perch (*Perca fluviatilis*) and pike perch (*Stizostedion luciperca*) in various waters of the U.S.S.R. *J. Fish. Res. Bd Can. 34*, 1559

Raghuraman, R. (1973), Rate and efficiency of conversion in the fish *Polyacanthus cupanus* fed different quantities of food. *Curr. Sci. 42*, 24

Ricker, W.E. (1946), Production and utilization of fish populations. *Ecol. Monogr. 16*, 373

Rosenthal, H. and Hempel, G. (1973), Experimental studies in feeding and food requirements of herring larvae (*Clupea harengus* L.) In: Steele, J.H. (ed.). *Marine Food Chains*, pp. 344–64. Edinburgh, Oliver & Boyd

Rumsey, G.L. and Ketola, H.G. (1975), Amino acid supplementation of casein diets of Atlantic salmon (*Salmo salar*) fry and of soyabean meal for rainbow trout (*Salmo gairdneri*) fingerlings. *J. Fish. Res. Bd Can. 32*, 422

Sabaut, J.J. and Luquet, P. (1973), Nutritional requirements of the gilthead bream *Chrysophrys aurata*. Quantitative protein requirements. *Mar. Biol. 18*, 50

Sarbahi, D.S. (1951), Studies of the digestive tracts and the digestive enzymes of the goldfish *Carassius auratus* (Linn.) and the largemouth blackbass, *Micropterus salmoides* (Lacepede). *Biol. Bull. mar. Biol. Lab., Woods Hole 100*, 244

Satia, B.P. (1974), Quantitative protein requirements of rainbow trout. *Prog. Fish Cult. 36*, 80

Singh, C.S. and Bhanot, K.K. (1970), Nutritive food values of algal feeds for common carp, *Cyprinus carpio* (Linnaeus). *J. Inland Fish Soc. India 2*, 121

Stilwell, C.E. and Konler, N.E. (1982), Food, feeding habits, and estimates of daily ration of the shortfin mako (*Isurus oxyrinchus*) in the Northwest Atlantic. *Can. J. Fish Ag. Sci. 39*, 407

Stroganov, N.G. and Buzinova, N.S. (1969), Enzymatic activity of the grass carp (*Ctenopharyngodon idella*) intestinal tract. II. Proteolytic enzymes. *Chem. Abs. 72*, 1078

Sumitra Vijayaraghavan, Royan, J.P. and Rao, T.S.S. (1982), Growth and food conversion efficiency in the fish *Etroplus suratensis* in relation to different feeding levels. *Indian J. Mar. Sci. 11*, 350

Swenson, W.A. and Smith, L.L. (1973), Gastric digestion, food consumption, feeding

124 Fish Energetics

periodicity and food conversion efficiency in walleye *Stizostedion vitreum vitreum. J. Fish. Res. Bd Can. 30*, 1327

Swenson, W.A. and Smith, L.L. (1976), Influence of food competition, predation and cannibalism on walleye (*Stizostedion vitreum vitreum*) and sauger (*S. canadense*) populations in Lake of the Woods, Minnesota. *J. Fish. Res. Bd Can. 33*, 1946

Tang, Y.A. and Hwang, T.L. (1966), Evaluation of the relative suitability of various groups of algae as food of milkfish produced in brackish water ponds. *Proc. World Symp. Warmwater Pond Fish Cult. 3*, 365

Tyler, A.V. (1970), Rates of gastric emptying in the young cod. *J. Fish. Res. Bd Can. 27*, 1177

Tyler, A.V. and Dunn, R.S. (1976), Ration, growth, and measures of somatic and organ condition in relation to meal frequency in winter flounder, *Pseudopleuronectes americanus*, with hypothesis regarding population homeostasis. *J. Fish. Res. Bd Can. 33*, 63

Vallet, F., Berhant, J., Leray, C., Bennet, B. and Pie, C. (1970), Preliminary experiments on the artificial feeding of Mugilidae. *Helgolander wiss. Meeresunters. 20*, 610

Van Dyke, J.M. and Sutton, D.L. (1977), Digestion of duckweed (*Lemna* spp.) by the grass carp (*Ctenopharyngodon idella*). *J. Fish. Biol. 11*, 273

Vivekanandan, E. (1976), Effects of feeding on the swimming activity and growth of *Ophiocephalus striatus. J. Fish Biol. 8*, 321

Vivekanandan, E. (1977), Surfacing activity and food utilization in the obligatory airbreathing fish *Ophiocephalus striatus* as a function of body weight. *Hydrobiologia 55*, 99

Vivekanandan, E., Pandian, T.J. and Visalam, C.N. (1977), Effects of algal and animal food combinations on surfacing activity and food utilization in the climbing perch *Anabas scandens. Polish Arch. Hydrobiol. 24*, 555

Wallace, J.C. (1973), Observations on the relationship between the food consumption and metabolic rate of *Blennius pholis. L. Comp. Biochem. Physiol. 45*, 293

Walters, V. (1966), On the dynamics of filter feeding by the wavyback tuna (*Euthynnus affinis*). *Bull. Mar. Sci. 92*, 245

Ware, D.M. (1972), Predation by rainbow trout (*Salmo gairdneri*): the influence of hunger, prey density, and prey size. *J. Fish. Res. Bd Can. 29*, 1193

Warren, C.E. and Davies, G.E. (1967), Laboratory studies on the feeding bioenergetics and growth of fishes. In: Gerking, S.D. (ed.). *The Biological Basis for Freshwater Fish Production*, pp. 175–214, Oxford, Blackwell

Werner, E.E. and Hall, D.J. (1974), Optimal foraging and the size selection of prey by the bluegill sunfish (*Lepomis macrochirus*). *Ecology 55*, 1

Williamson, G. and Payne, W.J.A. (1980), *An Introduction to Animal Husbandry in the Tropics*, p. 755. London, Longman

Winberg, G.G. (1961), Rate of metabolism and food requirements of fishes. Belorussian State Univ., Minsk (*Fish. Res. Bd Can. Trans. Ser. 1941*)

Windell, J.T (1967), Rates of digestion in fishes. In: Gerking, S.D. (ed.). *The Biological Basis for Freshwater Fish Production*, pp. 151–73. Oxford, Blackwell

Yoklavich, M. (1981), Growth, food consumption and conversion efficiency of juvenile English sole (*Parophrys vetulus*). In: Cailliet, G.H. and Simenstad, C.A. (eds). *Gutshop '81*, pp. 97–105. Seattle, Washington Sea Grant Publication

Zeitoun, I.H., Ullrey, D.E., Halver, J.E., Tack, P.K. and Magee, W.T. (1974), Influence of salinity on protein requirements of coho salmon (*Oncorhynchus kisutch*) smolts. *J. Fish. Res. Bd Can. 31*, 1145

5 LABORATORY METHODS IN FISH FEEDING AND NUTRITIONAL STUDIES

Clive Talbot

5.1 Introduction

The nutritional components and gross energy made available for cell maintenance, growth, locomotion and reproduction are determined by the amount of food consumed, the fraction that is assimilated and the nutrient content of the food. The relationship between fish and their food is affected by a complex interaction between a number of factors which include temperature, light, salinity, fish size, activity and behaviour, appetite, feeding regime, starvation, stress and type of food. These factors have been reviewed by Webb (1978), Braaten (1979), Brett (1979), Brett & Groves (1979) and Fänge & Grove (1979).

Most of our knowledge of the trophic dynamics of fish and the physiology of the gut and the digestive process has involved laboratory experiments. Mann (1978) and Windell (1978a,b) have reviewed the various general approaches which have been used to quantify the food requirements of fish per unit of time and to investigate the physiology of feeding and digestion. The three most important are:

(1) Direct measurement of the amount consumed in a specific meal or over a period of time.
(2) Measurement of the rate of gastric evacuation. Gastric evacuation rate (in terms of unit weight of food translocated in unit time) is a frequently measured parameter in physiological studies on the

factors affecting feeding and digestion. Daily food consumption may be estimated from such data on the assumption that the average rate at which a fish passes food out of its foregut must equal the average rate of consumption. Various mathematical models for estimating daily food consumption by this method have been reviewed by Elliott & Persson (1978), and Jobling (1981b).

(3) Balanced energy or nitrogen budgets (Chapter 12). Food consumption of fish may be estimated from their energy or nitrogen requirements, or the relationship between food consumption and growth. These methods assume that the nutrients ingested equal those lost in egestion and excretion, those retained in growth and those utilized in all metabolic processes. These various parameters are usually determined in the laboratory.

The consumption, evacuation and absorption of food are among the most important parameters measured in laboratory feeding experiments and are fundamental to the development of trophic models. The adequacy of these models depends on the adequacy of the experimental methods used. Although the biotic and abiotic factors that determine fish feeding and digestion have been extensively studied, little attention appears to have been paid to the way in which experimental design influences the results obtained. Experiments are often stressful to fish or impose unnatural or unrealistic feeding regimes. There is increasing evidence to suggest that different experimental procedures will produce widely differing results. The variety of methods used makes it difficult to compare the results obtained by different authors, even for the same species.

This chapter is largely concerned with a review of these methods, their limitations and their effects on fish, in an attempt to evaluate some of the implications of experimental design in fish feeding studies. Windell (1978b) provides a more detailed description of the computations associated with many of the methods considered here.

5.2 Food Consumption and Evacuation

The basis of almost all laboratory experiments designed to measure food consumption and gastric evacuation rates lies with following the fate of an *identifiable* meal. Numerous methods have been described.

5.2.1 Direct Observation of Feeding Activity

Many studies of consumption have been made by offering items of food of known weight and number to fish and observing how many are eaten. This measurement has been made by direct visual observation (Windell, 1966; Beukema, 1968; Windell *et al.*, 1969; Elliott, 1972; Grove *et al.*, 1978) or by recording feeding activity on film for subsequent analysis (Braum, 1978; Godin, 1981). The main disadvantage of direct observation is that only a small number of fish can be watched at any one time and each fish requires an individual tank. This problem can be overcome to some extent by filming a group of fish in the same tank, although the analysis of the films can be very laborious. When food consumption is measured by observation, the food is usually dispensed by hand. Fish may require a conditioning period before feeding regularly, and precautions must be taken to avoid startling the fish during the feeding period. Observation of food items eaten may be facilitated by using floating food.

An alternative to observing the amount eaten is to measure the difference between the quantity offered and that remaining at the end of the feeding period (Pentelow, 1939; Baldwin, 1957; Hunt, 1960; Brett *et al.*, 1969; Brett, 1971; Wallace, 1973). For this type of experiment, tanks either for groups of fish or for individuals require filters to trap uneaten food in the effluent water or a false mesh bottom under which the uneaten food collects. Tanks of this design also make it easier to count the number of food items eaten using direct observation.

These various methods are reliable and accurate, especially where natural food is used. Pelleted food is susceptible to breaking up in the water or in the mouths of fish, with subsequent loss of material. They do not, however, allow for the measurement of evacuation rate without further features of experimental design.

5.2.2 Examination of Gut Contents

Food consumption, evacuation rate and type of food eaten are frequently studied by quantifying the gut contents. Hyslop (1980) has reviewed the various methods of stomach contents analysis.

Serial Slaughter. A population of fish is fed to satiation and then sub-samples are killed immediately (to measure food consumption) and at intervals thereafter (to measure evacuation rate) (Magnuson, 1969; Brett & Higgs, 1970; Tyler, 1970; Jones, 1974; El-Shamy,

1976; Persson, 1979; Talbot & Higgins, 1982; De Silva & Owoyemi, 1983; Ryer & Boehlert, 1983). The fish are usually starved both pre- and post-prandially so that food eaten outside the experimental feeding period does not confuse the result. The removal of gut contents can be facilitated by freezing the fish beforehand. After partial thawing, the ingesta can be removed as a single mass with virtually no contamination by material derived from the gut itself (e.g. sloughed epithelium, mucus). The dissected gut can be divided into various regions and the evacuation of the stomach and subsequent filling and emptying of the hindgut can be followed (Talbot & Higgins, 1982).

The analysis of gut contents at autopsy suffers from the problem that it is wasteful of fish and the data are obviously collected from different fish at each sampling, which is less desirable than serial measurements on individual fish. The method is not suitable for very small fish due to the difficulty of reliable dissection. Corazza & Nickum (1983) refined the analysis of gut contents retrieved at autopsy by feeding walleye fingerlings (*Stizostedion vitreum vitreum*) with daphnids which had been coloured by immersion in an aqueous solution of National Fast Blue (2 g litre^{-1}). The use of dyed food enables the rate of evacuation of the test meal to be determined without the need to starve fish before or after the test meal is fed.

Sampling from Live Fish. The problem of having to kill fish can be overcome by removing stomach contents from live fish after they have ingested a meal. One of the earliest reported methods for removal of stomach contents involves inserting a glass tube into the oesophagus and forcing food into the tube by pressing the side of the fish (White, 1930). Seaburg (1957) described a stomach pump which consists of two copper tubes soldered together. Water is pumped through one tube by a rubber suction bulb fitted with one-way valves and the stomach contents are flushed down the other and collected in a specimen bottle. A smaller, simpler version is described by Strange & Kennedy (1981), which has only one tube inserted into the stomach. A similar one-tube apparatus is described by Foster (1977) and Light *et al.* (1983), but they used a pulsed flow of pressurized water from a hand-pumped compression reservoir rather than a suction bulb, and claim greater success with this modification. Giles (1980) described a stomach sampler which uses two syringes strapped together. Water is alternately forced into the stomach through a tube from one syringe and sucked out, along with food items, into the other syringe. Baker &

Fraser (1976) sampled the entire gut contents of small live fish (*Fundulus* sp.) by injecting water from a syringe through the alimentary tract by means of a tube inserted into the intestine at the anus. This method is not successful when used with fish which have distinct stomachs with strong sphincter muscles, or long and coiled intestines.

With appropriate regard for the anatomy of the gut, most of the flushing methods claim to recover at least 95% of the food items and with little effect on subsequent survival and growth. Meehan & Miller (1978), however, found differences in the efficiencies of removal of different types of food item and decreases in the efficiency with increasing fish size for three salmonid species. Hard-bodied prey, e.g. small crayfish (*Astacus* sp.) and cased caddis flies (Trichoptera), are not removed efficiently by stomach flushing.

Other alternatives to sacrificing fish in order to examine gut contents include injecting emetics such as arsenous acid and tartar emetic (local irritants) or apomorphine (which stimulates the vomiting centre) into the stomachs of fish (Markus, 1932; Jernejcic, 1969). The efficiency of recovery of stomach contents using emetics appears rather poor and is obviously stressful to the fish. Foster (1977) recorded significantly higher mortalities and delays in the resumption of feeding in fish given emetics compared with control fish and those subjected to pulsed gastric lavage. Wales (1962) reported the use of forceps to remove the stomach contents of live trout; 4 or 5 withdrawals were usually needed to empty the stomach. The use of forceps to assist the complete removal of stomach contents has been reported by several authors.

These various methods for collecting stomach contents of live fish are capable of measuring the amount of food consumed in a meal and the evacuation of food from the stomach. However, they do not permit the study of food passage along the intestine. As with serial slaughter, determination of gastric evacuation rate by serially sampling stomach contents using mechanical methods involves sub-sampling fish from a population. The short-term stress and possible injury caused to fish by these methods means that an individual fish should not be used more than once in a laboratory study if at all possible. Foster (1977) reported instances of tears at the junction of the stomach and oesophagus of grass pickerel (*Esox americanus vermiculatus*) sampled by stomach pump.

5.2.3 Radioisotope and Radiographic Methods

It is clear that all of the methods for sampling gut contents, from both live and dead fish, present some major difficulties and disadvantages. These problems have led several workers to develop techniques which allow the quantification of ingesta without the need to remove gut contents from the fish.

Food consumption has been measured by techniques based on the accumulation in the body of radioisotopes which are ingested as an integral part of the food (Davis & Foster, 1958; Kevern, 1966; Kolehmainen, 1974; Hakonson *et al.*, 1975). Food consumption is estimated from the amount of isotope (usually ^{137}Cs) that must be ingested to maintain a measured equilibrium body burden, given the concentration of isotope in the food, the fraction of consumed isotope that is absorbed and the fraction of the body burden eliminated per day. This method has been employed with fish from small ponds receiving low-level radioactive effluent which precludes its general practice. Although methods involving the turnover of radioisotopes have been mainly applied to field studies, they can be used for laboratory experiments, especially where food items are difficult to quantify. Farmer *et al.* (1975) measured food consumption of sea lampreys (*Petromyzon marinus*) by allowing them to feed on salmonids whose red blood cells had been labelled with ^{51}Cr. The loss of radioactivity from the host fish provided an estimate of the quantity of blood consumed by the parasite.

A major disadvantage of these radioisotope methods is that they measure food intake indirectly and depend on an extensive knowledge of the dynamics of the flow of isotope through the food chain under different environmental and physiological conditions. Fish have to be killed to measure the body burden of isotope. These methods are not generally suitable for short-term measurement of either food consumption or evacuation rate on individual fish.

Food consumption and evacuation rate have been measured more directly using food items labelled with an isotope that is biologically inert and poorly absorbed across the gut wall. Peters & Hoss (1974) labelled small fish and shrimps with radioactive cerium ($^{144}CeCl_3$). Pinfish (*Lagodon rhomboides*) were fed with the labelled diet and the gut of each fish was then repeatedly measured for radioactivity in a whole-animal scintillation counter. Cerium evacuation and serial slaughter methods provided similar estimates of evacuation time. A similar technique was reported by Storebakken *et al.* (1981).

Rainbow trout (*Salmo gairdneri*) were offered food labelled with ^{131}I (as Ag^{131}I) or ^{51}Cr (as ^{51}Cr$_2$O$_3$) and the radioactivity in the stomach and hindgut was measured separately using a gamma scintillation counter with a sodium iodide crystal detector. As the radioactive concentration of the food is known, the amount consumed and its subsequent evacuation can be measured directly from outside the fish.

These radioisotope methods have the advantage that fish do not need to be killed, and both food consumption and evacuation can be measured on the same fish although serial monitoring will cause stress to the fish. The food is virtually unaltered by the addition of isotope and will be eaten voluntarily. Fish need not be starved before or after the labelled meal is offered. The major disadvantage of all methods involving radioisotopes lies with the attendant hazards of handling radioactive substances. These hazards arise both with the preparation of the diet and with the disposal of radioactive waste water, food and fish.

X-Radiation presents fewer health and safety problems than do unsealed sources of gamma-radiation, and several radiographic methods have been reported for feeding studies of fish. Molnár & Tölg (1960) described a method for determining the evacuation rate of piscivorous fish which involves the radiographic visualization of structures (e.g. otoliths) of force-fed prey items in the intestine of the predator. More recently, several workers have fed fish with diets containing the radio-opaque compound barium sulphate. The rate of passage of this diet through the gut can be observed by taking radiographs serially (Edwards, 1971, 1973; Goddard, 1974; Jobling *et al.*, 1977; Grove *et al.*, 1978; Flowerdew & Grove, 1979; Gwyther & Grove, 1981; Ross & Jauncey, 1981). Barium sulphate is radio-opaque only when incorporated into food at high concentrations (e.g. 20% w/w) and the diet is usually force-fed. Barium sulphate itself does not appear to affect evacuation rate (Jobling *et al.*, 1977) and this method removes the need to kill fish. Fish are starved before and after the barium meal is fed. Force-feeding and serial X-raying, however, are both stressful procedures.

A radiographic method for studying trophic dynamics of fish has been described by Talbot & Higgins (1983) which incorporates the advantages of direct radioisotope methods but avoids the problem of handling isotopes. This method involves feeding fish on a diet labelled with metallic iron powder. The iron powder (particle size 100–200 μm) is added to finely ground commercial fish food at a concentration of 5% w/w. After thorough mixing, water is added (250 ml per kilogram of dry food) and the resulting paste is repelleted and dried. As the relationship between iron particle number and food weight is known, the quantity of ingesta in the gut of anaesthetized fish can be estimated accurately by counting the number of particles from the X-ray plate. This method obviates the need to force feed or sacrifice fish. As the labelled ingesta is readily distinguishable on the X-ray plate from other gut contents, fish do not need to be starved before or after the labelled food is presented, i.e. the normal feeding regime need not be changed. The position of the iron particles in the gut can be clearly seen on the X-ray plate, enabling the ingesta in different regions to be quantified (Figure 5.1). There is no evidence that the iron particles are evacuated differently from the other ingesta.

Figure 5.1: Typical Radiographs of Fish Showing Iron Particles in the Intestine. Four different fish are shown at various times after eating labelled food. Iron particles can be seen mainly in the stomach (A) and at different stages of distribution along the hindgut (B and D). E shows a fish which has evacuated all labelled food. The four fish shown ranged from 10.0 to 10.5 cm in length.

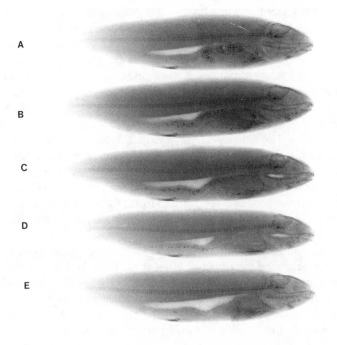

Assuming that the feeding period is less than the evacuation time, the number of iron particles present immediately after the feeding period provides an estimate of voluntary food consumption, and the rate of decrease in the number of iron particles provides a measure of evacuation rate. Large numbers of the fish can be fed at the same time in the same tank, and serial subsamples need not involve handling any fish more than once. Preparation of the labelled diet is easy and does not present the health hazards associated with radioisotope methods. The diet is readily eaten by fish and appears harmless to them. Juvenile Atlantic salmon (*Salmo salar*) fed once weekly over an 8-month period with labelled food did not show any significant difference in growth rate compared with control populations (P.J. Higgins, unpublished results). Fish of a wide range of sizes can be studied and the method has been used successfully on salmon from first feeding (0.3 g), to 2+ smolts (60 g).

5.2.4 Demand Feeders

A number of feeding studies have been made in which fish are self-trained to obtain food on demand by pressing a lever. This technique has been used with the stomachless goldfish, *Carassius auratus* (Rozin & Mayer, 1961, 1964) and blenny *Blennius pholis* (Grove & Crawford, 1980); flatfish, *Limanda limanda* (Gwyther & Grove, 1981) and rainbow trout, *Salmo gairdneri* (Adron *et al.*, 1973; Landless, 1976; Grove *et al.*, 1978). The number of actuations of the trigger per unit time can be recorded and this enables the voluntary feeding behaviour of fish to be studied in relation to environmental variables such as light, temperature, food quality (energy content, palatability, etc.) and feeding regime (trigger availability). Under conditions of constant light and temperature, free-running feeding rhythms are established which are related to the time required for digestive and absorptive processes (Adron *et al.*, 1973).

Statistical analysis of the actuations of the demand feeder can reveal the periodicity of the feeding rhythm. Each actuation of the demand feeder delivers a known quantity of food, enabling the quantity eaten in unit time to be measured. Any uneaten food must be removed and the record of the amount dispensed adjusted accordingly. The proportion of spurious actuations of the demand feeder by fish movement must be assessed as this may account for up to 10% of the actuations.

5.2.5 *Production of Faeces*

Several investigators have measured the evacuation time for the whole gut by monitoring the production of faeces. In an extensive study of 23 fish species, Lane & Jackson (1969) placed satiated fish into small aquaria and noted the presence or absence of faeces. For each trial, observations were made twice daily and voiding was judged to be complete when at least 2 of 4 aquaria were free of faecal matter. Using a demand feeder, Rozin & Mayer (1964) provided goldfish with food pellets dyed with carmine. Tanks were examined regularly and the colour of the faeces was noted. The observation of a piece of bright-red faeces indicated that the carmine previously ingested was being excreted in significant amounts. Blaxter (1963) fed herring larvae (*Clupea harengus*) with squid flesh and single meals of an orange-coloured copepod *Tigriopus*. Gut evacuation time was measured by observing the first signs of orange faeces. Laurence (1971) observed the passage of dyed plankters along the gut of transparent largemouth bass larvae (*Micropterus salmoides*).

5.3 Digestibility

Knowledge of the digestibility of dietary nutrients (i.e. protein, fat, carbohydrate, vitamins and minerals) is essential for the study of fish energetics and for the evaluation of different foodstuffs. During the time that food is in the gut of the fish, the nutrients are broken down (digested) by various enzymes into forms which can be absorbed across the gut wall and enter the blood system. Fänge & Grove (1979) have reviewed the biochemistry of enzyme digestion and nutrient absorption in fish. This usage of the term 'digestion' should not be confused with the process of food passage along the gut, which is more correctly described by the term 'evacuation'.

Digestibility coefficients for the various nutrients in food and for gross energy are determined by either a direct or an indirect quantitative measurement of the difference between the amounts ingested and subsequently egested. Since the digested portion itself is not determined for any of the nutrients, the term 'coefficient of apparent digestibility' is often used. The proximate composition of food and faeces is determined by chemical analysis, usually according to the prescribed methods of the Association of Official Analytical Chemists (AOAC, 1980). Direct methods involve measuring the total quantity of nutrient ingested and the amount of corresponding nutrient egested. Digestibility coefficient (D) is calculated as follows:

$$D\% = \frac{\text{Amount of nutrient ingested} - \text{amount of nutrient egested}}{\text{Amount of nutrient ingested}} \times 100$$

Direct digestibility measurements are difficult and time consuming to perform, largely because of the problems associated with the quantitative measurements of food consumption and faeces collection. Nevertheless, digestion coefficients for gross energy have been measured directly for perch, *Perca fluviatilis* (Solomon & Brafield, 1972); rainbow trout, *Salmo gairdneri* (Smith, 1971); brown trout, *Salmo trutta* (Elliott, 1976b) and salmon, *Oncorhynchus* spp. (Brett *et al.*, 1971).

Due to the problems of quantitative faeces collection, most digestibility studies on fish have used an indirect method which utilizes an indicator or reference substance in the diet. Indicators are referred to as external or internal, depending on whether the substance is added to the diet or is an integral naturally occurring component. Ideally, indicators should be totally indigestible and be evacuated at the same rate as the other gut contents. It is assumed that the amount of indicator ingested and egested is identical over equal periods of time. In addition, indicator substances should be readily determined chemically, should not alter the palatability of the test diet and should preferably be internal indicators (Kolb & Luckey, 1972; Maynard *et al.*, 1979). Internal indicator substances enable the digestibility of natural foodstuffs to be studied in both laboratory and field situations.

The digestibility of any given nutrient or nutrients is determined indirectly by calculating the ratio of the indicator concentration to nutrient concentration in the dry food and the same ratio in the dry faeces resulting from the test food.

$$\text{Digestibility} = 100 - \left(100 \; \frac{\%\ \text{indicator in food}}{\%\ \text{indicator in faeces}} \times \frac{\%\ \text{nutrient in faeces}}{\%\ \text{nutrient in food}} \right)$$

Total digestibility is calculated according to the formula:

$$\text{Total digestibility} = 100 - \left(100 \; \frac{\%\ \text{indicator in food}}{\%\ \text{indicator in faeces}} \right)$$

The advantage of the indirect method is that nutrient digestibility can be determined without the need to quantify either food intake or faeces output. Fish can be maintained under normal culture conditions and allowed to feed voluntarily. Large numbers of fish can be studied.

5.3.1 Indicators

Chromic oxide (Cr_2O_3) is the most commonly used external indicator substance for fish digestibility trials (Nose, 1960, 1967; Inaba *et al.*, 1962; Smith & Lovell, 1973; Cho *et al.*, 1974, 1976; Austreng, 1978; Lall & Bishop, 1979; Flowerdew & Grove, 1980; Jobling, 1981a). Chromic oxide is incorporated into formulated diets at a concentration of 2% or less and is usually assayed colorimetrically following acid digestion (Furukawa & Tsukahara, 1966), or by atomic absorption spectrophotometry in the same Kjeldahl digest as protein (Lied *et al.*, 1982). Bowen (1978) found that *Tilapia mossambica* is capable of selectively rejecting Cr_2O_3 from a diet of periphytic detrital aggregate.

Other dietary markers which have been used in fish digestibility studies include: ^{32}P as water-insoluble ammonium molybdate (Hirao *et al.*, 1960); silica (Hickling, 1966); hydrolysis-resistant organic matter, HROM (Buddington, 1979, 1980); hydrolysis-resistant ash, HRA (Bowen, 1981); crude fibre, CF (Tacon *et al.*, 1983); titanium (IV)-oxide and mineral elements (Lied *et al.*, 1982). Metallic iron powder has been suggested by Talbot & Higgins (1983) in conjunction with consumption and evacuation studies. Digestibility trials with mammals have also utilized polyethylene (Chandler *et al.*, 1964), $^{51}CrCl_3$ (Wightman, 1975), magnesium ferrite (Neumark *et al.*, 1974) and $^{144}CeCl$ (Ellis & Huston, 1968). De Silva & Perera (1983) investigated the digestibility of an aquatic macrophyte by the Asian cichlid *Etroplus suratensis* using the three internal markers, HROM, HRA and CF. They suggest that HROM is a better indicator substance than either CF or HRA.

5.3.2. Faeces Collection

For both direct and indirect digestibility studies of fish, it is essential that the faeces collected should represent quantitatively the undigested residue of the food consumed. This is difficult to achieve because of the aquatic environment. Faeces voided naturally into the aquarium water can be collected by fine dip net (Windell *et al.*, 1974), syphoning (Buddington, 1980), filtration columns (Ogino *et al.*, 1973) settling columns (Cho *et al.*, 1982) and mechanically rotating filter screens (Choubert *et al.*, 1982). Aquarium tanks are either flushed clear of uneaten food before faeces are collected or fish are fed in one aquarium and transferred to another for faecal collection. The main problem with collection of naturally voided faeces is leaching of nutrients leading to an overestimation of digestibility and, in small closed volumes of water, contamination by nitrogen compounds excreted across the fish's gills and in urine. Windell *et al.* (1978) found that storage of faeces in aquarium water for 1 h increased digestibility estimates for dry matter, protein and lipid by 11.5, 10.0 and 3.7%, respectively. Cho *et al.* (1982) point out that the major cause of leaching of naturally voided faeces is due to the break-up of faeces by physical handling. The settling column method described by Cho *et al.* (1982) claims to trap faeces within about 2 min of being voided by the fish with minimum break-up or leaching of faeces.

The general problem of leaching has led several investigators to develop methods for collecting faeces from the intestine before it is naturally expelled from the fish. Faeces can be collected by killing fish and dissecting the gut contents or by manual stripping from live fish where faeces are expelled by applying pressure to the abdominal cavity between the pelvic fins and the vent (Nose, 1960; Singh & Nose, 1967; Austreng, 1978; Windell *et al.*, 1978; Tacon *et al.*, 1983). Manual stripping may lead to the collection of incompletely digested food, resulting in an underestimation of digestibility, because of the difficulty of controlling the amount of faeces stripped. Austreng (1978) compared digestibility coefficients from faeces taken by dissection from the anterior and posterior areas of the rectum of rainbow trout. Higher digestibility coefficients were obtained in samples from the hindmost part of the rectum for all nutrients. Digestibility may be underestimated if stripped faeces are contaminated by body fluids or intestinal epithelium. Windell *et al.* (1978) described a suction method for removing the lowermost-formed faecal pellet from rainbow trout that allows more controllable sampling and reduces the possibility of contamination. These authors recommend intestinal dissection or anal suction as being preferable to manual stripping or collection of faeces from the aquarium water. Anal suction has the additional advantage that fish do not need to be killed; however, as with all manual faeces collection methods, some handling stress is inevitable. Handling may cause regurgitation or defaecation. Stripping faeces will unnaturally decrease hindgut contents, which in turn will affect gastric evacuation rate and subsequent food intake. Some appreciable time should elapse before fish are stripped again.

Table 5.1 shows a comparison of apparent digestibility coefficients in faeces collected by different methods from rainbow trout. Netting faeces from the aquarium water gave the highest digestibility coefficients, suggesting an overestimation due to leaching, whereas stripping of faeces gave the lowest values, suggesting contamination by body fluids and intestinal epithelium. There is a close agreement between the values obtained by the intestinal dissection, anal suction and settling column methods. The final choice of method will include factors such as the size of fish and ease of handling.

Irrespective of the method of faeces collection, it is important to feed the fish with the test diet containing indicator substances for a period which is longer than the evacuation time. Lied *et al.* (1982), for example, found that the level of indicator and the apparent digestibility of protein indicated that faeces derived from the test meal could not be obtained from the rectum of cod (*Gadus morhua*) until 72 h after the first feeding (water temperature 4–5°C).

5.4 Effects of Experimental Design on Feeding

5.4.1 Stress

Laboratory feeding experiments expose fish to both acute and chronic stress. Acute stress is usually short term, e.g. handling or force feeding, whereas chronic stress refers to long-term effects such as crowding or confinement. Regurgitation and defaecation have been mentioned previously as a reported response by some fish to acute stress. In addition, both acute and chronic stress have been shown to produce a variety of physiological changes. These include hyper-glycaemia; elevation in muscle and blood lactate, serum cortisol, skin mucus secretion and oxygen consumption; bradycardia; decrease in blood free fatty acid concentrations, protein synthesis and T_4 levels (for reviews see Love, 1980; Pickering, 1981). Many responses to acute stress occur within the first hour or so following the stressful experience, although the magnitude of the response and its duration are dependent on the species, nutritional status, severity of the stress and environmental temperature.

The effects of acute handling stress on feeding are variable. Transferring fish between tanks by dipnet abolished the feeding response in young coho salmon (*Oncorhynchus kisutch*) and rainbow trout for 4 to 7 days and 1 day, respectively (Wedemeyer, 1976) and brown trout for 3 days (Pickering *et al.*, 1982). Haddock (*Melano-*

Table 5.1: Comparison of Apparent Digestibility Coefficients in Faeces Collected by Different Methods from Rainbow Trout (*Salmo gairdneri*).

Method of faeces collection	Authors	Digestibility coefficient (% mean ± S.E.)		
		Dry matter	Crude protein	Crude lipid
Herring meal				
1 Dissection	Windell *et al.*, 1978	80.3 ± 1.0	90.2 ± 0.4	96.7 ± 0.5
2 Anal suction		79.1 ± 0.4	90.4 ± 0.1	97.4 ± 0.3
3 Netting		84.4 ± 0.1	94.6 ± 0.3	96.8 ± 0.2
4 Stripping		73.3 ± 1.6	77.5 ± 1.0	62.2 ± 5.1
5 Settling column	Cho *et al.*, 1982	80.2 ± 2.4	91.0 ± 0.8	97.3 ± 1.0
Dry pellet combination diet				
1 Dissection	Windell *et al.*, 1978	54.7 ± 1.0	80.5 ± 1.0	83.9 ± 0.5
2 Netting after:				
1 h		66.2 ± 0.7	90.5 ± 0.5	87.6 ± 0.3
8 h		69.7 ± 0.4	92.4 ± 0.5	90.2 ± 0.3
16 h		71.8 ± 0.5	90.5 ± 0.9	92.1 ± 0.2

grammus aeglefinus), cod (*Gadus morhua*) and whiting (*Merlangius merlangus*) would occasionally take food offered by hand within 1 h of their transfer to the experimental tank; however, most would not come to the surface to be hand-fed until the following day (Jones, 1974). Swenson & Smith (1973) found that the stomach evacuation rate of force-fed walleye (*Stizostedion vitreum vitreum*) was approximately half that for voluntarily feeding fish. Windell (1966) found that force-feeding bluegill sunfish (*Lepomis macrochirus*) increased the variation in evacuation rates, with both increases and decreases observed. Removal of stomach contents of largemouth bass (*Micropterus salmoides*) by stomach pumping did not appear to affect subsequent (once daily) feeding (Foster, 1977). Conflict for social dominance serves as a major stressor in the European eel, *Anguilla anguilla* (Peters 1982). The unavoidable confrontation with a dominant fish in a tank produced changes in a number of physiological and haematological parameters in subordinant fish. Stressed eels have soft, shrunken and translucent stomachs. The mucous epithelium of the stomach atrophies and cell-to-cell contact loosens. The gastric glands degenerate and sub-mucosal blood vessels contract.

It is clear that food consumption and evacuation in fish are affected by stress. Obviously, fish must be allowed to adapt to new holding or experimental facilities and every precaution must be taken to reduce chronic stress. Techniques that involve acutely stressful procedures, e.g. force-feeding or repeated handling, must be less desirable than more stress-free methods.

5.4.2 Feeding Regimes

The food intake of fish is controlled by systemic need (metabolic debt) and by the fullness of the foregut (Colgan, 1973). Systemic need rises with food deprivation but at a progressively decreasing rate as the fish reacts physiologically and behaviourally to conserve its resources. The amount of food in the foregut of a fish at any instant in time varies as a function of the rates of food ingestion and evacuation and these rates are simultaneous and interdependent. Voluntary food intake (appetite) is presumed to be zero when the foregut is full, regardless of systemic need, but is greater than zero with decreasing foregut content. Various authors have shown that the appetite of a fish is inversely related to foregut fullness (Brett & Higgs, 1970; Brett, 1971; Ware, 1972; Jones, 1974; Elliott, 1975a, b; Grove *et al.*, 1978; Grove & Crawford, 1980; Godin, 1981).

The relationship between foregut filling and emptying is obviously dependent on the feeding regime, including periods of starvation. For the purposes of this discussion, starvation is defined as a period of time following ingestion of the last meal which is in excess of the evacuation time for the whole gut. The choice of feeding regime has considerable bearing on the conclusion drawn from a feeding study. Many laboratory studies of ingestion and evacuation have employed starvation periods before and after the test meal is offered. In some cases this is a necessary consequence of the methods used, ensuring that the measured gut contents are not confused with ingesta originating from outside the test meal. This procedure is commonly adopted where meal size and evacuation rate are determined by an examination of gut contents from sacrificed or live fish. Periods of pre-prandial starvation may be variable and imposed in order to study the temporal development of appetite. Some experiments only employ post-prandial starvation. In this case, actively feeding fish are placed in food-free enclosures. Stomach contents are sampled at intervals by serial slaughter (Staples, 1975; Thorpe, 1977) or stomach pump (Swenson & Smith, 1973) and the data obtained for evacuation rate are used to estimate food consumption. Using a similar post-prandial starvation approach, Lane & Jackson (1969) measured the evacuation time from the whole gut of 23 species of fish by observing the production of faeces.

Starvation periods are known to affect the subsequent consumption of food and evacuation rate. A 7-day and 25-day fast was found to decrease the evacuation rate of bluegill sunfish (*Lepomis macrochirus*) by 22 and 51% respectively when compared to 2-day fasted fish (Windell, 1966). Elliott (1972) found that starvation periods up to 7 days prior to feeding brown trout (*Salmo trutta*) did not affect subsequent evacuation rate although periods of 10 days or more did significantly reduce the rate. However, Sarokon (1975, cited in Windell, 1978a) found that rainbow trout (*Salmo gairdneri*) starved for 3 and 6 days showed significantly slower evacuation rates than did fish starved for 18 h. A period of 4 and 6 days of resumed daily feeding was required before the rate returned to 'normal' in fish which had been starved for 3 and 6 days respectively. Jones (1974) found that the stomach evacuation rate of three species of gadoid was consistently slower in pre-prandially starved fish given a single meal than in fish given regular feeding. In contrast, long-term starvation (10–30 days) of plaice (*Pleuronectes platessa*) resulted in an increase in evacuation rate compared with control fish starved for 48 h (Goddard, 1974).

Persson (1979) acknowledged the effect of starvation on evacuation rate and attempted to minimize these effects by adopting a flexible pre-prandial starvation period related to temperature in a study of gastric evacuation in perch (*Perca fluviatilis*). Fish held at 4°C were starved for 96 h and fish at 21.7°C were starved for 12 h.

Post-prandial starvation also markedly affects the evacuation rate of a previously fed test meal. Largemouth bass larvae, *Micropterus salmoides* (Laurence, 1971), young yellow perch *Perca flavescens* (Noble, 1973) and fingerling walleye *Stizostedion vitreum vitreum* (Corazza & Nickum, 1983) have all been shown to evacuate a single meal of zooplankton almost twice as fast when the meal is followed immediately by an excess of food than when the fish are post-prandially starved. In the goldfish (*Carassius auratus*), ingestion of a second meal was found to stimulate egestion of the first meal (Rozin & Mayer, 1964). Elliott (1972), however, found that the rate of gastric evacuation was similar in brown trout (*Salmo trutta*) given one or three meals per day.

Pre-prandial starvation may lead to gorging when the test meal is fed. Brown (1951) compared young brown trout fry fed an excess ration 2, 4, 7 and 13 times per week. Fry fed only 2 or 4 times per week ate so much on each occasion that they bulged visibly for several hours afterwards. This was not observed with fish fed more frequently. Goldfish cannot consume a normal 24-h food intake when the duration of feeding is restricted to 1 h. However, the fish eat considerably more than they would normally eat in 1 h on a 24-h feeding regime. The food intake in 1 h is approximately the same when the fish have been deprived of food for between 4 and 47 h. This indicates that the capacity of the foregut may limit the short-term food intake (Rozin & Mayer, 1964). Brett (1971) fed juvenile sockeye salmon (*Oncorhynchus nerka*) a single meal every 11 or 22 h at 15°C. The stomach was virtually empty after each 22-h period, whereas approximately 10% of the previous meal was present in the stomach after each 11-h period. The mean amount of food consumed per 24 h (as per cent body weight) for the 22 h regime was 4.83 and 6.52% for the 11-h regime. The higher daily intake for the 11-h regime is made possible because it allows the greatest degree of appetite to develop for the least deprivation time. Fänge & Grove (1979) reviewed the reported effects of increasing meal size on stomach evacuation rate for a given size of fish. Both the time to evacuate a meal and the rate of evacuation (grams per hour) generally increase with increasing meal size.

Table 5.2: Feeding Regimes Used to Test the Effect (on Juvenile Atlantic Salmon) of Pre- and/or Post-prandial Starvation on the Amount of Food Consumed During a Test Meal and its Subsequent Evacuation.

Treatment	Pre-prandial period (48 h)[a]	Prandium (3 h)[b]	Post-prandial period (duration of evacuation)[a]	Abbreviation
A	Starve	Feed	Starve	SPS
B	Starve	Feed	Feed	SPF
C	Feed	Feed	Starve	FPS
D	Feed	Feed	Feed	FPF

[a] Normal food.
[b] Labelled food.

The complex interrelationships between feeding regime and fish size with food intake and evacuation rates were demonstrated by Talbot *et al.* (1984) for juvenile Atlantic salmon. Approximately 250 0+ sibling salmon parr were placed in each of four 1-m radial flow tanks (for tank design see Thorpe, 1981). The frequency distribution of fork lengths was bimodal and each tank contained approximately equal numbers of 'small' and 'large' fish (0.8–4.7 and 4.9–11.7 g, respectively). Four feeding regimes were involved, all of which had in common a 3-hour period (0900 to 1200) where iron particle labelled food (the prandium) was fed simultaneously to all treatments. The treatments, illustrated in Table 5.2, differed in respect of the presence or absence of pre- and/or post-prandial starvation. During all feeding periods, the fish were given food *ad libitum* (1 g every 10 min). At intervals after the end of feeding with labelled food, approximately 6 large and 12 small fish were taken from each treatment, anaesthetized, then X-rayed. The dry weight of the entire gut contents of each fish was determined by counting the number of particles from the X-ray plate and expressing this as a percentage of the dry fish weight. The square roots of the median gut contents for each size group at each sampling were regressed against time by the method of least squares, and the regressions were compared by analysis of variance. A square root transformation was used to linearize the data. The results of this experiment are shown in Figure 5.2 (A – D) and summarized briefly below:

(1) In fish of all sizes, pre-prandial starvation for 48 h causes a larger meal to be consumed when compared with pre-prandially fed fish. In addition, pre-prandial starvation results in relatively

Figure 5.2: The Effect of Starvation Periods on the Amount of a Test Meal Consumed by Juvenile Atlantic Salmon and its Evacuation Time. The various feeding regimes are shown in Table 5.2. The relationship between labelled gut contents (W) and time (T) was linearized by a square-root model ($\sqrt{W} = a + bT$) and regression parameters were fitted by the method of least squares. Regression lines shown separately for 'small' (———) and 'large' (– – – –) fish in A and B, and a common line (———) for both in C and D. Each data point represents the square root of the median value of the individual estimates, expressed as dry food per dry fish x 100 ($\sqrt{\% \ bw}$) (●, 0.8 to 4.7 g fish; ○, 4.9 to 11.7 g fish).

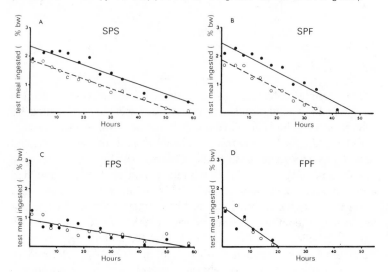

larger meals (as a percentage of body weight) being taken by smaller fish as compared with larger fish. This latter result was not evident for pre-prandially fed fish.

(2) Evacuation rate is unrelated to body size over the range of 0.8–11.7 g, irrespective of feeding history.

(3) Post-prandial starvation decreases evacuation rate but the magnitude of this effect is inversely related to stomach contents at the start of the starvation period.

(4) Test meals of different sizes are evacuated at similar rates as feeding occurs after the test meal is ingested, irrespective of pre-prandial starvation.

The results of this experiment confirm the effects of feeding history which have been reported by other authors. For all fish species, the results obtained for consumption and evacuation of a test meal and the effects of fish size may be very different, depending on whether the fish are deprived of food at any stage. Feeding regime will also affect the process of digestion and absorption of nutrients. Given supra-maximal rations, some fish species ingest an amount of food in excess of their energy requirements (Cowey, 1981). Much undigested food passes out in the faeces because its passage through the gastro-intestinal tract is too rapid to allow efficient digestion. In this situation, it is not possible to conduct experiments to determine the requirement for a given nutrient by a dose–response curve, as much of the nutrient may be lost in the faeces. Several authors have reported that the percentage loss of nutrients in the faeces increases as the ration size increases (Kinne, 1960; Pandian, 1967; Solomon & Brafield, 1972). Both temperature and ration size affect the energy losses in the faeces of brown trout (*Salmo trutta*). Values vary from 31% of the energy intake for fish on maximal rations at 3°C to 11% for fish at 22°C on a ration restricted to 10% of the maximal ration (Elliott, 1976a). However, many fish species appear to restrict their food intake in order to regulate gross energy intake (Jobling, 1981) and there are much published data for a variety of species which show that digestibility coefficients are independent of daily food intake, feeding frequency or evacuation rate (Kelso, 1972; Charles, 1984; Sampath, 1984).

5.5 Concluding Remarks

Laboratory studies of fish feeding are carried out for a variety of purposes, including understanding the physiology of the feeding and digestive process, fish husbandry, diet formulation and extrapolation to natural populations. The choice of experimental method is of critical importance because of its effects on food consumption, evacuation and digestion. Different species of fish from different habitats, environmental conditions and availability and type of food will exhibit feeding patterns ranging from almost continuous feeding to long periods of starvation. The occurrence of empty stomachs is frequently reported for wild fish in northern temperate regions, especially in winter (Seaburg & Moyle, 1964; Swenson & Smith, 1973; Doble & Eggers, 1978). Some species appear to be more or less continuous feeders on a diurnal basis, e.g. goldfish, *Carassius auratus* (Rozin & Mayer, 1964), pink salmon, *Oncorhynchus gorbuscha* (Godin, 1981), anchovy, *Engraulis mordax* (Leong & O'Connell, 1969), and herbivorous fishes such as the cichlids,

Sarotherodon niloticus and *Haplochromis nigripinus* (Moriarty & Moriarty, 1973). Other species, e.g. rainbow trout, *Salmo gairdneri* (Adron *et al.*, 1973; Landless, 1976), brown trout, *Salmo trutta* (Elliott, 1975 a, b) and adult pike, *Esox lucius* (Diana, 1979), appear to be discontinuous feeders and feed at a higher hunger threshold than continuous feeders.

For any given species of fish, a chosen experimental feeding regime may reflect a realistic feeding pattern but this will be just one of many possible feeding strategies. In feeding experiments involving starvation periods (or acutely stressful procedures), the results obtained may not be applicable to other trophic situations. Diana (1983) fitted parameters to a bioenergetics model to determine the energy budget of pike (*Esox lucius*). A poor fit was obtained between the predicted ration and growth (from the model) and the estimated ration and growth in the field. Food rations were estimated in the field by analysing stomach contents (as kilocalories ingested) divided by the evacuation time in days (measured by force-feeding pike, followed by stomach analysis using serial slaughter). Food ration estimates deviated at least 20% less than predicted in 8 of 18 time periods, and they were more than 20% above predicted values in only 4 of 18 periods. Diana (1983) recognized that ration estimates could potentially introduce substantial errors into the energy budget analysis, and attributed part of the discrepancy between predicted and estimated growth to poor ration data, possibly due to the effects of force-feeding on evacuation rate (cf. Swenson & Smith, 1973), and stress-induced regurgitation of stomach contents at capture. Similarly, estimates by Thorpe (1977) of daily food consumption by perch (*Perca fluviatilis*) may have been underestimated, due to a prolonged evacuation time caused by post-prandial food deprivation if the fish held in food-free enclosures would have otherwise fed normally during this period (cf. Noble, 1973; Corazza & Nickum, 1983; Figure 5.2, C and D). Paradoxically, the effects of post-prandial starvation on estimates of daily food consumption from evacuation rate data might be mitigated if pre-prandial starvation is also employed, as this will maximize the amount consumed during the test meal and increase evacuation rate (cf. Figure 5.2, A and C). Elliott & Persson (1978) and Jobling (1981b) discuss further the various mathematical descriptions of gastric emptying curves used to estimate daily food consumption and some effects on evacuation of meal size and frequency, fish size, and sampling procedures. The possible compounding effects of stress, feeding regimes and starvation on gastric evacuation deserve further consideration. Jobling & Davies (1979) employed a force-fed test meal and both pre- and post-prandial starvation in a study of gastric evacuation in plaice (*Pleuronectes platessa*). Radiographic methods using barium sulphate additionally employ repeated handling of individual fish during the post-prandial period. Generalized models of trophic dynamics of fish based on results obtained under only one experimental feeding regime cannot be wholly adequate, especially if acutely stressful procedures are used. Other aspects of the design of feeding experiments that deserve attention relate to fish holding facilities (see Hawkins, 1981) and feeding behaviour (e.g. intra-specific factors such as aggression: Peter (1979)).

As our knowledge of fish feeding strategies increases, it is possible to develop more precise theories of optimal foraging (see Chapter 3). Laboratory feeding experiments provide important data both for the development of these theories and for assessing the agreement between prediction and actual observed performance. Windell 1978b, quoting Beukema, 1968) points out that voluntary food consumption (appetite) is relevant only in its expression in feeding behaviour. Appetite is usually studied by measuring food consumption during *ad libitum* feeding following varying periods of starvation. The laboratory methods available to study appetite have evolved from the relatively simple 'starve–feed–starve' type to more sophisticated methods which use demand feeders or food which is labelled in some way, e.g. particulate markers, isotopes or dyes. These latter methods provide for greater flexibility in experimental feeding regimes and allow appetite to be studied in a more direct way under more natural circumstances, including studies where food is available to fish either constantly or at intervals more frequent than the gastric evacuation time.

Of particular interest to both physiologists and ecologists is a greater understanding of feeding and digestion in individual fish within a population. There is considerable variation in the time at which individual fish begin and stop feeding, the rate at which they feed and the rate at which food is evacuated from the foregut while they are feeding. There are frequent reports in the literature of fish within the same population that have empty stomachs in the presence of abundant food. Most investigations of food consumption and evacuation have concentrated on quantifying the content of the foregut. It would be profitable to investigate further the role of the hindgut in the regulation of feeding. These studies would require the use of labelled food. It is likely that digestibility increases with increased retention time of digesta in the gut. This aspect of digestive physiology deserves further study in view of the variability in evacuation rates found under different environmental conditions in nature and caused by different feeding regimes in laboratory experiments.

Many conclusions have been drawn relating to all aspects of fish feeding which are based on data whose value is limited in one way or another by the inadequacies of experimental method. The advent of more sophisticated techniques (which are by no means perfected yet) has removed to a large extent the constraints imposed by the need to starve fish and the need for force-feeding or repeated handling of individuals. The relationships between fish and their food require many types of data relating to population density, fish size, food availability and diel and seasonal effects of temperature, light, and fish and prey behaviour. The work of J.M. Elliott on brown trout (*Salmo trutta*) (summarized by Elliott, 1977) is notable in this respect. The choice of experimental method in the study of fish feeding will depend on the questions being asked. In all cases, however, the method chosen should be as biologically realistic as possible under any given environmental conditions with respect to feeding pattern and stress.

References

Adron, J.W., Grant, P.T. and Cowey, C.B. (1973), A system for quantitative study of the learning capacity of trout, *Salmo gairdneri*, and its application to the study of feeding preferences and behaviour. *J. Fish Biol. 5*, 625

A.O.A.C. (1980), *Official Methods of Analysis of the Association of Official Analytical Chemists*, 13th edn. Washington

Austreng, E. (1978), Digestibility determination in fish using chromic oxide marking and analysis of contents from different segments of the gastro-intestinal tract. *Aquaculture 13*, 265

Baker, A.M. and Fraser, D.F. (1976), A method for securing the gut contents of small, live fish. *Trans. Am. Fish. Soc. 105*, 520

Baldwin, N.S. (1957), Food consumption and growth of brook trout at different temperatures. *Trans. Am. Fish. Soc. 86*, 323

Beukema, J.J. (1968), Predation by the three-spined stickleback (*Gasterosteus aculeatus*): the influence of hunger and experience. *Behaviour, 31*, 1

Blaxter, J.H.S. (1963), The feeding of herring larvae and their ecology in relation to feeding. *Calif. Coop. Oceanogr. Fish. Invest. Rep. 10*, 79

Bowen, S.H. (1978), Chromic acid in assimilation studies — a caution. *Trans. Am. Fish. Soc. 107*, 755

Bowen, S.H. (1981), Digestion and assimilation of periphytic detrital aggregate by *Tilapia mossambica*. *Trans. Am. Fish. Soc. 110*, 239

Braaten, B.R. (1979), Bioenergetics — a review on methodology. In: Halver, J.E. and

Tiews, K. (eds). *Finfish Nutrition and Fishfeed Technology*, vol. II, pp. 461–504. Berlin, Heenemann

Braum, E. (1978), Ecological aspects of the survival of fish eggs, embryos and larvae. In: Gerking, S.D. (ed.). *Ecology of Freshwater Fish Production*, pp. 102–31. Oxford, Blackwell

Brett, J.R. (1971), Starvation time, appetite and maximum food intake of sockeye salmon, *Oncorhynchus nerka. J. Fish. Res. Bd Can. 28*, 409

Brett, J.R. (1979), Environmental factors and growth. In: Hoar, W.S., Randal, D.J. and Brett, J.R. (eds). *Fish Physiology*, vol. 8, pp. 599–675. New York, Academic Press

Brett, J.R. and Groves, T.D.D. (1979), Physiological energetics. In: Hoar, W.S., Randall, D.J. and Brett, J.R. (eds). *Fish Physiology*, vol. 8, pp. 279–352. New York, Academic Press

Brett, J.R. and Higgs, D.A. (1970), Effects of temperature on the rate of gastric digestion in fingerling sockeye salmon, *Oncorhynchus nerka. J. Fish. Res. Bd Can. 27*, 1767

Brett, J.R., Shelbourn, J.E. and Shoop, C.T. (1969), Growth rate and body composition of fingerling sockeye salmon, *Oncorhynchus nerka*, in relation to temperature and ration size. *J. Fish. Res. Bd Can. 26*, 2363

Brett, J.R., Sutherland, D.B. and Heritage, G.D. (1971), An environmental-control tank for the synchronous study of growth and metabolism of young salmon. *Fish. Res. Bd Can. Tech. Rep. 283*, 1

Brown, M.E. (1951), The growth of brown trout (*Salmo trutta* L.). IV. The effect of food and temperature on the survival and growth of fry. *J. exp. Biol. 23*, 473

Buddington, R.K. (1979), Digestion of an aquatic macrophyte by *Tilapia zilli* (Gervais). *J. Fish Biol. 15*, 449

Buddington, R.K. (1980), Hydrolysis-resistant organic matter as a reference for measurement of fish digestive efficiency. *Trans. Am. Fish. Soc. 109*, 653

Chandler, P.T., Kesler, E.M. and McCarthy, J.J. (1964), Polyethylene as a reference substance for digestion studies with young ruminants. *J. Dairy Sci. 47*, 1426

Charles, P.M., Sebastian, S.M., Raj, M.C.V. and Marian, M.P. (1984), Effect of feeding frequency on growth and food conversion of *Cyprinus carpio* fry. *Aquaculture, 40*, 293–300.

Cho, C.Y., Bayley, H.S. and Slinger, S.J. (1974), Partial replacement of herring meal with soybean and other changes in a diet for rainbow trout (*Salmo gairdneri*). *J. Fish. Res. Bd Can. 31*, 1523

Cho, C.Y., Slinger, S.J. and Bayley, H.S. (1976), Influence of level and type of dietary protein, and level of feeding on food utilization by rainbow trout. *J. Nutr. 106*, 1547

Cho., C.Y., Slinger, S.J. and Bayley, H.S. (1982), Bioenergetics of salmonid fishes: energy intake, expenditure and productivity. *Comp. Biochem. Physiol. 73B*, 25

Choubert, G., De La Noiie, J. and Luquet, P. (1982), Digestibility in fish: improved device for the automatic collection of faeces. *Aquaculture 29*, 185

Colgan, P. (1973), Motivational analysis of fish feeding. *Behaviour 45*, 38

Corazza, L. and Nickum, J.G. (1983), Rate of food passage through the gastro-intestinal tract of fingerling walleyes. *Prog. Fish Cult. 45*, 183

Cowey, C.B. (1981), Food and feeding of captive fish. In: Hawkins, A.D. (ed.) *Aquarium Systems*, pp. 223–46. London, Academic Press

Davis, J.J. and Foster, R.F. (1958), Bioaccumulation of radioisotopes through aquatic food chains. *Ecology 39*, 530

De Silva, S.S. and Owoyemi, A.A. (1983), Effect of dietary quality on the gastric evacuation and intestinal passage in *Sarotherodon mossambicus* (Peters) fry. *J. Fish Biol. 23*, 347

De Silva, S.S. and Perera, M.K. (1983), Digestibility of an aquatic macrophyte by the

150 Fish Energetics

cichlid *Etroplus suratensis* (Bloch) with observations on the relative merits of three indigenous components as markers and daily changes in protein digestibility. *J. Fish Biol. 23*, 675

Diana, J.S. (1979), The feeding pattern and daily ration of a top carnivore, the northern pike (*Esox lucius*). *Can. J. Zool. 57*, 2121

Diana, J.S. (1983), An energy budget for northern pike (*Esox lucius*). *Can. J. Zool. 61*, 1968

Doble, B.D. and Eggers, D.M. (1978), Diel feeding chronology, rate of gastric evacuation, daily ration, and prey selectivity in Lake Washington, juvenile sockeye salmon (*Oncorhynchus nerka*). *Trans. Am. Fish. Soc. 107*, 36

Edwards, D.J. (1971), Effect of temperature on rate of passage of food through the alimentary canal of plaice *Pleuronectes platessa* L. *J. Fish Biol. 3*, 433

Edwards, D.J. (1973), The effect of drugs and nerve section on the rate of passage of food through the gut of the plaice, *Pleuronectes platessa*. *J. Fish Biol. 5*, 441

Elliott, J.M. (1972), Rates of gastric evacuation in brown trout, *Salmo trutta* L. *Freshwat. Biol. 2*, 1

Elliott, J.M. (1975a), Weight of food and time required to satiate brown trout, *Salmo trutta* L. *Freshwat. Biol. 5*, 51

Elliott, J.M. (1975b), Number of meals in a day, maximum weight of food consumed in a day and maximum rate of feeding for brown trout, *Salmo trutta* L. *Freshwat. Biol. 5*, 287

Elliott, J.M. (1976a), Energy losses in the waste products of brown trout (*Salmo trutta* L.). *J. Anim. Ecol. 45*, 561

Elliott, J.M. (1976b), The energetics of feeding, metabolism and growth of brown trout (*Salmo trutta* L.) in relation to body weight, water temperature and ration size. *J. Anim. Ecol. 45*, 923

Elliott, J.M. (1977), Feeding, metabolism and growth of brown trout. *Freshwater Biological Association 45th Annual Report*, p. 70

Elliott, J.M. and Persson, L. (1978), The estimation of daily rates of food consumption for fish. *J. Anim. Ecol. 47*, 977

Ellis, W.C. and Huston, J.E. (1968), $^{144}Ce - ^{144}Pr$ as a particulate digesta flow marker in ruminants. *J. Nutr. 95*, 67

El-Shamy, F.M. (1976), Analyses of gastric emptying in bluegill (*Lepomis macrochirus*). *J. Fish. Res. Bd Can. 33*, 1630

Fänge, R. and Grove, D.J. (1979), Digestion. In: Hoar, W.S., Randall, D.J. and Brett, J.R. (eds). *Fish Physiology* vol. VIII, pp. 161–260. New York, Academic Press

Farmer, G.J., Beamish, F.W.H. and Robinson, G.A. (1975), Food consumption of the adult landlocked sea lamprey, *Petromyzon marinus* L. *Comp. Biochem. Physiol. 50A*, 753

Flowerdew, M. and Grove, D.J. (1979), Some observations of the effects of body weight, meal size and quality on gastric emptying time in the turbot, *Scophthalmus maximus* (L.) using radiography. *J. Fish Biol. 14*, 229

Flowerdew, M.W. and Grove, D.J. (1980), An energy budget for juvenile thick-lipped mullet, *Crenimugil labrosus* (Risso). *J. Fish Biol. 17*, 395

Foster, J.R. (1977), Pulsed gastric lavage: an efficient method of removing the stomach contents of live fish. *Prog. Fish Cult. 39*, 166

Furukawa, A. and Tsukahara, H. (1966), On the acid digestion method for the determination of chromic oxide as an index substance in the study of digestibility of fish feed. *Bull. Jpn Soc. Sci. Fish. 32*, 502

Giles, N. (1980), A stomach sampler for use on live fish. *J. Fish Biol. 16*, 441

Goddard, J.S. (1974), An X-ray investigation of the effects of starvation and drugs on intestinal mobility in the plaice, *Pleuronectes platessa* L. *Ichthyologica 6*, 49

Godin, J.J. (1981), Effect of hunger on the daily pattern of feeding rates in juvenile pink salmon, *Oncorhyncus gorbuscha* Walbaum. *J. Fish Biol. 19*, 63

Grove, D.J. and Crawford, C. (1980), Correlation between digestion rate and feeding frequency in the stomachless teleost, *Blennius pholis* L. *J. Fish Biol. 16*, 235

Grove, D.J., Lozoides, L. and Nott, J. (1978), Satiation amount, frequency of feeding and gastric emptying rate in *Salmo gairdneri*. *J. Fish Biol. 12*, 507

Gwyther, D. and Grove, D.J. (1981), Gastric emptying in *Limanda limanda* (L.) and the return of appetite. *J. Fish Biol. 18*, 245

Hakonson, T.E., Gallegos, A.F. and Whicker, F.W. (1975), Cesium kinetics data for estimating food consumption rates of trout. *Health Phys. 29*, 301

Hawkins, A.D. (ed.) (1981), *Aquarium Systems*. New York, Academic Press

Hickling, C.F. (1966), On the feeding process in the white amur (*Ctenopharyngodon idella*). *J. Zool. 148*, 408

Hirao, S., Yamada, J. and Kikuchi, R. (1960), On improving efficiency of feed for fish culture. I. Transit and digestibility of diet in eel and rainbow trout observed by use of 32P. *Bull. Tokai Regional Fish Res. Lab. 7*, 67

Hunt, B.P. (1960), Digestion rate and consumption of Florida gar, warmouth, and largemouth bass. *Trans. Am. Fish. Soc. 89*, 206

Hyslop, E.J. (1980), Stomach contents analysis — a review of methods and their application. *J. Fish Biol. 17*, 411

Inaba, D., Ogino, C., Takamatsu, C., Sugano, S. and Hata, H. (1962), Digestibility of dietary components in fishes. I. Digestibility of dietary proteins in rainbow trout. *Bull. Jpn Soc. Sci. Fish. 28*, 367

Jernejcic, F. (1969), Use of emetics to collect stomach contents of walleye and largemouth bass. *Trans. Am. Fish. Soc. 98*, 698

Jobling, M. (1981a), Dietary digestibility and the influence of food components on gastric evacuation in plaice *Pleuronectes platessa* L. *J. Fish Biol. 19*, 29

Jobling, M. (1981b), Mathematical models of gastric emptying and the estimation of daily rates of food consumption for fish. *J. Fish Biol. 19*, 245

Jobling, M. and Davies, P.S. (1979), Gastric evacuation in plaice, *Pleuronectes platessa* L.: effects of temperature and size. *J. Fish Biol. 14*, 539

Jobling, M., Gwyther, D. and Grove, D.J. (1977), Some effects of temperature, meal size and body weight on gastric evacuation time in the dab, *Limanda limanda* (L.). *J. Fish Biol. 10*, 291

Jones, R. (1974), The rate of elimination of food from the stomachs of haddock *Melanogrammus aeglefinus*, cod *Gadus morhua* and whiting *Merlangius merlangus*. *J. Cons. perm. int. Explor. Mer. 35*, 225

Kelso, J.R.M. (1972), Conversion, maintenance and assimilation for walleye, *Stizostedion vitreum vitreum*, as affected by size, diet and temperature. *J. Fish. Res. Bd Can., 29*, 1181–92.

Kevern, N.R. (1966), Feeding rate of carp estimated by a radioisotopic method. *Trans. Am. Fish. Soc. 95*, 363

Kinne, O. (1960), Growth, food intake and food conversion in a euryplastic fish exposed to different temperatures and salinities. *Physiol. Zool. 33*, 288

Kolb, A.R. and Luckey, T.D. (1972), Markers in nutrition. *Nutr. Abstr. Rev. 42*, 813

Kolehmainen, S.E. (1974), Daily feeding rates of bluegill (*Lepomis macrochirus*) determined by a refined radioisotope method. *J. Fish Res. Bd Can. 31*, 67

Lall, S.P. and Bishop, F.J. (1979), Studies on the nutrient requirements of rainbow trout, *Salmo gairdneri*, grown in seawater and freshwater. In: Pillay, T.V.A. and Dill, W.A. (eds). *Advances In Aquaculture*. Farnham, Fishing News Books

Landless, P.G. (1976), Demand-feeding behaviour of rainbow trout. *Aquaculture 7*, 11

Lane, T.H. and Jackson, H.M. (1969), Voidance time for 23 species of fish. *Invest. Fish Control No. 33*

Laurence, G.C. (1971), Digestion rate of larval largemouth bass. *N.Y. Fish and Game*

152 Fish Energetics

Journal 18, 52

Leong, R.J. and O'Connell, C.P. (1969), A laboratory study of particulate and filter feeding of the northern anchovy (*Engraulis mordax*). *J. Fish. Res. Bd Can. 26*, 557

Lied, E., Julshamn, K. and Braekkan, O.R. (1982), Determination of protein digestibility in Atlantic cod (*Gadus morhua*) with internal and external indicators. *Can. J. Fish. Aquat. Sci. 39*, 854

Light, R.W., Adler, P.H. and Arnold, D.E. (1983), Evaluation of gastric lavage for stomach analyses. *North Am. J. Fish. Mgmt 3*, 81

Love, R.M. (1980), *The Chemical Biology of Fishes*, vol. II: *Advances 1968–1977*. New York, Academic Press

Magnuson, J.J. (1969), Digestion and food consumption by skipjack tuna *Katsuwonus pelamis*. *Trans. Am. Fish. Soc. 98*, 379

Mann, K.H. (1978), Estimating the food consumption of fish in nature. In: Gerking, S.D. (ed.) *Ecology of Freshwater Fish Production*, pp. 250–73. Oxford, Blackwell

Markus, H.C. (1932), The extent to which temperature changes influence food consumption in largemouth bass (*Huro floridana*). *Trans. Am. Fish Soc. 62*, 202

Maynard, L.A., Loosli, J.K., Hintz, H.F. and Warner, R.G. (1979), *Animal Nutrition*, 7th edn. New York, McGraw-Hill

Meehan, W.R. and Miller, R.A. (1978), Stomach flushing: effectiveness and influence on survival and condition of juvenile salmonids. *J. Fish. Res. Bd Can. 35*, 1359

Molnár, G. and Tölg, I. (1960), Roentgenologic investigation of the duration of gastric digestion in the pike perch (*Lucioperca lucioperca*). *Acta. Biol. Hung. 11*, 103

Moriarty, C.M. and Moriarty, D.J.W. (1973), Quantitative estimation of the daily ingestion of phytoplankton by *Tilapia nilotica* and *Haplochromis nigripinus* in Lake George, Uganda. *J. Zool. 171*, 15

Neumark, H., Helavi, A., Amir, S. and Yerushalmi, S (1974), Assay and use of magnesium ferrite as a reference in absorption trials with cattle. *J. Dairy Sci. 58*, 1476

Noble, R.L. (1973), Evacuation rates of young yellow perch *Perca flavescens* (Mitchill). *Trans. Am. Fish. Soc. 102*, 759

Nose, T. (1960), On the effective value of freshwater green algae *Chlorella ellipsoidea*, as a nutritive source to goldfish. *Bull. Freshwater Fish. Res. Lab. Tokyo 10*, 12

Nose, T. (1967), On the metabolic faecal nitrogen in young rainbow trout. *Bull. Freshwater Fish. Res. Lab. Tokyo 17*, 97

Ogino, C., Kakino, J. and Chem, M.S. (1973), Determination of metabolic fecal nitrogen and endogenous nitrogen excretion of carp. *Bull. Jpn Soc. Sci. Fish. 39*, 519

Pandian, T.J. (1967), Transformation of food in the fish *Megalops cyprinoides*. II. Influence of quality of food. *Marine Biol. 1*, 107

Pentelow, F.T.K. (1939), The relation between growth and food consumption in the brown trout (*Salmo trutta*). *J. exp. Biol. 16*, 446

Persson, L. (1979), The effects of temperature and different food organisms on the rate of gastric evacuation in perch (*Perca fluviatilis*). *Freshwat. Biol. 9*, 99

Peter, R.E. (1979), The brain and feeding behaviour. In: Hoar, W.S., Randall, D.J. and Bretts, J.R. (eds). *Fish Physiology*, vol. VIII, pp. 121–59. New York, Academic Press

Peters, D.S. and Hoss, D.E. (1974), A radioisotopic method of measuring food evacuation time in fish. *Trans. Am. Fish. Soc. 103*, 626

Peters, G. (1982). The effect of stress on the stomach of the European eel, *Anguilla*

anguilla L. *J. Fish Biol. 21*, 497.

Pickering, A.D. (ed.) (1981), *Stress and Fish*. New York, Academic Press

Pickering, A.D., Pottinger, T.G. and Christie, P. (1982), Recovery of the brown trout, *Salmo trutta* L., from acute handling stress: a time-course study. *J. Fish Biol. 20*, 229

Ross, B. and Jauncey, K. (1981), A radiographic estimation of the effect of temperature on gastric emptying time in *Sarotherodon niloticus* (L.) × *S. auvens* (Steindachner) hybrids. *J. Fish Biol. 19*, 333

Rozin, P. and Mayer, J. (1961), Regulation of food intake in the goldfish. *Am. J. Physiol. 201*, 968

Rozin, P. and Mayer, J. (1964), Some factors influencing short-term food intake in goldfish. *Am. J. Physiol. 206*, 1430

Ryer, C.H. and Boehlert, G.W. (1983), Feeding chronology, daily ration, and the effects of temperature upon gastric evacuation in the pipefish, *Syngnathus fuscus*. *Env. Biol Fish. 9*, 301

Sampath, K. (1984), Preliminary report on the effects of feeding frequency in *Channa striatus*. *Aquaculture, 40*, 301–6.

Seaburg, K.G. (1957), A stomach sampler for live fish. *Prog. Fish. Cult. 19*, 137

Seaburg K.G. and Moyle, J.B. (1964), Feeding habits, digestion rates and growth of some Minnesota warm water fishes. *Trans. Am. Fish. Soc. 93*, 269

Singh, R.P. and Nose, T. (1967), Digestibility of carbohydrates in young rainbow trout. *Bull. Freshwater Fish. Res. Lab. Tokyo 17*, 21

Smith, R.R. (1971), A method for measuring digestibility and metabolizable energy of feeds. *Prog. Fish Cult. 33*, 132

Smith, B.W. and Lovell, R.T. (1973), Determination of apparent protein digestibility in feeds for channel catfish. *Trans. Am. Fish. Soc. 102*, 831

Solomon, D.J. and Brafield, A.E. (1972), The energetics of feeding metabolism and growth of perch (*Perca fluviatilis* L.). *J. Anim. Ecol. 41*, 699

Staples, D.J. (1975), Production biology of the upland bully *Philypnodon breviceps* Stokell in a small New Zealand lake. III. Production, food consumption and efficiency of food utilization. *J. Fish Biol. 7*, 47

Storebakken, T., Austreng, E. and Stoenberg, K. (1981), A method for determination of feed intake in salmonids using radioactive isotopes. *Aquaculture 24*, 133

Strange, C.D. and Kennedy, G.J.A. (1981), Stomach flushing of salmonids: a simple and effective technique for the removal of stomach contents. *Fish. Mgmt 12*, 9

Swenson, W.A. and Smith, L.L. Jr (1973), Gastric digestion, food consumption, feeding periodicity and food conversion efficiency in walleye (*Stizostedion vitreum vitreum*). *J. Fish. Res. Bd Can. 30*, 1327

Tacon, A.G.J., Haaster, J.V., Featherstone, P.B., Kerr, K. and Jackson, A.J. (1983), Studies on the utilization of full-fat soybean and solvent-extracted soybean meal in a complete diet for rainbow trout. *Bull. Jpn Soc. Sci. Fish. 49*, 1437

Talbot, C. and Higgins, P.G. (1982), Observations on the gall bladder of juvenile Atlantic salmon, *Salmo salar* L., in relation to feeding. *J. Fish Biol. 21*, 663

Talbot, C. and Higgins, P.J. (1983), A radiographic method for feeding studies on fish using metallic iron powder as a marker. *J. Fish Biol. 23*, 211

Talbot, C., Higgins, P.J. and Shanks, A.M. (1984), Effects of pre- and post-prandial starvation on meal size and evacuation rate of juvenile Atlantic salmon (*Salmo salar* L.). *J. Fish Biol. 25*, 551

Thorpe, J.E. (1977), Daily ration of adult perch, *Perca fluviatilis* L. during summer in Loch Leven, Scotland. *J. Fish Biol. 11*, 55

Thorpe, J.E. (1981), Rearing salmonids in freshwater. In: Hawkins, A.D. (ed.). *Aquarium Systems*. New York, Academic Press

Tyler, A.V. (1970), Rates of gastric emptying in young cod. *J. Fish Res. Bd Can. 27*, 1177

Wales, J.H. (1962), Forceps for removal of trout stomach content. *Prog. Fish Cult.* *24*, 171

Wallace, J.C. (1973), Observations on the relationship between food consumption and metabolic rate of *Blennius pholis* L. *Comp. Biochem. Physiol.* *45A*, 293

Ware, D.M. (1972), Predation by rainbow trout (*Salmo gairdneri*): the influence of hunger, prey density and prey size. *J. Fish. Res. Bd Can.* *29*, 1191

Webb, P.W. (1978), Partitioning of energy into metabolism and growth. In: Gerking, S.D. (ed.) *Ecology of Freshwater Fish Production*, pp. 184–214. Oxford, Blackwell

Wedemeyer, G.A. (1976), Physiological response of juvenile coho salmon (*Oncorhynchus kisutch*) and rainbow trout (*Salmo gairdneri*) to handling and crowding stress in intensive fish culture. *J. Fish. Res. Bd Can.* *33*, 2699

White, H.C. (1930), Some observations on Eastern brook trout, *Salvelinus fontinalis*, of Prince Edward Island. *Trans. Am. Fish. Soc.* *60*, 101

Wightman, J.A. (1975), An improved technique for measuring assimilation efficiency by the ^{51}Ce $-^{14}C$ twin tracer method. *Oecologia (Berl.)* *19*, 273

Windell, J.T. (1966), Rate of digestion in the bluegull sunfish. *Invest. Indiana Lakes Streams* *7*, 185

Windell, J.T. (1978a), Digestion and the daily ration of fishes. In: Gerking, S.D. (ed.). *The Ecology of Freshwater Fish Production*, pp. 159–83. Oxford, Blackwell

Windell, J.T. (1978b), Estimating food consumption rates of fish populations. In: Bagenal, T. (ed.). *Methods for Assessment of Fish Production in Fresh Waters.* I.B.P. Handbook No. 3, third edn, pp. 227–54. Oxford, Blackwell

Windell, J.T., Norris, D.O., Kitchell, J.F. and Norris, J.S. (1969), Digestive response of rainbow trout, *Salmo gairdneri*, to pellet diets. *J. Fish. Res. Bd Can.* *26*, 1801

Windell, J.T., Armstrong, R.D. and Clinebell, J.R. (1974), Substitution of brewer's single cell protein (BSCP) into pelleted fish feed. *Feedstuffs* *46*, 16

Windell, J.T., Foltz, J.W. and Sarakon, J.A. (1978), Methods of fecal collection and nutrient leaching in digestibility studies. *Prog. Fish Cult.* *40*, 51

6 PROTEIN AND AMINO ACID REQUIREMENTS

A.G.J. Tacon and C.B. Cowey

6.1 Introduction

Over the past 25 years, considerable progress has been made in the study of the dietary nutrient requirements of fishes (for reviews see Cowey & Sargent, 1972, 1979; Halver, 1972; National Research Council, 1981, 1983; and Millikin, 1982). Despite some obvious similarities between fishes and other vertebrates in basic qualitative nutrient needs, the two groups have markedly different quantitative nutrient requirements. For example, the optimal dietary protein level required for maximal growth in farmed fishes is reported to be 50–300% higher than that of terrestrial farm animals (Cowey, 1975). In the main, these quantitative differences have been attributed to the carnivorous/omnivorous feeding habit of fishes and their apparent preferential use of protein over carbohydrate as a dietary energy source. However, the common expression by nutritionists (including major review authors) of nutrient requirements solely in terms of a 'dietary percentage' has itself limited value unless it is related to the feed intake and subsequent growth of the animal. This chapter attempts to relate the protein and amino acid requirements of fishes to the 'growing animal' with respect to its dietary feeding regime, developmental status, position in the aquatic food chain, and its physical environment. In addition, this chapter critically assesses the methodology employed by researchers for the measurement of nutrient requirements.

6.2 Metabolic Considerations

The study of dietary requirements in fishes has been almost entirely based on studies comparable to those conducted with terrestrial farm animals. At least two important metabolic differences between these vertebrate groups have been noted by other reviewers (Chapter 2) and are now re-emphasized. First, in contrast to warm-blooded animals, fish are aquatic ectotherms and consequently do not have to expend a large proportion of energy in maintaining body temperature (Nijkamp *et al.*, 1974). Secondly, because they live in water, the primary end-product of nitrogen metabolism, i.e. ammonia, can rapidly be disposed of by passive diffusion through permeable surfaces (gills). Toxic levels of ammonia do not build up in tissues and there is no necessity to convert ammonia to innocuous molecules such as urea or uric acid. Consequently, fish derive more metabolizable energy from catabolism of proteins than do terrestrial animals, which must convert ammonia to non-toxic substances (Brett & Groves, 1979). The 'efficient mechanism possessed by fish for protein catabolism and excretion of nitrogen' is seen by Smith *et al.* (1978) as one of the factors that contribute to the high energetic efficiency of fish. Others include cold-blooded existence, low energy cost of voluntary activity in water, and low energy cost of reproduction. These authors have produced values for the energetic efficiency of rainbow trout (*Salmo gairdneri*) over all phases of production (9.6 g body protein produced per megajoule of digestible energy consumed) that are 2 to 20-fold better than comparable values for chickens, pigs and cattle. However, although these values are partly based on direct measurements of heat increment, they have not found universal acceptance in the field of fish energetics, where the metabolizable energy value of protein is a controversial matter. Consequently, the overall figures for energetic efficiency in fish may require modification as more data become available.

Although there are several facets of the nutrition of fish that contrast with their terrestrial counterparts, that which has the greatest significance, at least from the viewpoint of fish cultivation, is their high requirement for dietary protein. The point is also of academic interest in that the amounts of essential amino acids required by fish seem vastly to exceed their requirements for protein synthesis. It is therefore necessary to ascertain with confidence the level of essential amino acid requirement. Consequently this chapter has concentrated on a careful evaluation of methods employed and results obtained in the field of protein nutrition rather than attempting to review requirements of many species of fish for 40 or more essential nutrients.

6.3 Protein Requirement

Based on feeding techniques pioneered and developed for terrestrial animals, the dietary protein requirements of fish were first investigated in the Chinook salmon (*Oncorhynchus tshawytscha*) by DeLong *et al.* (1958). Fish were fed a balanced diet containing graded levels of a high-quality protein (casein: gelatin mixture supplemented with crystalline amino acids to simulate the amino acid profile of whole hen's-egg protein) over a 10-week period, and the observed protein level giving optimum growth was taken as the requirement. Since these early studies, the approach used by workers today has changed very little, if at all, with the possible exception of the use by some researchers of maximum tissue protein retention or nitrogen balance in preference to weight gain as the criterion of requirement (Ogino, 1980a). Dietary protein requirements are normally expressed in terms of a fixed dietary percentage or as a ratio of protein to dietary energy (calculated using either gross, digestible or metabolizable energy values). To date, over 20 different fish species have been examined in this manner and the results show a uniformly high dietary protein requirement in the range 35–55%, or equivalent to 45–70% of the gross energy content of the diet in the form of protein (Table 6.1). Although a high protein requirement might have been expected for carnivorous fish species such as plaice *Pleuronectes platessa* (50%; Cowey *et al.*, 1972) or snakehead *Channa micropeltes* (52%; Wee & Tacon, 1982), the fact that a relatively high protein requirement was also observed in the herbivorous grass carp *Ctenopharyngodon idella* (41–43%; Dabrowski, 1977) suggests that the requirement may in part be a function of the methodology used for its determination. The use by different workers of different dietary protein sources, non-protein energy substitutes, feeding regimes, fish age classes and methods for the determination of dietary energy content and dietary requirement leaves little common ground for direct comparisons to be made within or between fish species. However, using the information calculated in Table 6.1, some general comments can be made with respect to diet formulation and consumption; feeding regime and growth; abiotic factors (temperature and salinity).

Table 6.1: Quantitative Dietary Protein Requirement of Several Fish Species.

Species	Temperature range (0°C)	Dietary protein source	Study period (days)	Initial body weight (g)	Feeding regime employed
Oreochromis mossambicus	26–28	Fish meal	40	1.8	6% bw d^{-1}
O. niloticus	24–28	Fish meal	56	0.0128	15% bw d^{-1}
O.aureus	27.6	Fish/soy meal	84	0.37	*c.* 8.83% bw d^{-1}
O. aureus	31	Casein/albumin	77	0.016	10–20% + bw d^{-1}
Tilapia zilli	24–26	Casein	21	1.8	5% bw d^{-1}
Channa micropeltes	27–29	Fish meal	56	125	2% bw d^{-1}
Fugu rubripes	25–26	Casein	21	2	10% bw d^{-1}
Chanos chanos	25–28	Casein	30	0.040	10% bw d^{-1}
Ctenopharyngodon idella	22–23	Casein	40	0.15–0.2	Fixed feed intake (?)
Chrysophrys aurata	21	Casein/amino acids	112	2.8	Ad lib. (*c.* 2.4% d^{-1})
Morone saxatilis	20	Fish/soy meal	70	1.4	Ad lib. (*c.* 3.78– d^{-1})
M. saxatilis	24.5	Fish/soy meal	42	2.5	Ad lib. (?)
Cyprinus carpio	23	Casein	30	6.04	Ad lib. (?)
C. carpio	23–28	Fish meal	35	64.2	5% bw d^{-1}
Anguilla japonica	25	Casein	56	3.03	Ad lib. (?)
Micropterus dolomieui	20–21	Fish/gelatin/amino acids	26–68	0.65–4.03	Ad lib. (*c.* 4.5–5.2% d^{-1})
M. salmoides	23	Fish/gelatin/amino acids	14–59	2.05–5.8	Ad lib. (*c.* 3.6–2.4% d^{-1})
Pleuronectes platessa	15	Cod muscle	84	14	Ad lib. (*c.* 1.50% d^{-1})
Salvelinus alpinus	10	Fish meal	56	18.5	Ad lib. (?)
Salmo gairdneri	8–12	Fish meal composite	153	61	Fixed feed intake (?)
S. gairdneri	16–27	Fish meal	70	3.5	Fixed feed intake (*c.* 4.5% d^{-1})
S. gairdneri	9–12.5	Casein/gelatin	70	6.25	Ad lib. (?)

Table 6.1: continued

Species	Dietary crude protein requirement (%)	Dietary gross energy (TE, kJ g⁻¹)[a]	Dietary gross protein: TE	Growth and feed utilization			Protein requirement g
				Final body weight (g)	Specific growth rate[b] (% day)	Food conversion ratio[c]	
Oreochromis mossambicus	42(40)	17.83	0.55	8.4	3.85	1.46	25.5
O. niloticus	35	17.08	0.48	0.61	6.90	1.78	52.5
O. aureus	36	?	?	8.51	3.73	1.59	31.8
O. aureus	34–56	13.23–17.91	0.60–0.73	c. 8.0	7.9–7.75	?	?
Tilapia zilli	35	16.06	0.51	3.43	3.07	1.20	17.5
Channa micropeltes	52	19.0	0.64	297	1.55	1.25	10.40
Fugu rubripes	47(50)	20.08	0.55	4.43	3.79	0.88	?
Chanos chanos	39.1(40)	16.87	0.54	0.175	4.91	1.96	39.1
Ctenopharyngodon idella	41–43	?	?	0.4–0.6	2.3–2.9	?	?
Chrysophrys aurata	38.4	19.01	0.47	18.0	1.66	1.7	9.21
Morone saxatilis	48.8(47)	18.29	0.62	6.62	2.22	1.25	18.44
M. saxatilis	55.3(55)	21.10	0.61	9.4	3.15	1.0	?
Cyprinus carpio	38	18.64	0.48	13.52	2.68	?	?
C. carpio	35	18.30	0.45	208	3.36	1.53	17.50
Anguilla japonica	44.5	17.16	0.61	5.85	2.19	?	?
Micropterus dolomieui	45.25	18.17	0.58	2.55–7.0	2.72–0.90	1.4–2.0	20.4–9.96
M. salmoides	40–41	18.13	0.52	4.95–11.4	2.30–1.53	1.55–1.45	14.4–9.84
Pleuronectes platessa	50	16.91	0.69	30.7	0.93	1.33	7.50
Salvelinus alpinus	36–43.6	18.36–19.07	0.56–0.45	55.75–53.5	1.9–2.0	?	?
Salmo gairdneri	42	20.8	0.47	387	1.21	?	?
S. gairdneri	40	18.78	0.50	25.5	2.84	1.4	18.0
S. gairdneri	40–45	19.0–19.25	0.50–0.55	18.5–19.2	1.5–1.6	1.19–1.11	?

[a] Calculated using gross energy values of 23.4 kJ g⁻¹ protein; 39.2 kJ g⁻¹ lipid and 17.2 kJ g⁻¹ carbohydrate (Cho, 1982), and not including crude fibre.

[b] Specific growth rate (SGR) = [(log$_e$ final body weight − log$_e$ initial body weight)/time in days] × 100.

[c] Food fed/weight gain (dry weight basis).

[d] Protein requirement, grams per kilogram body weight per day = [dietary protein requirement (%) × feeding rate (%)]/10.

[e] Protein requirement, grams per kilogram live weight gain = dietary protein requirement (%) × FCR × 10.

Table 6.1: continued

Species	Protein requirement	Remarks	Reference
Oreochromis mossambicus	613	Reports optimal dietary protein level of 40%; actual value is 42% from diet analyses. Best FCR observed with fish fed 57.4% protein diet.	Jauncey, 1982
O. niloticus	623	Growth and FCR became poorer above dietary protein concentration of 35%. FCR value quoted is suspect in view of feeding rate employed and SGR observed.	Santiago *et al.*, 1982
O. aureus	572	FCR improved with increasing dietary protein level. Maximum protein level tested only 36%. FCR value quoted is suspect in view of feeding rate employed and SGR observed.	Davis & Stickney, 1978
O. aureus	?	Diet composition not related to reported diet analyses. No accurate data given on feed intake or FCR. Diets low in lipid, no plant oil used.	Winfree & Stickney, 1981
Tilapia zilli	420	Reduced growth above 35% dietary protein level. Maximum carcass protein deposition observed at 30% protein level. FCR value quoted is suspect in view of feeding rate employed and SGR observed.	Mazid *et al.*, 1979
Channa micropeltes	650	FCR decreased and SGR increased with increasing dietary protein level. Feed intake fixed at 2% d^{-1} thus feed intake on low protein diets restricted.	Wee & Tacon, 1982
Fugu rubripes	?	Reports optimal dietary protein content of 50%, actual value c. 47%. Author states feeding regime of 10% bw d^{-1}, with SGR of 3.79% d^{-1} and FCR of 0.88. Error in calculation, therefore analysis of data not possible.	Kanazawa *et al.*, 1980
Chanos chanos	766	Reduced growth with 60% protein diet. Poor FCR observed with 40% protein diets suggests under feeding of fry.	Lim *et al.*, 1979
Ctenopharyngodon idella	?	No data given on feed intake and FCR. Feed intake restricted to lowest ad lib. intake observed for all treatments. Fish underfed, low SGR.	Dabrowski, 1977

Species	Value	Comments	Reference
Chrysophrys aurata	653	Feed intake decreased with increasing dietary protein and energy content. No reduced growth observed at high protein levels.	Sabaut & Luquet, 1973
Morone saxatilis	610	Alpha-cellulose used within high protein diets to replace high dietary oil levels used within low protein diets.	Millikin, 1983
M. saxatilis	553	Increased growth with increasing dietary protein level. Maximum level of dietary protein tested was 55%.	Millikin, 1982
Cyprinus carpio	?	Maximum carcass protein deposition and growth observed with 38% and 55% protein diets, respectively. Food intake increased with increasing protein levels — suggests underfeeding. No fish oil used in diets.	Ogino & Saito, 1970
C. carpio	536	No reduced growth observed at high protein levels. No plant oil used in diets.	Jauncey, 1981
Anguilla japonica	?	No weight increase observed above 44.5% protein level. Poor growth.	Nose & Arai, 1973
Micropterus dolomieui	633	FCR improved with increasing dietary protein content up to 68.2%. Feed intake decreased with increasing dietary protein and energy content.	Anderson et al., 1981
M. salmoides	620–695	FCR improved with increasing dietary protein level up to 67% protein.	Anderson et al., 1981
Pleuronectes platessa	665	Maximum weight gain observed with 50% protein diet. Low feed intake and SGR.	Cowey et al., 1972
Salvelinus alpinus	?		Jobling & Wandsvik, 1983
Salmo gairdneri	?	No data on feed intake and FCR. Poor growth.	Austreng & Refstie, 1979
S. gairdneri	560	Growth and FCR improved with increasing dietary protein content. Feed intake fixed within all groups to lowest recorded ad lib. feed intake observed.	Satia, 1974
S. gairdneri	476–500	Protein requirement increased from 40 to 45% with increasing salinity. No data on feed intake.	Zeitoun et al., 1973

6.3.1 Diet Formulation and Composition

Over nine different protein sources have been used for the determination of the dietary protein requirement in fishes. These range from purified proteins (casein, gelatin, albumin) to whole protein composites (fish meal, soybean meal, cod muscle) used either alone or in combination with free amino acid supplements (often so as to simulate the amino acid profile of whole hen's-egg protein). In view of the different essential amino acid (EAA) compositions of these protein sources and the hypothesis that the nutritive value of a food protein is generally governed by how close its available EAA profile meets the animal's own requirement profile, one would expect these different protein sources, if not supplemented to a common amino acid pattern, to produce different dietary protein requirement optima within the same species. Similarly, depending on the feeding regime employed, fishes vary in their ability to utilize free amino acids (Yamada *et al.*, 1981a), and consequently when certain fish species are fed amino acid supplemented diets they often display sub-optimal growth and feed conversion efficiency compared with the same fish species fed a 'whole' protein ration (DeLong *et al.*, 1958; Winfree & Stickney, 1981). Coupled with the low palatability of purified proteins to carnivorous fishes (particularly those species that rely heavily on olfaction to detect their food), and the necessity to achieve optimum feed intake and growth, it is imperative that a standard reference protein be used by all workers which is both equally palatable to all species and which would supply an optimal EAA profile. In view of the close similarity between the dietary EAA requirement of fishes and the EAA profile of fish muscle/carcass (Cowey & Tacon, 1983; Ogata *et al.*, 1983), it is recommended that a lipid-extracted fish meal or lipid-extracted fish muscle be used as the standard reference protein.

In addition to using different dietary protein sources, workers also differ in their choice of an energy source to substitute for protein in the lower protein diets, and in their choice of calorific equivalents assigned to the major nutrients in the formulation of so-called 'isoenergetic diets'. Not only do individual protein substitutes (commonly either dextrin, starch, cellulose or lipid) differ in their respective gross (total) energy content and digestible energy value to individual fish species (Cho *et al.*, 1982; Jobling, 1983a) and in their ability to 'spare' dietary protein (Page & Andrews, 1973; Takeuchi *et al.*, 1978; Gropp *et al.*, 1982), but fishes also vary in their metabolic response to certain protein substitutes when used at high dietary concentrations (Hilton *et al.*, 1983). For example, the dependence of carnivorous fishes on a high protein diet may in fact derive not only from a highly developed capacity to metabolize protein but also from a limited ability to catabolize carbohydrates such as dextrin or starch which are commonly used as dietary protein substitutes (Cowey & Luquet, 1983). Similarly, the use of dietary protein substitutes with differing dietary energy densities has generally led to the formulation of high protein rations having a higher digestible energy value than the corresponding low-protein diets (Jobling, 1983a).

6.3.2 Feeding Regime and Growth

Since 'optimal growth' is generally used as the criterion for estimating dietary protein requirement, it is essential that food supply is not limiting. In studies with terrestrial farm animals, ad lib. feeding is relatively easily attained by allowing the confined animal continuous access to food and by weighing any remaining uneaten food. However, with fishes the situation is complicated by the aquatic environment due to the rapid deterioration of uneaten food in water, and by the different feeding habits of individual fish species. For example, the frequency of food presentation required to promote optimum growth and feed efficiency has been shown to vary from juvenile rainbow trout (*Salmo gairdneri*) which require 'satiation' feeding only twice daily (Grayton & Beamish, 1977) to the first feeding larval phase of common carp (*Cyprinus carpio*), which requires continuous or very regular feeding throughout a 24-h cycle (Bryant & Matty, 1981). Feeding regimes which have been employed have generally been related to the convenience of the researcher during his or her working day, rather than to the fish species' own requirement. For example, the feeding regimes used by individual workers listed in Table 6.1 included the following:

(a) All fish fed a restricted ration at a fixed percentage of body weight per day. Feeding rates employed varied from 2 to 20% per day (depending on the fish species and age class) with a feeding frequency of between 2 and 4 feeds over a 12-h working day (Davis & Stickney, 1978; Lim *et al.*, 1979; Mazid *et al.*, 1979; Kanazawa *et al.*, 1980; Jauncey, 1981, 1982; Winfree & Stickney, 1981; Santiago *et al.*, 1982; Wee & Tacon, 1982).

(b) All fish fed a restricted ration at a fixed rate based on the lowest ad lib. feed intake observed within any group, with a feeding frequency of between 2 and 3 feeds over a 12-h working day (Satia, 1974; Dabrowski, 1977).

(c) All fish fed to 'excess' at a fixed undetermined rate by automatic feeder (Austreng & Refstie, 1979).

(d) All fish fed to 'satiation' with a feeding frequency of between 2 and 4 feeds over a 12-h working day (Cowey *et al.*, 1972; Nose & Arai, 1973; Sabaut and Luquet, 1973; Zeitoun *et al.*, 1973; Anderson *et al.*, 1981; Millikin, 1982, 1983; Jobling & Wandsvik, 1983).

By far the most common feeding method employed has been the use of a fixed feeding regime in which the feeding level has been set arbitrarily. In contrast to the preferred feeding method where fish are fed to satiation at an optimum feeding frequency, a fixed feeding regime directly influences the outcome of the observed dietary requirement. Thus, Ogino (1980a) reported a decrease in the dietary protein requirement of juvenile carp and rainbow trout from 60–65% to 30–32% when the feeding level was increased from 2 to 4% body weight per day in both species (Figure 6.1). In view of the above relationship between feeding level and dietary protein requirement, and the arbitrary way in which the feeding levels/frequency are set by most researchers, one may well question the value of requirements obtained in this manner. Coupled with the higher digestible energy content of high-protein diets (particularly when using non-digestible fibre sources as the dietary protein substitute – Mazid *et al.*, 1979; Winfree & Stickney, 1981; Millikin, 1983), and the belief that fish eat (if offered palatable feed) to satisfy their energy requirement (Page & Andrews, 1973; Sabaut & Luquet, 1973; Grove *et al.*, 1978; Hunt, 1980), it is not surprising, therefore, that a restricted/fixed feeding regime will favour the growth of those fish fed the higher protein diets (Ogino & Saito, 1970; Wee & Tacon, 1982; Table 6.1). For example, the high protein requirement observed for grass carp fry (41–43%; Dabrowski, 1977) almost certainly arose from all experimental fish being fed a restricted ration (fed only twice daily, and fixed on the lowest recorded ad lib. feed intake), and consequently fish fed the lower protein diets not being able to consume sufficient feed to meet their dietary protein and energy requirements.

Although a fixed feeding regime does allow direct comparisons between different laboratories employing similar feeding levels, no absolute estimate of dietary requirement can be made unless a series

Figure 6.1: Relationship Between Feeding Rate and the Dietary Crude Protein Level Needed to Satisfy the Protein Requirement of Common Carp and Rainbow Trout (12–13 g protein per kilogram body weight per day; Ogino, 1980a). Each vertical line represents the combined limits of protein requirement for both species at each feeding level.

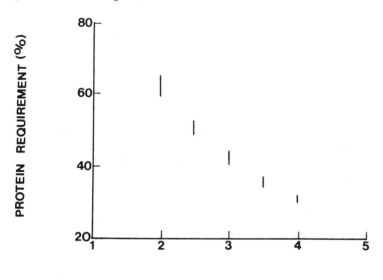

FEEDING RATE ($\%$ bw day^{-1})

of different feeding levels are tested so as to elicit a maximum growth response. Experiments with a number of fish species, including rainbow trout, have shown that maximum daily food intake can occur with as few as two satiation feedings per day (Grayton & Beamish, 1977), but these experiments have usually involved only one diet, and the results cannot necessarily be regarded as valid for an experiment in which a series of diets (of different protein and digestible energy contents) are being tested. If a feeding programme involving different feeding levels is not possible, then a feeding regime involving satiation feeding at an optimal feeding frequency is recommended. However, it should be emphasized that optimal feeding frequency is itself dependent on a variety of factors including fish size, meal size, the anatomy of the gastro-intestinal tract (gut length, stomach fullness), the digestible energy content of the diet, and temperature — all of which have been shown to effect the return of appetite in fish (Elliott, 1975; Grove *et al.*, 1978; Gwyther & Grove, 1981).

In view of the different dietary formulations employed by individual workers and the association between food consumption and dietary energy density (Sabaut & Luquet, 1973; Cho *et al.*, 1976; Hunt, 1980), it may be more meaningful to express protein requirement as digestible protein energy relative to the digestible energy content of the diet. Although this method of expression remains to be tested (since no direct measurements of nutrient and diet digestibility have been made in the studies presented in Table 6.1), the common practice of calculating nutrient digestibility using 'standard' digestibility coefficients, often obtained from other fish species under widely different culture conditions, should be strongly discouraged (for review see Jobling, 1983a).

To allow valid comparisons between results from different laboratories, it is clear that the protein requirements of fish can no longer be simply expressed solely in percentage terms or as a protein: energy ratio. It is in fact necessary to express dietary nutrient requirements in terms of feed intake (grams or kilojoules of nutrient required per kilogram body weight per day) and to a lesser extent to weight gain (grams or kilojoules of nutrient required per kilogram live weight gain). Table 6.1 therefore also shows the protein requirement of the fish species examined as grams of protein required per kilogram body weight per day, and as grams of protein required per kilogram live weight gain (where available information has allowed this calculation). To be strictly correct these requirements should be expressed on a digestible protein intake basis. However, despite this, the following tentative conclusions can be drawn from the gross requirement values calculated in Table 6.1.

(a) There exists an almost linear relationship between daily protein requirement (grams of protein per kilogram body weight per day) and the specific growth rate (SGR, per cent per day) of the different fish species examined (Figure 6.2). This correlation is perhaps not surprising since fish growth (protein synthesis) and metabolic activity are paced at gradually declining rates with increasing fish size and/or decreasing water temperature (Brett & Groves, 1979). For example, the highest recorded daily protein requirement and SGR were observed with the rapidly growing first feeding fry of the tropical omnivorous cichlid *Oreochromis niloticus* (52.5 g protein per kilogram body weight per day, and SGR 6.90% d[1]; Table 6.1). At present no information is available on the quantitative protein requirements of the first feeding fry of carnivorous fish species.

(b) Despite the large differences observed in daily protein requirement, the near-linear relationship between protein requirement
and SGR implies that the utilization of dietary protein for new
tissue growth is relatively constant within and between the
individual fish species examined (Figure 6.2). This is surprising,
since it might be expected that carnivorous fishes as compared
with omnivorous or herbivorous fish species would derive a

Figure 6.2: Relationship Between Optimal Dietary Protein Requirement (Grams
of Protein Per Kilogram Body Weight Per Day) and the Observed Specific Growth
Rate (Per Cent Per Day) of Several Fish Species of Different Age Classes.

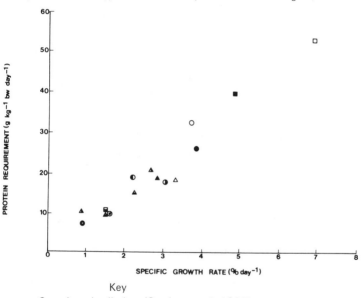

Key

□ - *Oreochromis niloticus* (Santiago *et al.*, 1982)

■ - *Chanos chanos* (Lim *et al.*, 1979)

○ - *Tilapia aurea* (Davis & Stickney, 1978)

● - *Oreochromis mossambicus* (Jauncey, 1982)

△ - *Cyprinus carpio* (Jauncey, 1981)

▲ - *Salmo gairdneri* (Satia, 1974)

◑ - *Pleuronectes platessa* (Cowey *et al.*, 1972)

▫ - *Channa micropeltes* (Wee & Tacon, 1982)

◐ - *Tilapia zilli* (Mazid *et al.*, 1979)

◐ - *Morone saxatilis* (Millikin, 1983)

▲ - *Micropterus dolomieui* (Anderson *et al.*, 1981)

▲ - *Micropterus salmoides* (Anderson *et al.*, 1981)

⊙ - *Chrysophrys aurata* (Sabaut & Luquet, 1973)

greater proportion of their metabolic. energy through protein catabolism. However, it may be that the high dietary protein requirement of carnivorous fish species is offset by the higher post-prandial oxygen consumption rate (specific dynamic action, SDA) and the lower tissue protein/energy deposition efficiency observed with fishes fed such rations (Cho *et al.*, 1976, 1982).

(c) Finally, the dietary protein requirements of fishes are not dissimilar from those of terrestrial farm animals when expressed relative to feed intake (grams of protein per kilogram body weight per day and live weight gain (grams of protein per kilogram live weight gain). However, since terrestrial farm animals are maintaining body temperature well above ambient, they have a higher total energy requirement for maintenance. Consequently, at the dietary level the relationship between protein requirement and growth is masked by the higher proportion of non-protein calories present. For example, the maintenance requirement per unit body weight of the chick is reported to be about five times greater than that of common carp growing at 23°C (Nijkamp *et al.*, 1974).

6.3.3 Abiotic Factors — Temperature and Salinity

The influence of water temperature on protein requirement and fish growth has been the subject of numerous investigations. The early study of DeLong and co-workers (1958) with fingerling Chinook salmon (*O. tshawytscha*) was said to show an increase in the dietary protein requirement from 40 to 55% with an increase in water temperature from 8.3 to 14.4°C (DeLong *et al.*, 1958). More recently a similar rise in dietary protein requirement was reported in fingerling striped bass (*Morone saxatilis*) from 47 to 55% with an increase in water temperature from 20.5 to 24.5°C (Millikin, 1983; Table 6.1). In contrast, fingerling rainbow trout (*Salmo gairdneri*) showed no difference in growth at dietary protein levels of 35, 40 and 45% at temperatures of 9, 12, 15 and 18°C in one study (Slinger *et al.*, 1977) or in another study with temperatures of 9, 15 and 18°C (Cho & Slinger, 1978). Although distinct temperature effects were observed in terms of growth, the greater absolute need for protein at the higher water temperatures was apparently satisfied through the increased consumption of the lower protein diets. These latter studies are in line with the observations made in Figure 6.2, and the hypothesis that an increase in water temperature (up to an optimum level) is accompanied by an increased feed intake (Brett *et al.*, 1969; Choubert *et al.*,

1982), increased growth rate and metabolic rate (Jobling, 1983b) and a faster gastro-intestinal transit time (Ross & Jauncey, 1981; Fauconneau *et al.*, 1983) under conditions where food supply is not limiting. The weight of evidence is that increased water temperature does not lead to increased protein requirement. In both cases where such a requirement was claimed, the effect of water temperature on dietary protein requirement was investigated by comparing the results obtained in successive experiments at different water temperatures. In addition, the sub-optimal growth and increased feed intake observed with fish fed the higher protein diets suggests that the ad lib. feeding regime employed in fact led to a restricted feed intake.

Very few studies have been undertaken concerning the effect of salinity on protein requirement. Experiments conducted with fingerling rainbow trout (a euryhaline fish) are reported to show an increase in the absolute dietary requirement for protein from 40 to 45% with a salinity increase from 10 to 20 parts per thousand (Zeitoun *et al.*, 1973; Table 6.1). However, no increase in dietary protein requirement was observed in a similar experiment conducted with coho salmon fingerlings (*O. kisutch*; Zeitoun *et al.*, 1974). In view of the speculative method for arriving at dietary requirement from the dose–response curve (Zeitoun *et al.*, 1973), and the lack of information on the protein requirement of these fish species in full strength sea water (35 parts per thousand), there are no firm data demonstrating that the protein requirements of fish are elevated with increased salinity.

6.4 Amino Acid Requirement

In common with terrestrial farm animals, fish require the same ten indispensable or essential amino acids (EAA) within their diet: namely threonine, valine, methionine, isoleucine, leucine, phenylalanine, lysine, histidine, arginine and tryptophan. The quantitative EAA requirements of fish have traditionally been determined by feeding graded levels of each amino acid within an amino acid test diet so as to elicit a dose–response curve (for reviews see Ketola, 1982; Cowey & Luquet, 1983). Dietary requirement is then usually taken at 'break-point' on the basis of the observed growth response. Several workers have also used free amino acid levels within specific tissue pools (whole blood, plasma or muscle; Kaushik, 1979) or the oxidation of radioactively labelled amino acids (administered orally or by injection; Walton *et al.*, 1982) as the criteria for estimating dietary requirement.

Within these experimental test diets the protein component is supplied almost entirely in the form of crystalline amino acids or in combination with selected 'whole' protein sources (commonly either casein, gelatin, zein, gluten or fish meal). Although more recent studies have favoured the use of a higher proportion of 'whole' protein sources within amino acid test diets (Kaushik, 1979; Jackson & Capper, 1982), the overall amino acid profile of the total protein component of the diet is usually carefully controlled so as to simulate the amino acid profile of a specific reference protein (with the exception of the amino acid under test). For example, amino acid patterns commonly simulated within experimental test diets include: casein/gelatin mixtures (3:2 or 38:12, Halver, 1957; Nose, 1979; Walton *et al.*, 1982), whole hen's-egg protein (Wilson *et al.*, 1978; Robinson *et al.*, 1981; Kim *et al.*, 1983), fish meal/soybean/groundnut meal composites (Jackson & Capper, 1982), fish meal (Rumsey *et al.*, 1983), whole trout egg (Ketola, 1983), and zein/fish meal (1:1, Kaushik, 1979).

In contrast to the above standard method in which fishes are fed graded levels of crystalline amino acids, Ogino (1980b) determined the quantitative EAA requirement of fishes simultaneously on the basis of the daily increase (retention) of individual amino acids within the fish carcass. Fishes were fed a diet containing a 'whole' protein source of high biological value, and using the observed daily retention values for each amino acid within the carcass of growing fish, the dietary EAA requirement was then computed, dietary requirements being based on a diet containing 35 or 40% protein, having a protein digestibilty of 80 or 90%, and assuming a feeding rate of 3% of the body weight per day.

Table 6.2 summarizes the quantitative EAA requirements of the fish species which have been studied to date using the above-mentioned techniques. To allow comparisons to be made between individual fish species for specific amino acids, the dietary requirements are also expressed as a percentage of the total dietary protein and relative to feed intake milligrams of amino acid required per kilogram body weight per day).

6.4.1 EAA Requirements Based on Feeding Graded Levels of Amino Acids

Quantitative dietary requirements for all 10 EAAs have been established for only four fish species (common carp, Japanese eel (*Anguilla japonica*), channel catfish (*Ictalurus punctatus*) and the

Table 6.2: Quantitative Essential Amino Acid (EAA) Requirements of Selected Fish Species. Values are expressed in order as a percentage of the dietary protein, a percentage of the dry diet (the denominator being the percentage protein in the diet), and as milligrams EAA per kilogram body weight per day (where available data exist).

Species	Temperature (°C)	Simulated amino acid (AA) profile of protein source	Feeding regime[1]	Initial body weight (g)	Arginine
C. carpio	25±0.5	Casein: gelatin (38:12)	Ad lib., 4f d⁻¹	0.5–4.0	3.3(1.3/38.5);—
I. punctatus	26.7±1	Whole hen's egg	3% bw d⁻¹, 3f d⁻¹	2–10	4.29(1.03/24);309
O. tshawytscha	10	Whole hen's egg	Ad lib., 3f d⁻¹	2–4	6.0(2.4/40);—
A. japonica	?	?	?	?	3.9(1.7/42);—
S. gairdneri	?	Whole hen's egg	?	12–14	>4.0(1.4/35);—
S. gairdneri	9.4	Whole trout egg	Fixed, ?	1–2	5.4–5.9(2.5–2.8/47);—
S. gairdneri	14	Fish meal	4.5% bw d⁻¹, 3f d⁻¹	1.5–9	3.43(1.2/35);210
S. gairdneri	12	Zein: fish meal (1:1)	Ad lib., 4f d⁻¹	20–30	—
S. gairdneri	15	Casein: gelatin (3.2)	2% bw d⁻¹, 3f d⁻¹	27	3.5–4.0(1.6–1.8/45); 800–900
O. kisutch	8–15	White cod muscle	2–5bw d⁻¹, 4f d⁻¹	5–14	6.0(2.4/40);—
O. mossambicus	10	Whole hen's egg	Ad lib.,3f d⁻¹	2–4	<4.0(1.59/40); 636
C. carpio	25±0.5	Fish meal composite	4% bw d⁻¹, 3f d⁻¹	1.7	3.8(1.52/40);456
S. gairdneri	?				3.5(1.4/40);420

Calculated on the basis of tissue deposition of EAA, with fish fed a whole protein source of high biological value having a protein digestibility of 80%, and a feeding rate of 3% bw d⁻¹ for both species (carp 62–74 g, 20–25°C; trout 68–127 g, 15–18°C)

[1] Feeding regime: indicates feeding level and number of feedings per day.

Table 6.2: continued

Species	Histidine	Isoleucine	Leucine	Lysine	Methionine[1]
C. carpio	2.1(0.8/38.5);—	2.5(0.9/38.5);—	3.3(1.3/38.5);—	5.7(2.2/38.5);—	2.1(0.8/38.5);—[a]
I. punctatus	1.54(0.37/24);111	2.58(0.62/24);186	3.5(0.84/24);252	5.1(1.5/30);450	1.34(0.32/24);96[b]
O. tshawytscha	1.8(0.7/40);—	2.2(0.9/41);—	3.9(1.6/41);—	5.0(2.0/40);—	1.5(0.6/40);—[c]
A. japonica	1.9(0.8/42);—	3.6(1.5/42);—	4.8(2.0/42);—	4.8(2.0/42);—	2.1(0.9/42);[d]
S. gairdneri				3.7(1.3/35);—	—
S. gairdneri				6.1(2.9/47);—	1.57–2.14 (0.55–0.75/35)[e]; 247–337
S. gairdneri				—	
S. gairdneri				—	
S. gairdneri				—	
S. gairdneri				4.3(1.95/45);430	1.0(0.5/50);100[f]
O. kisutch	1.7(0.7/40);—			—	
O. mossambicus	—	2.3(0.92/40);276	4.1(1.64/40);492	4.1(1.62/40);648	<1.33(0.53/40); 212[g]
C. carpio	1.4(0.56/40);168	2.4(0.96/40);288	4.4(1.76/40);528	5.3(2.12/40);636	1.6(0.64/40);192
s. gairdneri	1.6(0.64/40);192			5.3(2.12/40);636	1.8(0.72/40);216

[2] In the presence of dietary cystine (a, 2%; b, 0.24%; c, 1%; d, 1%; e, 0.3%; f, 2%; g, 0.74%).

Table 6.2: continued

Species	Methionine[3]	Phenylalanine[4]	Phenylalanine[5]	Threonine
C. carpio	3.1(1.2/38.5);—	3.4(1.3/38.5);—[h]	6.5(2.5/38.5);—	3.9(1.5/38.5);—
I. punctatus	2.34(0.56/24);168	2.04(0.49/24);147[i]	5.0(1.2/24);360	2.21(0.53/24);159
O. tshawytscha	—	4.1(1.7/41);—[j]	—	2.2(0.9/40);—
A. japonica	2.9(1.2/42);—	2.9(1.2/42);—[k]	5.2(2.2/42);—	3.6(1.5/42);—
S. gairdneri	—	—	—	—
S. gairdneri	—	—	—	—
S. gairdneri	—	—	—	—
S. gairdneri	1–2(0.5–1/50);100–200	—	—	—
S. gairdneri	—	—	—	—
O. kisutch	—	—	—	—
O. mossambicus	—	2.9(1.16/40);348	—	—
C. carpio	—	3.1(1.24/40);372	—	3.3(1.32/40);396
S. gairdneri	—	—	—	3.4(1.36/40);408

3 In the absence of dietary cystine.
4 In the presence of dietary tyrosine (h, 1%; i, 1%; j, 0.4%; k, 2%).
5 In the absence of dietary tyrosine.

Table 6.2: continued

Species	Tryptophan	Valine	Reference
C. carpio	0.8(0.3/38.5);—	3.6(1.4/38.5);—	Nose, 1979
I. punctatus	0.5(0.12/24);36	2.96(0.71/24);213	N.R.C., 1983
O. tshawytscha	0.5(0.2/40);—	3.2(1.3/40);—	N.R.C., 1983
A. japonica	1.0(0.4/42);—	3.6(1.5/42);—	Nose, 1979
S. gairdneri	—	—	Kim et al., 1983
S. gairdneri	—	—	Ketola, 1983
S. gairdneri	—	—	Rumsey et al., 1983
S. gairdneri	—	—	Kaushik, 1979
S. gairdneri	0.45(0.25/55);50	—	Walton et al., 1982
O. kisutch	0.5(0.2/40);—	—	Klein & Halver, 1970
O. mossambicus	—	—	Jackson & Capper, 1982
C. carpio	0.6(0.24/40);72	2.9(1.16/40);348	Ogino, 1980b
S. gairdneri	0.5(0.2/40);60	3.1(1.24/40);372	Ogino, 1980b

Chinook salmon). In the main these requirements have been based on the results obtained from a single laboratory and only one experiment. Although numerous independent studies have recently been performed on the amino acid requirements of rainbow trout, significant differences in requirement (grams of amino acid per 100 g protein) exist within and between individual fish species (Table 6.2). For example, differences of the order of 65, 72 and 114% were observed between independent laboratories for the lysine, arginine and methionine requirement of rainbow trout (*c.* 1–30 g body weight). Similarly, inter-species variations ranged between 22% (for valine) to as high as 122% (for tryptophan). Whereas one would have expected the quantitative EAA requirements of fish to decrease with age and decreasing tissue protein synthesis (growth), one may well question whether or not the observed variations in requirement are real or merely an artifact of the method employed. The lack of precision of this feeding method can be viewed at the following levels.

(a) The formulation of the amino acid test diet. At present, widely different dietary protein sources are used and there is equally wide variation in the reference protein whose amino acid pattern is being mimicked. A more meaningful comparison would be possible if the same standard reference protein found general use, preferably a low-lipid fish muscle or fish meal (Table 6.3).

(b) Fishes fed rations in which a significant proportion of the dietary protein is supplied in the form of free amino acids generally display sub-optimal growth and feed conversion efficiency compared with those fed protein-bound amino acids (Wilson *et al.*, 1978; Robinson *et al.*, 1981; Yamada *et al.*, 1981a; Walton *et al.*, 1982). Amino acid requirements are therefore ascertained under conditions of sub-optimal growth.

(c) Dietary free amino acids are more rapidly assimilated in fish than protein-bound amino acids. Experiments with rainbow trout (Yamada *et al.*, 1981b), common carp (Plakas *et al.*, 1980) and tilapia (*Oreochromis niloticus*; Yamada *et al.*, 1982) fed free amino acid diets showed that peak plasma amino acid concentrations occurred sooner (12–24 h, 2–4 h, 2 h, respectively) than with an equivalent casein-based diet (24–36 h, 4 h, 4 h, respectively). Furthermore, in carp, individual free amino acids appear to be absorbed at varying rates from the gastro-intestinal tract, and consequently peak plasma concentrations for individual amino acids do not occur simultaneously (Plakas *et al.*, 1980).

Table 6.3: Mean Fish EAA Requirement Ratio (%)[a] in Selected Fish Tissues and Commonly Used Dietary Reference Protein Sources.

Amino acid	Fish carcass Ogino (1980b)	Casein: gelatin (38:12) Arai et al., 1971	Casein Plakas et al., 1980	Whole hen's-egg protein Cowey & Sargent, 1972	Fish eggs Ketola, 1982	Fish meal Jackson et al., 1982	Cod muscle Cowey & Sargent, 1972
Threonine	100	67[b]	79[b]	79[b]	89	65[b]	80
Valine	100	116	121[b]	125[b]	123[b]	99	99
Methionine	100	99	100	102	93	120	111
Isoleucine	101	121[b]	130[b]	134[b]	136[b]	111	114
Leucine	100	119	125[b]	109	113	112	113
Phenylalanine	99	102	94	97	83	82	79[b]
Lysine	101	86	89	79[b]	83	94	107
Histidine	100	114	113	95	104	110	94
Arginine	100	83	61[b]	95	94	117	98
Tryptophan	100	175[b]	149[b]	129[b]	91	127[b]	114

[a] Based on the mean quantitative EAA requirements of rainbow trout and common carp (Ogino, 1980b).
[b] Variation from fish requirement ratio greater than 20%.

For optimal protein synthesis to occur, it is essential that all amino acids (whether they be derived from whole proteins or amino acid supplements) are presented simultaneously to the tissue. If such an equilibrium is not achieved, then amino acid catabolism ensues with consequent loss of growth and feed efficiency. For those warm-water fish species which display a rapid uptake and assimilation of free amino acids, it is therefore essential that either:

(i) the release or absorption of free amino acids from the diet is reduced so as to minimize the variations in absorption rate observed between free and protein-bound amino acids (achieved by coating individual amino acids with casein; Murai *et al.*, 1982); or

(ii) that the frequency of feed presentation is increased from 2 or 3 feeds per day to up to 18 feeds per day so as to minimize the variations observed in plasma amino acid concentration

(Yamada *et al.*, 1981a).

(d) Amino acid requirements are influenced by interactions among the EAAs themselves, between essential and non-essential amino acids, and between amino acids and other nutrients. For example, if cystine is deficient in the diet, it can be synthesized by fish from methionine; the requirement for methionine is therefore partially dependent on the cystine content of the diet. A similar relationship also exists between phenylalanine and the non-essential amino acid tyrosine (for review see Ketola, 1982; Table 6.2). Particular consideration must therefore be given to the dietary concentration of these non-essential amino acids.

(e) Finally, interpretation of the dose–response curve itself is very often suspect, since 'breakpoint' is not always clearly discernible.

6.4.2 EAA Requirements Based on Carcass Deposition

Ogino (1980b) determined the EAA requirement of rainbow trout and common carp on the basis of the observed daily carcass deposition of individual amino acids. In contrast to the variations in requirement observed for the same fish species fed conventional amino acid test diets, there was no significant difference in the EAA requirement of carp and trout on the basis of the carcass deposition method (Table 6.2). However, although quantitative differences in requirement (milligrams of amino acid per kilogram body weight per day) do exist within fishes of different age classes with respect to growth, the dietary requirements observed are within the range reported for fish fed amino acid test diets (Table 6.2).

Interestingly, recalculation of the data obtained by Ogino (1980b) shows that there is no difference between the relative proportions of individual EAAs required in the diet and the relative proportions of the same 10 EAAs present within the fish carcass (Table 6.3). A similar relationship has also been seen in the growing pig and chick (Boorman, 1980), and to a lesser extent within the 4 fish species for which EAA requirements have been determined using amino acid test diets (Figure 6.3). Since the EAA profile of fish muscle protein does not differ greatly (if at all) between individual fish species (Connell & Howgate, 1959; Njaa & Utne, 1982), it follows, therefore, that the pattern of requirement for individual fish species will also be similar (Cowey & Tacon, 1983).

Compared with the conventional method of feeding graded levels of individual amino acids, the carcass deposition method of Ogino (1980b) offers numerous advantages:

Figure 6.3: Relationship Between Pattern of EAA Requirements Found by Feeding Experiments Using Amino Acid Test Diets With Carp (●), Japanese Eel (■), Channel Catfish (□) and Chinook Salmon (○) and the Pattern of the Same Amino Acids in Fish Carcass. The level of each amino acid is represented as a percentage of the sum of all 10 EAAs in each pattern. The line represents coincidence of requirement and tissue patterns.

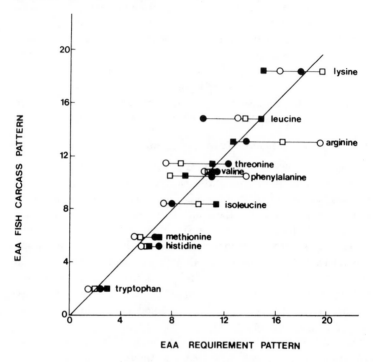

(a) Fish are fed rations in which the protein component is supplied in the form of a 'whole' protein of high biological value. Amino acid requirements can therefore be ascertained in fish displaying optimal growth.

(b) The dietary requirement for all 10 EAAs can be determined simultaneously in one single experiment. Using conventional amino acid test diets, up to 10 separate experiments have to be performed, each experiment involving the use of up to 6 dietary regimes employing varying dietary concentrations of the single EAA under test.

(c) Quantitative EAA requirements can equally be established for first-feeding fry and brood-stock fish with no loss of precision.

Concluding Remarks

Examination of the methods used to identify minimal dietary protein level for optimal growth in many species of fishes has led to doubts that it has yet been possible to devise a series of diets ideally suited for the purpose. Particular problems relate to certainty or otherwise that all diets in any one experiment are more or less isoenergetic, and whether or not there might be marked differences in palatability of the low- and high-protein diets within a given experiment. For many species the oft-quoted view that fishes eat to meet an energy requirement holds true only if the food provided is palatable.

The general dietary protein requirement of an animal is recognized as the requirement for essential amino acids together with some requirement for non-specific nitrogen. In view of the difficulties involved in a satisfactory experimental design for evaluating optimal protein requirement, progress might be more rapid if the effort expended were concentrated on the quantification of essential amino acid requirement. In this area, problems of palatability and of constancy of dietary energy density between treatments need not arise. However, there are other problems, of which the main one is the utilization of free as compared with protein-bound amino acids. This may be brought within acceptable limits by adopting an appropriate feeding regime.

Large differences in requirements for essential amino acids have previously been noted between species, and these have been treated with caution. Agreement on quantitative requirements for certain amino acids between different laboratories for a given species (rainbow trout) has been little better. Confidence in the results obtained will only be realized when there is a much greater measure of agreement between different laboratories.

References

Anderson, R.J., Kienholz, E.W. and Flickinger, S.A. (1981), Protein requirements of smallmouth bass and largemouth bass. *J. Nutr. 111*, 1085

Arai, S., Nose, T. and Hashimoto, Y. (1971), A purified test diet for the eel, *Anguilla japonica*. *Tansuiku Suisan Kenkyusho Kenkyu Hokoku 21*, 161

Austreng, E. and Refstie, T. (1979), Effect of varying dietary protein level in different families of rainbow trout. *Aquaculture 18*, 145

Boorman, K.N. (1980), Dietary constraints on nitrogen retention. In: Buttery, P.J. and Lindsay, D.B. (eds). *Protein Deposition in Animals*, pp. 147–66. London, Butterworth

Brett, J.R. and Groves, T.D.D. (1979), Physiological Energetics. In: Hoar, W.S., Randall, D.J. and Brett, J.R. (eds). *Fish Physiology*, vol. 8, pp. 279–352. New York, Academic Press

Brett, J.R., Shelbourn, J.E. and Shoop, C.T. (1969), Growth rate and body composition of fingerling sockeye salmon, *Oncorhynchus nerka*, in relation to temperature and ration size. *J. Fish. Res. Bd Can. 26*, 2363

Bryant, P.L. and Matty, A.J. (1981), Adaptation of carp (*Cyprinus carpio*) larvae to artificial diets. 1. Optimum feeding rate and adaptation age for a commercial diet. *Aquaculture 23*, 275

Cho, C.Y. and Slinger, S.J. (1978), Effect of ambient temperature on the protein requirements of rainbow trout and on the fatty acid composition of gill phospholipids. In: *1977 Annual Report, Fish Nutrition Laboratory*, University of Guelph, Ontario, Canada

Cho, C.Y., Slinger, S.J. and Bayley, H.S. (1976), Influence of level and type of dietary protein, and of level of feeding on feed utilization by rainbow trout. *J. Nutr. 106*, 1547

Cho, C.Y., Slinger, S.J. and Bayley, H.S. (1982), Bioenergetics of salmonid fishes: energy intake, expenditure and productivity. *Comp. Biochem. Physiol. 73B*, 25

Choubert, G., Fauconneau, B. and Luquet, P. (1982), Influence of a rise in water temperature on food dry matter, nitrogen and energy digestibility in rainbow trout (*Salmo gairdneri*). *Reprod. Nutr. Devel. 22*, 941

Connell, J.J. and Howgate, P.F. (1959), The amino acid composition of some British food fishes. *J. Sci. Fd Agric. 10*, 241

Cowey, C.B. (1975), Aspects of protein utilization by fish. *Proc. Nutr. Soc. 34*, 57

Cowey, C.B. and Luquet, P. (1983), Physiological basis of protein requirement of fishes. Critical analysis of allowances. In: Arnal, M., Pion, R. & Bonin, D. (eds). *Protein Metabolism and Nutrition*, vol. I, pp. 365–84. INRA, Paris

Cowey, C.B. and Sargent, J.R. (1972), Fish nutrition. *Adv. Mar. Biol. 10*, 383

Cowey, C.B. and Sargent, J.R. (1979), Nutrition. In: Hoar, W.S., Randall, D.J. and Brett, J.R. (eds). *Fish Physiology*, vol. VIII, pp. 1–69. New York, Academic Press

Cowey, C.B. and Tacon, A.G.J. (1983), Fish nutrition — relevance to marine invertebrates. In: Pruder, G.D., Langdon, C.J. and Conklin, D.E. (eds). *Proceedings of the Second International Conference on Aquaculture Nutrition: Biochemical and Physiological Approaches to Shellfish Nutrition*, pp. 13–30. Baton Rouge, Louisiana State University

Cowey, C.B., Pope, J.A., Adron, J.W. and Blair, A. (1972), Studies on the nutrition of marine flatfish. The protein requirement of plaice (*Pleuronectes platessa*). *Brit. J. Nutr. 28*, 447

Dabrowski, K. (1977), Protein requirements of grass carp fry (*Ctenopharyngodon idella* Val.). *Aquaculture 12*, 63

Davis, A.T. and Stickney, R.R. (1978), Growth responses of *Tilapia aurea* to dietary protein quality and quantity. *Trans. Am. Fish. Soc. 107*, 479

DeLong, D.C., Halver, J.E. and Mertz, E.T. (1958), Nutrition of salmonid fishes VI. Protein requirements of chinook salmon at two water temperatures. *J. Nutr. 65*, 589

Elliott, J.M. (1975), Number of meals in a day, maximum weight of food consumed in a day and maximum rate of feeding for brown trout, *Salmo trutta* L. *Freshwat. Biol. 5*, 287

Fauconneau, B., Choubert, G., Blanc, D., Breque, J. and Luquet, P. (1983), Influence of environmental temperature on flow rate of foodstuffs through the gastrointestinal tract of rainbow trout. *Aquaculture 34*, 27

Grayton, B.D. and Beamish, F.W.H. (1977), Effects of feeding frequency on food intake, growth and body composition of rainbow trout (*Salmo gairdneri*).

Aquaculture 11, 159

Gropp, J., Schwalb-Buhling, A., Koops, H. and Tiews, K. (1982), On the protein-sparing effect of dietary lipids in pellet feeds for rainbow trout (*Salmo gairdneri*). *Arch. FischWiss. 33*, 79

Grove, D.J., Loizides, L.G. and Nott, J. (1978), Satiation amount, frequency of feeding and gastric emptying rate in *Salmo gairdneri. J. Fish Biol. 12*, 507

Gwyther, D. and Grove, D.J. (1981), Gastric emptying in *Limanda limanda* (L) and the return of appetite. *J. Fish Biol. 18*, 245

Halver, J.E. (1957), Nutrition of salmonid fishes. III. Water-soluble vitamin requirements of chinook salmon. *J. Nutr. 62*, 225

Halver, J.E. (1972). *Fish Nutrition*. New York, Academic Press

Hilton, J.W., Atkinson, J.L. and Slinger, S.J. (1983), Effect of increased dietary fibre on the growth of rainbow trout (*Salmo gairdneri*). *J. Fish. Res. Bd Can. 40*, 81

Hunt, J.N. (1980), A possible relation between the regulation of gastric emptying and food intake. *Am. J. Physiol. 239*, G1

Jackson, A.J. and Capper, B.S. (1982), Investigations into the requirement of the tilapia, *Sarotherodon mossambicus* Licus, for dietary methionine, lysine and arginine in semi-synthetic diets. *Aquaculture 29*, 289

Jackson, A.J., Capper, B.S. and Matty, A.J. (1982), Evaluation of some plant proteins in complete diets for the tilapia (*Sarotherodon mossambicus*). *Aquaculture 27*, 97

Jauncey, K. (1981), The effects of varying dietary composition on mirror carp (*Cyprinus carpio*) maintained in thermal effluents and laboratory recycling systems. In: *Proceedings of World Symposium on Aquaculture in Heated Effluent and Recirculation Systems, vol. II*, pp. 247–61. Berlin, Heenemann

Jauncey, K. (1982), The effects of varying dietary protein level on the growth, food conversion, protein utilization and body composition of juvenile tilapias (*Sarotherodon mossambicus*). *Aquaculture 27*, 43

Jobling, M. (1983a), A short review and critique of methodology used in fish growth and nutrition studies. *J. Fish Biol. 23*, 685

Jobling, M. (1983b), Influence of body weight and temperature on growth rate of Arctic charr (*Salvelinus alpinus*). *J. Fish Biol. 22*, 471

Jobling, M. and Wandsvik, A. (1983), Quantitative protein requirements of Arctic charr, *Salvelinus alpinus* (L). *J. Fish Biol. 22*, 705

Kanazawa, A., Teshima, S., Sakamoto, M. and Shinomiya, A. (1980), Nutritional requirements of the puffer fish: purified test diet and optimum protein level. *Bull. Jpn Soc. Scient. Fish. 46*, 1357

Kaushik, S. (1979), Application of a biochemical method for the estimation of amino acid needs in fish: quantitative arginine requirements in different salinities. In: Halver, J.E. and Tiews, K. (eds). *Finfish Nutrition and Fishfeed Technology, vol. I*, pp. 197–207. Berlin, Heenemann

Ketola, H.G. (1982), Amino acid nutrition of fishes: requirements and supplementation of diets. *Comp. Biochem. Physiol. 73B*, 17

Ketola, H.G. (1983), Requirement for dietary lysine and arginine by fry of rainbow trout. *J. Anim. Sci. 56*, 101

Kim, K.I., Hayes, T.B. and Amundsen, C.H. (1983), Protein and arginine requirement of rainbow trout. *Fed. Proc. 42*, 2198 (Abstract)

Klein, R.G. and Halver, J.E. (1970), Nutrition of salmonid fishes: arginine and histidine requirement of chinook and coho salmon. *J. Nutr. 100*, 1105

Lim, C., Sukhawongs, S. and Pascual, F.P. (1979), A preliminary study on the protein requirement of *Chanos chanos* (Forskal) fry in a controlled environment. *Aquaculture 17*, 195

Mazid, M.A., Tanaka, Y., Katayama, T., Asadur Rahman, M., Simpson, K.L. and

Chichester, C.O. (1979), Growth response of *Tilapia zilli* fingerlings fed iso-calorific diets with variable protein levels. *Aquaculture 18*, 115

Millikin, M.R. (1982), Qualitative and quantitative nutrient requirements of fishes: a review. *Fish. Bull. 80*, 655

Millikin, M.R. (1983), Interactive effects of dietary protein and lipid on growth and protein utilization of age-0 striped bass. *Trans. Am. Fish. Soc. 112*, 185

Murai, T., Akiyama, T., Ogata, H., Hirasawa, Y. and Nose, T. (1982), Effect of coating amino acids with casein supplemented to gelatin diet on plasma free amino acids of carp. *Bull. Jpn Soc. Sci. Fish. 48*, 703

National Research Council (1981), *Nutrient Requirements of Coldwater Fishes*. National Academy Press, Washington, DC

National Research Council (1983), *Nutrient Requirements of Warmwater Fishes and Shellfishes*. National Academy Press, Washington, DC

Nijkamp, H.J., van Es, A.J.H. and Huisman, A.E. (1974), Retention of nitrogen, fat, ash, carbon, and energy in growing chickens and carp. *Eur. Assoc. Animal Prod. 14*, 277

Njaa, L.R. & Utne, F. (1982), A comparison of the amino acid composition of fifteen species of whole fish. *Fisk. Dir. Skr. Ernaering, 11*, 25

Nose, T. (1979), Summary report on the requirements of essential amino acids for carp. In: Halver, J.E. & Tiews, K. (eds). *Finfish Nutrition and Fishfeed Technology, vol. I*, pp. 145–56. Berlin, Heenemann

Nose, T. and Arai, S. (1973), Optimum level of protein in purified diet for eel, *Anguilla japonica*. *Bull. Freshwater Fish. Res. Lab. Tokyo 22*, 145

Ogata, H., Arai, S. and Nose, T. (1983), Growth responses of cherry salmon (*Oncorhynchus masou*) and Amago salmon (*O. rhodurus*) fry fed purified casein diets supplemented with amino acids. *Bull. Jpn Soc. Sci. Fish. 49*, 1381

Ogino, C. (1980a), Protein requirements of carp and rainbow trout. *Bull. Jpn Soc. Sci. Fish. 46*, 385

Ogino, C. (1980b), Requirements of carp and rainbow trout for essential amino acids. *Bull. Jpn Soc. Sci. Fish. 46*, 171

Ogino, C. and Saito, K. (1970), Protein nutrition in fish. 1. The utilization of dietary protein by young carp. *Bull. Jpn Soc. Sci. Fish. 36*, 250

Page, J.W. and Andrews, J.W. (1973), Interactions of dietary levels of protein and energy on channel catfish (*Ictalurus punctatus*). *J. Nutr. 103*, 1339

Plakas, S.M., Katayama, T., Tanaka, Y. and Deshimaru, O. (1980), Changes in the levels of circulating plasma free amino acids of carp (*Cyprinus carpio*) after feeding a protein and amino acid diet of similar composition. *Aquaculture 21*, 307

Robinson, E.H., Wilson, R.P. and Poe, W.E. (1981), Arginine requirement and apparent absence of a lysine–arginine antagonist in fingerling channel catfish. *J. Nutr. 111*, 46

Ross, B. and Jauncey, K. (1981), A radiographic estimation of the effect of temperature on gastric emptying time in *Sarotherodon niloticus* (L.) × *S. aurens* (Steindachner) hybrids. *J. Fish Biol. 19*, 333

Rumsey, G.L., Page, J.W. and Scott, M.L. (1983), Methionine and cystine requirement of rainbow trout. *Prog. Fish Cult. 45*, 139

Sabaut, J.J. and Luquet, P. (1973), Nutritional requirements of the gilthead bream *Chrysophrys aurata*. Quantitative protein requirements. *Mar. Biol. 18*, 50

Santiago, C.B., Banes-Aldaba, M. and Laron, M.A. (1982), Dietary crude protein requirement of *Tilapia nilotica* fry. *Philipp. J. Biol. 11*, 255

Satia, B.P. (1974), Quantitative protein requirements of rainbow trout. *Prog. Fish Cult. 36*, 80

Slinger, S.J., Cho, C.Y. and Holub, B.J. (1977), Effect of water temperature on protein and fat requirements of rainbow trout (*Salmo gairdneri*). In: *Proceedings 12th Annual Nutrition Conference for Feed Manufacturers*, pp. 1–5. Guelph, Ontario,

University of Guelph

Smith, R.R., Rumsey, G.L. and Scott, M.L. (1978), Heat increment associated with dietary protein, fat, carbohydrate, and complete diets in salmonids: Comparative energetic efficiency. *J. Nutr. 108*, 1025

Takeuchi, T., Yokoyama, M., Watanabe, T. and Ogino, C. (1978), Optimum ratio of dietary energy to protein for rainbow trout. *Bull. Jpn Soc. Sci. Fish. 44*, 727

Walton, M.J., Cowey, C.B. and Adron, J.W. (1982), Methionine metabolism in rainbow trout fed diets of differing methionine and cystine contents. *J. Nutr. 112*, 1525

Wee, K.L. and Tacon, A.G.J. (1982), A preliminary study on the dietary protein requirement of juvenile snakehead. *Bull. Jpn Soc. Sci. Fish. 48*, 1463

Wilson, R.P., Allen, O.W., Robinson, E.H. and Poe, W.E. (1978), Tryptophan and threonine requirements of fingerling channel catfish. *J. Nutr. 108*, 1595

Winfree, R.A. and Stickney, R.R. (1981), Effects of dietary protein and energy on growth, feed conversion efficiency and body composition of *Tilapia aurea. J. Nutr. 111*, 1001

Yamada, S., Tanaka, Y. and Katayama, T. (1981a), Feeding experiments with carp fry fed an amino acid diet by increasing the number of feedings per day. *Bull. Jpn Soc. Sci. Fish. 47*, 1247

Yamada, S., Simpson, K.L., Tanaka, Y. and Katayama, T. (1981b), Plasma amino acid changes in rainbow trout (*Salmo gairdneri*) force-fed casein and a corresponding amino acid mixture. *Bull. Jpn Soc. Sci. Fish. 47*, 1035

Yamada, S., Tanaka, Y., Katayama, T., Sameshima, M. and Simpson, K.L. (1982), Plasma amino acid changes in *Tilapia nilotica* fed a casein and a corresponding free amino acid diet. *Bull. Jpn Soc. Sci. Fish. 48*, 1783

Zeitoun, I.H., Halver, J.E., Ullrey, D.E. and Tack, P.I. (1973), Influence of salinity on protein requirements of rainbow trout (*Salmo gairdneri*) fingerlings. *J. Fish. Res. Bd Can. 30*, 1867

Zeitoun, I.H., Ullrey, D.E., Halver, J.E., Tack, P.I. and Magee, W.T. (1974), Influence of salinity on protein requirement of coho salmon (*Oncorhynchus kisutch*) smolts. *J. Fish. Res. Bd Can. 31*, 1145

7 THE HORMONAL CONTROL OF METABOLISM AND FEEDING

A.J. Matty and K.P. Lone

7.1 Introduction

Hormone in fishes, as in other animals, may be concerned with the regulation and control of many aspects of metabolism, ranging from mineral metabolism, osmoregulation and other physiological activities, including colour change, to behavioural changes. However, in keeping with the title of this book, *Fish Energetics — New Perspectives*, we propose in this chapter to deal only with those hormones that are known to be concerned with the control and regulation of energy metabolism and growth. The endocrine control of feeding in fishes is poorly understood and it is, therefore, inevitable that much of this chapter will be concerned with protein, carbohydrate and fat metabolism.

It is difficult to conceive of any metabolic process in a fish, as in any other vertebrate, taking place in the body of the animal without the involvement of hormones. Growth does not take place normally, nor do the digestion and storage of food to be used ultimately in energetic processes, without the catalytic intervention of endocrine gland secretions. The glands involved in fishes as in higher vertebrates, are the thyroid, the gonads, the 'adrenals' (inter-renal and chromaffin tissue), the endocrine pancreas and the pituitary gland.

7.2 The Pancreas

In mammals the regulation of nutrient homeostasis is largely the responsibility of the Islets of Langerhans. In fishes, although the hormones insulin and glucagon are present in the Islets of Langerhans (Brockmann bodies or principal islets as the discrete endocrine pancreas is often called in fishes), their roles in nutrient homeostasis do not parallel those of mammals. Nevertheless, these hormones, as in man, do appear in fishes to control the rates at which various fuels enter and leave the extracellular space and also the concentration of these fuels as they pass through this space.

7.2.1 Insulin

The most obvious response to exogenous insulin injections in fishes is hypoglycaemia (Ince, 1983). The insulin used to demonstrate this response has nearly always been of mammalian origin. This hypoglycaemic response is a slow one and is very variable from species to species of fish. However, whereas it appears that the blood glucose levels of teleosts are generally very resistant to the effects of mammalian insulins, the insulin obtained from teleosts brings about a marked hypoglycaemia within a few hours. Oral glucose tests, a diagnostic procedure useful in the detection of human diabetes, result in teleosts showing a persistent hyperglycaemia over many hours. This indicates that the rapid glucose metabolic response shown by mammals is not one that insulin is called upon to modulate in the normal physiology of fishes, at least in teleosts. It has been suggested that whereas in mammals rapid glucose homeostasis is required to maintain brain function, in teleosts, due to higher brain glycogen reserves, this rapid response is not required (Plisetskaya, 1968). Glycogen levels in both muscle and liver of fishes, however, appear to be substantially lower than those recorded for mammals. Glycogen stores in fish muscle are depleted very rapidly during exercise but blood glucose levels do not change significantly.

Changes in blood glucose do, in all probability in normal physiological conditions, stimulate insulin release. This has been demonstrated in the cannulated silver eel *Anguilla anguilla*, for whereas the release is dose dependent over a range of glucose loads from 10 to 100 mg kg^{-1}, higher doses produce no greater effect than 100 mg kg^{-1} (Ince & Thorpe, 1977a). This is good evidence that insulin is required to enable relatively low levels of glucose to leave the extracellular space to function as cellular fuel. This response may be fairly specific to the insulin molecule, for although there is great similarity in mammalian and fish insulins, differences do occur. For example, a feature of teleost insulin is that there is an extra residue at the beginning of the β chain.

In mammals insulin facilitates glycogen formation from glucose partly due to a more rapid accumulation of glucose but also by increasing activity of glycogen synthetase. In fishes the effect is variable but a more consistent glycogenic response is found *in vitro* (Tashima & Cahill, 1968). Ince (1983) has suggested that the paradoxical *decrease* in tissue glycogen found in fishes after insulin treatment *in vivo* may be due to the effect of a counter-regulating hormone released in response to hypoglycaemia. However, Otto-lenghi *et al.* (1982) observed in the catfish *Ictalurus melas* that after insulin treatment liver glycogen showed a decrease whereas in both white and dark muscle there was an increase. Also, hormone treatment was found not to change the heart glycogen levels in fed catfish but did so in starved animals.

The role of insulin in carbohydrate metabolism may in fishes be directed to oxidative clearance of glucose but not generally to glycogen deposition. However, whereas this statement needs further corroboration, the role of insulin in the regulation of protein metabolism is now well defined and this may be the major function of the hormone in fishes (Murat *et al.*, 1981).

Some years ago it was shown in the toadfish, *Opsanus tau*, that insulin accelerated the incorporation of radioactive glycine into skeletal muscle protein. Other amino acids have since been shown to be incorporated into the skeletal muscle after the treatment of other teleosts with mammalian insulin (Ahmad & Matty, 1975). Plasma amino nitrogen is reduced by treatment with both teleost and mammalian insulin. This is the expected result of increased uptake of amino acids by muscle and liver cells. There may also be, as in mammals, an inhibition by insulin of the mobilization of amino acids from protein but this has not yet been clearly demonstrated.

In the establishment of a control role for endocrine glands, the surgical removal of the gland and observed post-operative changes have played a key part. Isletectomy or removal of the Brockmann bodies can be performed in teleost fish, particularly in species such as *Cottus scorpius* and *Lophius piscatorius* where the bodies are concentrated into one or two discrete tissue masses. This operation has been made on a number of teleosts inducing a diabetic state with hyperglycaemia and glycosuria. Only in the trout *Salmo gairdneri* has isletectomy been attempted to demonstrate an increase in plasma amino acids (Matty & Kumar, unpublished). The levels of several plasma amino acids, e.g. glycine, lysine and proline, were elevated by over 100% 7 days following isletectomy.

Insulin, also in the rainbow trout, has been shown to inhibit gluconeogenesis from alanine in both fed and starved fishes (Cowey *et al.*, 1977). This is similar to the inhibition of gluconeogenesis from amino acids seen in mammals. In addition to incorporation of amino acids and the suppression of gluconeogenesis in the liver, insulin significantly reduces protein turnover by the liver of fishes. The half-life of protein in the carp liver is increased to twice the control values after insulin treatment.

Like glucose, amino acids stimulate insulin release. Arginine and lysine injections of 10, 25 and 100 mg kg^{-1} cause greater significant increases in the plasma levels of insulin in the eel than does glucose at the same dose level. Also lysine, leucine and phenylalanine stimulate a biphasic insulin release from the perfused eel pancreas (Ince & Thorpe, 1977b).

Whereas protein and amino acid conservation and tissue deposition may be the prime role of insulin in fishes, and therefore central to energy regulation (remembering that the major energy sources in fishes are derived from protein), there is evidence that lipid metabolism is also influenced by insulin. Injections of insulin have been shown to produce a lowered level of plasma free fatty acids (Leibson *et al.*, 1968; Ince & Thorpe, 1975; Lewander *et al.*, 1976). These results, obtained in carp, sea bass, goldfish, eel and pike, are in contrast to those obtained in toad fish, where there was no change in fatty acid levels (Tashima & Cahill, 1968), or in the catfish, where no effect of insulin on fatty acid synthetase was observed (Warman & Bottino, 1978). This latter observation stands in marked contrast to the situation in mammals, where the administration of insulin promotes the biosynthesis of fatty acid synthetase. In view of the relatively high dose of insulin injected in some of the above investigations, the question must be asked as to whether the dose is pharmacological. However, we must remember that some metabolic processes in fishes seem quite different from those of terrestrial animals. In mammals the activities of the lipogenic enzymes, acetyl-Co A carboxylase and fatty acid synthetase are affected by the nature and amount of dietary lipids, but in fishes the lipogenic enzymes appear to be insensitive to the presence or absence of fat in the diet. Again this is a feature of fishes which distinguishes them from mammals in their handling of energy sources. If there is a physiological role for insulin in modulating lipid metabolism, then it may be quite different from that seen in mammals.

7.2.2 Glucagon

The pancreatic hormone glucagon, when administered exogenously, brings about an increase in blood sugar mainly as a result of liver glycogenolysis (Birnbaum *et al.*, 1976; Ince & Thorpe, 1977a; Chan & Woo, 1978). However, glucagon also brings about gluconeogenesis (Inui & Yokote, 1977; Walton & Cowey, 1979). Recently, Morata *et al.* (1982) demonstrated that glucagon and adrenalin increase the release of glucose from liver slices of the rainbow trout into the incubation medium but that insulin had no effect. Glycogen even tended to increase in these slices in the presence of glucagon and adrenalin. Glucose release by these hormones is thus further shown to be mainly due to gluconeogenesis. Like insulin in fishes, glucagon reduces the plasma amino acid level, but unlike insulin (which affects amino acid mobilization from body protein) glucagon stimulates hepatic amino acid incorporation.

7.3 The Thyroid

The thyroid in fishes is ubiquitous though diffuse in most teleosts. The same type of iodoamino acids as in mammalian glands are present in fish thyroids. Thyroxine (T4) and tri-iodothyronine (T3) are the primary thyroid hormones (Gorbman, 1969, 1978; Fontaine, 1975; Eales, 1979; Leatherland, 1982). Control of T4 production is through pituitary TSH, although this control has not been demonstrated in cyclostomes (Peter, 1973; Matty *et al.*, 1976; Simpson, 1978). In blood, both T4 and T3 are bound to plasma proteins, and it is the free fraction of the hormones which bring the characteristic changes in metabolism. The free fraction is metabolized by deiodination or is eliminated from the body through the entero-hepatic system (biliary-faecal pathway) (Donaldson *et al.*, 1979). Both hormones (T4 and T3) have been measured in the blood of different fishes by gas–liquid chromatography, competitive protein binding or radioimmunoassay. These studies show that levels of the hormone are species dependent and also dependent on the physiological state of the fish (Higgs *et al.*, 1982). Marked seasonality has been shown in the plasma levels of T4 and T3 in different species (White & Henderson, 1977; Osborn &

Simpson, 1978; Osborn *et al.*, 1978; Eales & Fletcher, 1982).

7.3.1 Carbohydrate Metabolism

In mammals thyroid hormones have well marked effects on metabolism. They increase the oxygen consumption after a latent period of about 24 h. This increase is also seen *in vitro* with different tissue slices except in brain, spleen and testis. This has been shown to be due to the futile cycling of glucose-free fatty acids (Sestoft, 1980). Apart from this, the thyroid state of the animal affects the Cori-cycle, glucose–alanine cycle and other glucogenic and pentose phosphate cycle enzymes (Valtonen, 1974; Paul & Dhar, 1980; Huang & Lardy, 1981; Muller & Seitz, 1982).

The energy output of a fish, like that of any other animal, may be found by measuring its oxygen consumption during a measured time and calculating from the results the equivalent heat produced. In mammals thyroid hormones have a marked calorigenic effect and it was understandable that a similar effect should have been sought in fishes. Matty (1957) surgically removed the thyroid gland of the parrot fish but observed no depression of oxygen consumption. Numerous experimenters have injected and immersed fishes in thyroid hormone preparations and measured oxygen consumption. Equivocal and contradictory results have been reported but the general consensus is that a calorigenic effect cannot be determined.

Fontaine *et al.* (1953) claimed a decrease in hepatic glycogen content of eel after either interperitoneal injection or feeding of T4 or iodinated casein. Hochachka (1962) studied the effects of T3 and T4 on the carbon dioxide production from radioactive gluconate-1-C^{14}. His results show that the glucose metabolism is more directed via the pentose phosphate pathway than via the glycolytic pathway. This effect, also observed in mammals, is very important when seen in the perspective of NDP production which is used in anabolic processes and also for providing pentose for nucleic acid synthesis. Le Ray *et al.* (1970) administered *per os* T4 and T3 (500 mg kg^{-1}) and propyl thiouracil (PTU) (1 or 5 μg 100 g^{-1}) to *Mugil auratus* and noted that the liver glycogen stores were affected seriously. T4 lowered the glycogen whereas T3 had no effect. Low doses of PTU, like T4, lowered the liver glycogen (33%), whereas at higher doses it resulted in glycogen accumulation (143%). These effects of PTU confirmed the earlier hypothesis of Le Ray *et al.* (1969) that lower doses of PTU activated the thyroid while higher doses inhibit. In this same study, whereas cytochrome oxidase increased, the glucose-6-phosphate dehydrogenase decreased in response to T4 or T3 administration (Le Ray *et al.*, 1970).

Ray & Medda (1976) showed that immersion of *Ophicephalus punctatus* for 15–30 days in thyroxine (2.5 μg 100 ml^{-1}) increased both acid and alkaline phosphatase activities in the liver. Acid phosphatase but not alkaline phosphatase was increased 5 days after a single injection of 1–2 μg g^{-1} of T4. Alkaline phosphatase needed a higher (4 μg g^{-1}) dose to bring about this response. Immersion in the antithyroid drug thiourea (1 mg ml^{-1}) did not produce any change in activity of these liver enzymes in 15 days, but after 30 days the activities of both enzymes were decreased.

Falkmer & Matty (1966) reported a fall in blood sugar level in the hagfish (*Myxine glutinosa*) after injection of a pharmacological (5 mg kg^{-1}) dose of T4. Injection of T4 and T3 (1 and 0.5 mg kg^{-1} body weight) lowered the blood sugar (15–25%) with a concomitant increase in the glycogen of the heart and muscle of carp (Murat & Serfaty, 1971a,b). Hypoglycaemia with depletion of liver glycogen stores has been observed in *Clarias batrachus* 4 days after T4 administration. Muscle glycogen increased but no change in the brain glycogen was observed. Islets of fishes autopsied between 72 and 96 h after thyroxine administration contained severely damaged β-cells, but the α-cells remained normal (Bhatt & Khanna, 1976).

7.3.2 Lipid Metabolism

Several studies are available on the relationship between thyroid hormones and lipid metabolism in teleosts. Barrington *et al.* (1961) observed a decrease in abdominal fat in rainbow trout treated with thyroxine. Baker-Cohen (1961) postulated that an increase in the liver size of radiothyroidectomized platyfish was due to lipid deposition. La Roche *et al.* (1965) noted increased deposition of visceral lipids in radiothyroidectomized rainbow trout. Injection of mammalian thyroid powder (4 × 125 mg per fish) to 2-year-old (350 g) rainbow trout lowered plasma lipids to about one-half of the control values, but individual components, i.e. free fatty acids (FFA) and total cholesterol decreased more than triglycerides (Takashima *et al.*, 1972). Murat & Serfaty (1971a,b) administered carp with thyroxine (1 mg kg^{-1}) and 24 h after the injection, plasma FFA increased from 0.25 μeq ml^{-1} to 0.40 μeq ml^{-1}. Decreases in the hepatic and visceral lipid reserves regardless of the nutritional status, with a concomitant increase in the plasma FFA, were observed in *Salvelinus fontinalis* treated with thyroxine. Further, in this same study an increase in the FFA levels of the visceral adipose tissue was also observed but little effect was noted in muscle (Narayansingh & Eales, 1975b).

All the above findings indicate that thyroid hormones mobilize the lipid reserves. However, there are some studies available which do not support this conclusion. Le Ray *et al.* (1969) did not find any effect of desiccated thyroid powder, thyroid hormones or thiouracil on the deposition of abdominal fat and total lipid contents of *Astyanax* and *Mugil auratus*. No correlation could be found between the seasonal fluctuations of thyroid hormones with lipid deposits of *Salmo trutta* (Swift, 1955). Similarly, serum cholesterol remained unchanged with seasonal variations of serum thyroid hormones in *Oncorhynchus kisutch* (Leatherland & Sonstegard, 1980, 1981), and in some cases lower serum cholesterol and FFA have been correlated with hypothyroidism in this fish (Leatherland & Sonstegard, 1981). Further, Higgs and his co-workers (1976, 1977) found an increase in the muscle lipids of juvenile *Oncorhynchus kisutch* given injections of thyroxine. Oral administration of T3 or T4 (20 or 40 ppm) to the same species did not have any effect on muscle fat (Fagerlund *et al.*, 1980), but when T3 or T4 (20–100 ppm) was given *per os* to *Sarotherodon mossambica*, a decrease in muscle fat content was observed (Lone *et al.*, unpublished). Many of the differences in results regarding the effect of thyroid hormones and metabolism in teleosts may doubtless have arisen as a result of variation in factors such as nutritional state, levels of the hormones injected and route of administration, ambient temperature, photoperiod, salinity and size, etc. Among the factors discussed above, the nutritional state is important. There are some studies available on the food quality and endogenous hormone levels in fishes. Although protein (35 and 45%) had no effect on plasma T4 and T3 levels of rainbow trout held at 9, 15 and 18°C, starvation or restriction of protein or energy in the diet depressed the thyroid hormone levels (Simpson, 1978; Higgs & Eales, 1979). Also, an increase of unsaturated fatty acids or total fat in the diet of trout (*Salmo gairdneri*) increased the plasma T4 levels at low (7°C) temperature (Leatherland *et al.*, 1977).

7.3.3 Protein and Nucleic Acid Metabolism

Thyroid hormones have an extensive and diverse range of effects in most if not all cells of the vertebrate body, but one of the most important effects is on protein and nucleic acid turnover. In mammals this effect is seen as the so-called 'biphasic' response, meaning that lower doses (physiological or near physiological) of the hormones are protein anabolic whereas higher (pharmacological) doses are catabolic (Tata, 1974, 1980; Bernal & DeGroot, 1980).

Little is known of protein metabolism and thyroid hormones in fishes. Hoar (1958) and Thornburn & Matty (1963) noted that the ammonia excretion increased in thyroxine-treated *Carassius auratus*. Further, ammonia excretion increased at higher (20°C) but not at lower (14°C) temperatures (Thornburn & Matty, 1963), and the incorporation of ^{14}C-Leucine into muscle protein was also increased. Smith & Thorpe (1977) showed that thyroxine (0.25–5 mg d^{-1} kg^{-1}) decreased the nitrogen excretion of fed *Salmo gairdneri* at 12°C. Matty & Sheltaway (1967) showed that the incorporation of ^{14}C-Glycine into the skin guanine increased after intraperitoneal injection of T4. The incorporation of ^{14}C-leucine into liver and gill proteins of brook trout was stimulated by T3 but not by T4 (Narayansingh & Eales, 1975a).

In two fairly recent reports (Ray & Medda, 1976; Medda & Ray, 1979), the effect of thyroid hormones and their analogues on nitrogen excretion, protein and nucleic acid metabolism were shown to be 'biphasic' and temperature dependent. In lata fish at 25°C, T4 (0.5–4 g g^{-1}) had an anabolic effect with regard to the nitrogen excretion of starved fishes, but at 30°C nitrogen excretion increased. Thyroxine or its analogues (TRIAC and TETRAC) increased the protein and RNA levels in liver and muscle after 4 days, but higher doses were catabolic at the same temperature.

Mindful of the previous equivocal results, a detailed study was undertaken in the laboratory of the authors on the role of thyroid hormones in protein and nucleic acid metabolism (Matty *et al.*, 1982). We have seen that T4 does not support growth in tilapia or carp when given in the food. However, T3 is potent in this regard at a dose level of 20 mg per kilogram of food but only in tilapia. Carp do not respond to feeding of either T3 or T4 (Lone *et al.*, unpublished). On the other hand, both T3 and T4, when injected intraperitonally (0.5–1 g g^{-1}) increased the protein and RNA of the liver and muscle of both tilapia and carp (Matty *et al.*, 1982). Actinomycin D inhibits these effects in both tissues, when given alone or simultaneously with T4. The time course of this inhibition is tissue dependent. Further studies have pointed out that the effects of the activator (T4) or inhibitor (actinomycin D) are mediated through different loci (Lone *et al.*, unpublished). Temperature seems to mediate these effects of thyroid hormones, for, with an increase in the ambient temperature

(15°C to 30°C), there is a linear increase in both protein and RNA levels of the liver and muscle of tilapia given T3 and T4. Further experiments showed that there was no inhibition of the movement of T4 into sub-cellular fractions at lower temperature (Lone *et al.*, 1982). The metabolic responses of fishes to thyroid hormones have also been shown to be age dependent (Matty *et al.*, 1982).

7.4 Gonads

7.4.1 Androgens

In mammals, sex steroids, both androgens and œstrogens, have far-reaching effects on carbohydrate, lipid and protein metabolism (Young & Pluskal, 1977). Studies on the effect of steroids in fish have been restricted mainly to sex differentiation, sex reversal, inducement of sterility and development of mono-sex culture, spermatogenesis, oogenesis and ovulation. Also, many studies have been made on the seasonal variation of sex steroids in the plasma and gonads of fish (Idler & Truscott, 1966; Peter, 1979; Demski & Hornby, 1982; Lam, 1982). The metabolic effects of sex steroids have not been studied in detail and few reports are available on this aspect of hormone physiology. Hirose & Hibiya (1968) injected 1-year-old rainbow trout with 4-chlorotestosterone (1 and 2.5 mg kg^{-1}) and observed changes in the serum and liver constituents. Intramuscular injections of 4-chlorotestosterone brought no change in the haematocrit and serum proteins but non-protein nitrogen was reduced. Histochemical tests of the liver of treated fish showed an increased deposition of glycogen in this organ. Takashima *et al.* (1972) observed no changes in plasma lipids after intraperitoneal injections of methyltestosterone but lipids decreased significantly in the viscera. A decrease in visceral lipids and a slight increase in carcass lipid content have also been observed by Simpson (1976) in rainbow trout fed methyltestosterone (2.5 mg kg^{-1}) in food. An increase in muscle lipids and cholesterol after feeding methyltestosterone (1–10 mg kg^{-1}) to carp was again observed (Lone & Matty, 1980) but a detailed fatty acid analysis of the total lipids and phospholipids of the coho salmon fed methyltestosterone and testosterone (2.5 mg kg^{-1}) did not reveal any change (Yu *et al.*, 1979). Weigand & Peter (1980) studied the effect of testosterone on the total lipids and plasma free fatty acids (FFA) in goldfish. The hormone did not change total plasma lipids but increased FFA levels. The authors suggested that this increase in FFA was due to either hypothalamic or neural pathways but the exact mechanism is not clear. For detailed information on the body lipids after feeding androgenic steroids, the reader is referred to Higgs *et al.* (1982) and Lone & Matty (1980, 1982).

A seasonal study of sex steroids and plasma lipids and cholesterol in striped mullet (*Mugil cephalus*) has shown that whereas testosterone shows only one peak in November, the lipids and cholesterol exhibit two peaks. According to the authors, the summer peak (June–July) is due to active feeding and preparation for the development of gonads and spawning migration, and the second peak, coinciding with the November peak of the testosterone, is considered by them to reflect the fact that lipid had been mobilized by the steroid for gonad development (Dindo & MacGregor, 1981).

The androgen molecule has been shown to have inherent protein anabolic properties and this has been shown many times in mammals (Kochakian, 1976). In fishes also, these steroids have shown growth-promoting effects (Higgs *et al.*, 1982) but the effect on protein and nucleic acid metabolism is far from clear. When testosterone propionate $(0.5–4 \mu g\, g^{-1})$ was injected intraperitoneally into *Heteropneustes fossilis* (40–60 g) for 7 consecutive days, it failed to cause any change in protein metabolism (Medda *et al.*, 1980). In contrast to the above study, others undertaken on carp (Lone & Matty, 1980, 1983) and rainbow trout (Ince *et al.*, 1982; Lone *et al.*, 1982; Lone & Ince, 1983) have shown that feeding of natural or synthetic androgenic–anabolic steroids in both fish species bring about marked changes in protein and nucleic acid metabolism. In general, feeding of the steroid induces an increase in the RNA and protein contents of the liver, kidneys and muscle.

7.4.2 Oestrogens

More studies have been made on the effect of oestrogens on fat and carbohydrate metabolism in fishes than on the effect of androgens. All the studies show that oestrogens increase the plasma lipid, cholesterol, protein, calcium and phosphate levels in the many different teleost species studied (Plack *et al.*, 1971; Mugiya & Watanabe, 1977; Medda *et al.*, 1980; Weigand & Peter, 1980; Mugiya & Ichii, 1981; Dindo & MacGregor, 1981). Oestrogen treatment (10 and 100 μg per fish of 9–14 g) increased the body fat reserves of female *Notemegonus crystoleucas* when kept at long and intermediate photoperiods, whereas fishes kept at shorter photoperiods exhibited fat depletion. In males, lower doses (20–25 μg per fish) of oestrogen evoked body lipid depletion, and higher doses (50–100 μg per fish) stimulated fat deposition (de Vlaming *et al.*, 1977). Apart from body lipids, hepatic levels of fat also increase after œstrogen treatment, and this is said to be the main reason for high hepato-somatic indices in fishes (McBride & Overbeeke, 1971).

In addition to their acute effects on lipid metabolism, œstrogens also have far-reaching effects on carbohydrate metabolism in fishes. Injections of *Notemegonus*, goldfish and brook trout, with œstrogens decrease the liver glycogen content (de Vlaming *et al.*, 1977; Whiting & Wiggs, 1978). These changes are consistent with earlier reports. This decrease in glycogen content has been explained on the basis of production of energy for the synthesis of vitellogenic proteins and other ovarian changes. Œstrogen treatment also reduces tyrosine aminotransferase (TAT) in the livers of brook trout. This decrease of TAT is interesting in that there is an elevation of this enzyme in a tissue exhibiting proteolysis. Further, this enzyme has been shown to increase in physiological states that favour protein depletion, for example increased glucocorticoid levels. The decrease in TAT after œstrogen treatments means that amino acid precursors for gluconeo-genesis or glycogenesis will not be available during the vitellogenic stages of egg production, ultimately decreasing the glycogen stores in the liver.

Little is known about the effects of œstrogens on the protein and nucleic acid metabolism of fishes. The majority of studies made on vertebrates, including fishes, indicate that œstrogens activate the liver to secrete a lipophosphoprotein (vitellogenin) to be deposited in the developing ovary (Yaron *et al.*, 1977; Tata, 1980). Studies on fishes show that œstrogen ($0.5-4\ \mu g\ g^{-1}$) increases the proteins of both liver and plasma in male and female *Heteropneustes fossilis*. Also the liver RNA and hepato-somatic index of both male and female increased but no change in muscle gonads protein or RNA was noted (Medda *et al.*, 1980). Similar results were reported earlier for the flounder *Platichthys flesus* (Emmerson *et al.*, 1979) and cod (Plack *et al.*, 1971).

7.5 Interrenal Tissue

In mammals, adrenal steroids have distinct effects on metabolism, particularly on protein and carbohydrate metabolism. They promote gluconeogenesis from non-carbohydrate precursors such as amino acids and glycerol and also promote liver glycogen deposition, the elevation of blood glucose and the inhibition of glucose utilization by tissues. They have highly catabolic actions on muscles, which are characterized by protein wasting due to a reduction in synthesis and an increase in intracellular proteolysis, and cause retardation of overall growth (Mayer & Rosen, 1977; Williams, 1981).

In fishes, the occurrence and biosynthesis of corticosteroids produced by the interrenal or cortical tissue lying between the kidney have been discussed (Idler & Truscott, 1972). In lampreys, cortisol or other classical glucocorticoids have been reported to be absent in the plasma (Buus & Larsen, 1975) but very low concentrations of these compounds were identified after ACTH stimulation in *Myxine* (Fontaine, 1975). Cortisol, cortisone, corticosterone and aldosterone have been identified in teleosts, and Selachii have 1 α-hydroxy-corticosterone, which is not present in the other sub-class of Chondrichthyes, the Holocephali (Idler & Truscott, 1972). The detailed effects of these steroids in fishes have been presented by Chester-Jones *et al.* (1974) and Fontaine (1975). In the following paragraphs only intermediary metabolic effects will be discussed.

The majority of the reports that are available on the effects of corticoids deal with carbohydrate metabolism. Hypophysectomy does not induce any change in the blood sugar level of *Myxine glutinosa* (Falkmer & Matty, 1966). In contrast, hypophysectomy does affect the carbohydrate metabolism in teleosts. Hypophysectomy induces hypoglycaemia in *Anguilla anguilla* and *Anguilla rostrata*, and a decrease in hepatic glycogen of fasting fishes, but muscle glycogen is not changed (Butler, 1968). The decrease in glycogen was shown to be due to the lack of gluconeogenesis as the animals were capable of glycogenesis from injected glucose. Cortisol injection in these fishes increased both liver and muscle glycogen. Similar effects of hypophysectomy and administration of ACTH or corticoids were observed by Chidambaram *et al.* (1973). In contrast to the above effects, Ball *et al.* (1966) observed an increase in glycogenolysis after ACTH or corticosteroid injections. Hypophysectomy in *Poecilia latipinna* and *Fundulus heteroclitus* induced the storage of liver glycogen but ACTH or cortisol depleted these stores. This means that ACTH or corticoids can have both anabolic or catabolic effects in teleosts as far as glycogen stores are concerned. Similar effects of ACTH have been observed on glycogenolysis in fasting *Tilapia mossambica* (Swallow & Fleming, 1966).

The injection of corticoids to intact fishes increases gluconeogenesis, glycaemia and deposition of glycogen in the liver (Muller & Hanke,

1974). In lampreys, ACTH or cortisol increased the glycaemia and liver glycogen in short-term experiments. Prolonged administration of these hormones, however, had a glycogen-mobilizing effect (Bentley & Follett, 1965). Interrenalectomy of the skate *Raja erinacea* did not induce any change in the liver glycogen (Idler *et al.*, 1969). No study of exogenous 1α-hydroxycorticosterone on carbohydrate metabolism in these fish is available. It is argued that it may be possible that in these fish this hormone and other corticosteroids have more subtle effects on lipid metabolism.

Stress-induced changes in carbohydrate metabolism have also been attributed to the elevation of cortisol or other glucocorticoids (Leach & Taylor, 1980), and Storer (1967) and Pickford *et al.* (1970) have shown that exogenous cortisol decreases the parietal muscle and growth of goldfish and causes a reduction in body weight of *Fundulus heteroclitus*. Ball & Ensor (1969) demonstrated that cortisol not only inhibited the regeneration of amputated dorsal fins of hypophysecto-mized *Poecilia latipinna* but also decreased the weight and length in this fish, and Leatherland & Lam (1971) observed that injection of ACTH to *Gasterosteus aculeatus* decreased the body length. Smith & Thorpe (1977) also observed catabolic actions of cortisol in starved rainbow trout, but fed trout did not show any response. Finally, Chan & Woo (1978) have demonstrated the classical mammalian effects of cortisol in eels characterized by increased metabolic rate, liver transaminases and glycogen levels. This is also accompanied by the elevation of ammonia and potassium excretion, hyperglycaemia and hyperaminoacidaemia and a peripheral proteolysis.

Skeletal muscle represents a major proportion of the body mass in fishes, and its proteins may contribute up to 70–80% of the total fish mass. It is from this large reservoir of amino acids that the supply of nutrients and energy to other tissues of the body are regulated by corticosteroids, as explained in the previous paragraph. The control may be either short term, as when increased cortisol levels cause rapid hyperglycaemia, or long term, when corticosteroids increase gluconeogenesis from protein and this gradually results in muscle atrophy and cessation of growth. This gluconeogenesis in fishes is characterized, as in mammals, by an increase in transaminases such as tyrosine aminotransferase (Whiting & Wiggs, 1978).

7.6 Chromaffin Tissue

Chromaffin tissue producing adrenalin and noradrenalin occurs in all groups of fishes in the region of the postcardinal venous drainage of the kidney, dorsal aorta or heart. Adrenalin in mammals increases the concentration of both glucose and fatty acids in the plasma, the glucose response being extremely rapid in response to injection of the hormone. This hyperglycaemic effect of adrenalin is seen in all fish groups but the site of its action may differ.

In lampreys, adrenalin mobilizes glycogen in liver but not muscle (Bentley & Follett, 1965). In elasmobranchs, catecholamine hormones appear to have an established part to play in carbohydrate metabolism. Prolonged hyperglycaemia occurs when relatively small doses of adrenalin are injected. In this case, unlike the lampreys but similar to mammals, glycogen is mobilized from both liver and muscle (De Roos & De Roos, 1972). In the dogfish *Scyliorhinus stellaris*, plasma glucose levels are elevated within 15 min of an intra-aortic infusion of adrenalin. The metabolic and energy-mobilizing role of noradrenalin is not so clear in elasmobranchs, but lipolysis of adipose tissue is also stimulated by adrenalin as in mammals.

In bony fish, as in cartilaginous fish, adrenalin release or injection causes an immediate and marked hyperglycaemia. Noradrenalin also brings about almost immediate hyperglycaemia but is somewhat slower acting than adrenalin. The biochemical pathway of glycolysis in the liver is similar to that found in mammals with the enzyme glycogen phosphorylase existing in an active and inactive form and stimulated by glucose.

7.7 The Pituitary

There is evidence in fishes as in mammals that hormones produced by both the adenohypophysis and the neurohypophysis are involved in the mobilization and deposition of metabolites. This evidence is at the moment confined to the hormones prolactin, growth hormone and arginine vasotocin, but it would not be at all surprising if pituitary polypeptides, more usually associated with target organs such as the gonadotropins, TSH, ACTH and MSH, were found to have some direct effect on intermediate metabolism and energy shifts.

7.7.1 Growth Hormone

The understanding of the mechanisms by which growth hormone improves food conversion and stimulates tissue growth in fishes has not attracted much attention. This may largely be due to the difficulty in obtaining fish growth hormone preparations for metabolic investigations. Nevertheless it has been suggested that growth hormone of fishes may (a) affect rates of protein synthesis and/or breakdown, (b) stimulate fat mobilization and oxidation, (c) stimulate, either directly or indirectly, insulin synthesis and release (Markert *et al.*, 1977).

Matty (1962) first reported nitrogen retention in *Cottus scorpius* following mammalian growth hormone injections. Reduction in total plasma protein as well as plasma urea occurred. More recently, Higgs *et al.* (1975, 1976) estimated that the mean amount of protein in bovine growth hormone treated coho salmon was substantially higher than that of control fish. Matty & Cheema (unpublished) have shown that porcine and shark growth hormone treatment of rainbow trout can stimulate the *in vivo* incorporation of L-(^{14}C)-leucine into skeletal muscle protein. These observations indicate that growth hormone release may result in a positive nitrogen balance in fish, as in mammals, improving the uptake of amino acids into cells and influencing the rate of RNA synthesis and energy conservation.

In mammals, growth hormone exhibits a ketogenic effect since it can mobilize free fatty acids from lipid depots. The percentage of lipid in the muscle of coho salmon declines following injection of bovine growth hormone (Higgs and co-workers, 1975, 1976, 1977). Bovine growth hormone also brings about a decrease in whole body lipid expressed as a percentage of dry weight in sockeye salmon (Clarke, 1976). In salmon bovine growth hormone injections also cause a rise in muscle free fatty acid (McKeown *et al.*, 1975). All this suggests that more of the ingested amino acids are thus made available for growth as lipids are being preferentially used as an energy source.

A number of studies have implicated fish growth hormone in the modulation of carbohydrate metabolism and possibly in stimulating insulin activity. As long ago as 1940 Abramowity *et al.* found that growth hormone was diabetogenic in the dogfish, and in 1962 Matty demonstrated elevated plasma glucose levels in the teleost *Cottus scorpius*, and Enomoto (1964) found that bovine growth hormone injections resulted in a transient glucosuria in the rainbow trout. Growth hormone may thus inhibit peripheral glucose utilization, and the resulting hyperglycaemia results in insulin release. On the other hand, growth hormone may influence the pancreatic islets directly. Prolonged bovine growth hormone treatment increases the size and number of Islets of Langerhans in salmon but there is little other evidence for a direct growth hormone–islet pathway in fishes.

7.7.2 Prolactin

Prolactin has been related to a number of metabolic processes in fishes and, therefore, like growth hormone, may play a part in the energy movements of fishes. However, there is no evidence to indicate the involvement of prolactin in carbohydrate metabolism in fishes. Prolactin does, however, increase lipid deposition in *Fundulus*, and cause an increase in free fatty acids in the plasma and muscle of salmon.

7.7.3 Arginine Vasotocin

Arginine vasotocin injections have resulted in hyperglycaemia in lampreys (Bentley & Follett, 1965) and in increased plasma levels of glucose and free fatty acids in coho salmon (McKeown *et al.*, 1976).

7.8 The Control of Growth

Physiologically useful energy in excess of metabolic requirements is available for growth and therefore the hormonal control of growth in fishes must be briefly considered. Much of the above discussion relating to the effect of hormones on protein, lipid and carbohydrate metabolism and of the sparing action of lipids and carbohydrates on protein is obviously related to growth and cannot be easily separated. However, hormones and growth in fishes were reviewed in detail by Donaldson *et al.* in 1979, and therefore mention will only be made of the investigations of the past 4 years.

Insulin does not appear to play a major role in the control of growth in length and weight of fishes. However, using a high dose of bovine insulin, Ablett *et al.* (1981) obtained a significant increase in body weight of rainbow trout after a period of 56 days.

Recent experiments have continued to indicate that thyroid hormones may accelerate growth when given at a physiological level. Higgs *et al.* (1979) confirmed earlier reports that in salmonids tri-iodothyronine (T3) enhances growth. Thyroxine (T4) and tri-iodothyronine were fed to under-yearling coho salmon, and Higgs *et al.* found that T3 significantly increased weight and length; thyroxine, however, did not influence growth although it did increase food consumption. T3 improved food conversion but T4 did not. Lam (1980) immersed yolk-sac larvae of the tilapia *Saratherodon mossambicus* in a 0.1 ppm solution of thyroxine and observed markedly accelerated development and enhanced survival. Matty *et al.* (1982) immersed fry of the same species in thyroxine and observed weight gains. The optimum dose for promoting growth was lower for 1-week-old fry (10 μg T4 per 100 ml water) than for 3-week-old fry (25 μg T4 per 100 ml water). It is possible that thyroid hormones are more important in the early life of fishes when there is active growth, and once fishes are mature, these hormones shift the emphasis of their action to the control of carbohydrate and lipid metabolism.

There have been few reports on the role of fish growth hormone in growth and metabolism since the review of Donaldson *et al.* (1979), and the manner in which growth hormone improves food conversion and stimulates growth is still uncertain. Leatherland & Nuti (1981) have observed that bovine growth hormone decreases the plasma free fatty acid content of rainbow trout. Liver lipids were lowered but not muscle lipids. Houdebine *et al.* (1981) have also shown that highly purified tilapia growth hormone specifically stimulates casein synthesis in rabbit mammary gland explants. The growth hormone of a primitive bony fish, the sturgeon *Acipenser guildenstadti*, has recently been isolated and highly purified (Farmer *et al.*, 1981) but its role in metabolism has not yet been considered.

Naturally occurring androgen may play a part in the control of growth in fishes (Lone & Matty, 1980).

7.9 The Control of Feeding

The control of feeding in fishes, like so many other physiological processes, involves the nervous system and the secretion of several hormones. Peter (1979) has reviewed the role of the brain in fish feeding behaviour, Fänge & Grove (1979) the role of the autonomic nervous system and gastro-intestinal hormones, and Donaldson *et al.* (1979) the role of growth hormone, steroid hormones and thyroid hormones. In fishes, as in higher vertebrates, absorbed nutrients influence secretion of hormones and this response differs with diet.

Growth hormone may possibly control fish growth not only by regulating protein synthesis but also by stimulating appetite (that is, voluntary food intake) and in addition by improving food conversion. Evidence for these possible actions is presented by Donaldson *et al.* (1979). Other hormones may also affect appetite and food conversion; for example, tri-iodothyronine increases the appetite of operant-conditioned rainbow trout (Matty & Majid, unpublished). The hypothalamic control of feeding, such as exists in the 'hunger' and 'satiety' centres of the mammalian hypothalamus, has not yet been found in fishes (Peter *et al.*, 1976; Majid, 1979). Nevertheless, the lateral hypothalamus appears to play some role as a feeding centre (Peter, 1979). Much more research is needed in this area and, as Peter (1979) has remarked, 'Presentation of a (any) model to serve as a working hypothesis for the control of feeding behavior and food intake in teleosts is premature at this stage.' Surprisingly, as Peter (1979) has again pointed out, the effects of blood glucose levels on food intake in teleosts have not yet been studied *per se*. In mammals it is the hyperglycaemia resulting from feeding that stimulates enteroglucagon and insulin production and inhibits the release of adrenalin and growth hormone, and the hypoglycaemia resulting from fasting stimulates the release of corticosteroids, pancreatic glucagon, adrenalin and growth hormone. It is from this control mechanism that the changes in intermediate metabolism, i.e. glycogenolysis, lipolysis and gluconeogenesis from amino acids, take place, resulting in energy distribution and growth. In fishes the present evidence suggests that although some pathways of food intake–hormone release–metabolism may be similar to those seen in higher vertebrates, many are dissimilar. Perhaps (and there is considerable suggestive evidence, some of which has been presented in this chapter) it is the level of plasma amino acids that is the set-point that regulates energy balance in fishes rather than the level of blood glucose. This is readily understandable when one remembers that the major component of the food intake in fishes is protein, and that carbohydrates are but a minor energy source for most fish species. The gross efficiency of food energy utilization is similar in chicken and carp (Nijkamp *et al.*, 1974). Thus, if the metabolizable energy of carbohydrates and proteins is essentially similar in carp to that in higher vertebrates, then an amino acid indicator of energy level (controlling, via the endocrine system, the modulation of metabolic pathways) will be just as efficient as a glucose indicator.

References

Ablett, R.F., Sinnhuber, R.O. and Selivonic, D.P. (1981), Insulin and metabolism in trout. *Gen. Comp. Endocrinol. 44*, 418

Abramowity, A.A., Hisaw, F.L., Boettiger, E. and Papandrea, D.N. (1940), Diabetogenic hormone in the dogfish. *Biol. Bull. 78*, 189

Ahmad, M.M. and Matty, A.J. (1975), Effect of insulin on the incorporation of ^{14}C-leucine into goldfish muscle protein. *Biologia (Lahore) 21*, 119

Baker-Cohen, K.F. (1961), The role of the thyroid in the development of platyfish. *Zoologica 46*, 181

Ball, J.N. and Ensor, D.M. (1969), In: La specificité zoologique des hormones hypophysaires. *Coll. Int. C.N.R.S. 177*, 215

Ball, J.N., Giddings, M.R. and Hancock, M.P. (1966), Pituitary and liver glycogen in *Poecilia. Am. Zool. 6*, 595

Barrington, E.J.W., Barron, N. and Piggins, D.J. (1961), Thyroid and growth of rainbow trout. *Gen. Comp. Endocrinol. 1*, 170

Bentley, P.J. and Follett, B.K. (1965), Hormones and carbohydrate metabolism of the lamprey, *Lampetra fluviatilis. J. Endocrinol. 31*, 127

Bernal, J. and De Groot, L.J. (1980), Mode of action of thyroid hormones. In: De Vischer, M. (ed.) *The Thyroid Gland*, pp. 123–142. New York, Raven Press

Bhatt, S.D. and Khanna, S.S. (1976), Histopathology of endocrine pancreas of a freshwater fish, *Clarias batrachus. Acta Biol. Acad. Hung. 27*, 25

Birnbaum, M.J., Schulz, J. and Fain, J.N. (1976), Hormone stimulated glycogenolysis in isolated goldfish hepatocytes. *Am. J. Physiol. 231*, 191

Butler, D.G. (1968), Hormonal control of gluconeogenesis. *Gen. Comp. Endocrinol. 10*, 85

Buus, O. and Larsen, L.O. (1975), Lampreys and corticosteroids. *Gen. Comp. Endocrinol. 26*, 96

Chan, D.K.O. and Woo, N.Y.S. (1978), Cortisol on eel metabolism. *Gen. Comp. Endocrinol. 35*, 205

Chester-Jones, I., Ball, J.N., Henderson, I.W., Sandor, T. and Baker, B.I. (1974), Endocrinology of fishes. In: Florkin, M. and Sheer, B.T. (eds). *Chemical Zoology*, vol. VIII, pp. 523–93. New York, Academic Press

Chidambaram, S., Meyer, R.K. and Halser, A.D. (1973), Hormone control of glycemia and hematocrit. *J. Exp. Zool. 184*, 75

Clarke, W.C. (1976), Prolactin and growth hormone on lipid content of sockeye salmon. *Reg. Conf. Comp. Endocrinol. Oregan State Univ.* Corvallis Abst. No. 38

Cowey, C.B., Knox, D., Walton, M.J. and Adron, J.W. (1977), Regulation of gluconeogenesis by diet and insulin in rainbow trout (*Salmo gairdneri*). *Brit. J. Nutr. 38*, 463

De Roos, R. and De Roos, C.C. (1972), Plasma glucose in elasmobranchs. *Gen. Comp. Endocrinol.* Suppl. 3, 192

de Vlaming, V.L., Vodicnik, M.J., Bauer, G., Murphy, T. and Evans, D. (1977), Oestrogen effects on goldfish. *Life Sci. 20*, 273

Demski, L.S. and Hornby, P.J. (1982), Brain–gonadal steroid interactions. *Can. J. Fish. Aq. Sci. 39*, 36

Dindo, J.J. and MacGregor, R. (1981), Reproductive mobilization of steroids and lipids. *Trans. Am. Fish. Soc. 110*, 403

Donaldson, E.M., Fagerlund, U.H.M., Higgs, D.A. and McBride, J.R. (1979), Hormonal enhancement of growth. In: Hoar, W.S., Randall, D.J. and Brett, J.R. (eds). *Fish Physiology*, vol. VIII, pp. 455–597. New York, Academic Press

Eales, J.G. (1979), Thyroid function in cyclostomes and fishes. In: Barrington, E.J.W. (ed.), *Hormones and Evolution*, vol. I, pp. 341–436. New York, Academic Press

Eales, J.G. and Fletcher, G.L. (1982), Thyroid hormones of flounder. *Can. J.*

Zool. 60, 304

Emmersen, J., Korsgaard, B. and Petersen, I. (1979), Dose response kinetics in male flounders. *Comp. Biochem. Physiol. 63B*, 1

Enomoto, Y. (1964), Growth-promoting hormones and trout (*Salmo irideus*). *Nippon Suisan Gakkaishi 30*, 537

Fagerlund, U.H.M., Higgs, D.A., McBride, J.R., Plotnikoff, M.D. and Dosanjh, B.S. (1980), Hormones and Coho salmon. *Can. J. Zool. 58*, 1980

Falkmer, S. and Matty, A.J. (1966), Blood sugar regulation in hagfish. *Gen. Comp. Endocrinol. 6*, 334

Fänge, R. and Grove, D. (1979), Digestion. In: Hoar, W.S., Randall, D.J. and Brett, J.R. (eds). *Fish Physiology*, vol. VIII, pp. 162–260. New York, Academic Press

Farmer, S.W., Hayashida, T., Papkoff, H. and Polenov, A.L. (1981), Sturgeon growth hormone. *Endocrinol. 108*, 377

Fontaine, M., Leloup, J. and Olivereau, M. (1953), Étude histologique et biochimique. *Compr. Rend. Soc. Biol. 157*, 225

Fontaine, Y.A. (1975), Hormones in fishes. *Biochem. Biophys. Perspect. Mar. Biol. 2*, 139

Gorbman, A. (1969), Thyroid function and its control in fishes. In: Hoar, W.S. and Randall, D.J. (eds). *Fish Physiology*, vol. II, pp. 241–74. New York, Academic Press

Gorbman, A. (1978), In: Werner, S.C. and Ingbar, S.H. (eds). *The Thyroid Gland.* pp. 22–30. Maryland, Harper and Row

Higgs, D.A. and Eales, J.G. (1979), Diet and thyroid function of brook trout. *Can. J. Zool. 57*, 396

Higgs, D.A., Donaldson E.M., Dye, H.M. and McBride, J.R. (1975), Growth hormone and thyroxine and effects in coho salmon, *Oncorhynchus kisutch. Gen. Comp. Endocrinol. 27*, 240

Higgs, D.A., Donaldson, E.M., Dye, H.M. and McBride, J.R. (1976), Growth hormone and thyroxine on composition of coho salmon (*Oncorhynchus kisutch*). *J. Fish. Res. Bd Can. 33*, 1585

Higgs, D.A., Fagerlund, U.H.M., McBride, J.R., Dye, H.M. and Donaldson, E.M. (1977), Growth hormone, 17a-methyltestosterone and *L*-thyroxine on growth of coho salmon (*Oncorhynchus kisutch*). *Can. J. Zool. 55*, 1048

Higgs, D.A., Fagerlund, U.H.M., McBride, J.R. and Eales, J.G. (1979), Orally administered gyroid hormones on growth of salmon. *Can. J. Zool. 57*, 1974

Higgs, D.A., Fagerlund, U.H.M., Eales, J.G. and McBride, J.R. (1982), Thyroid and steroids as anabolic agents. *Comp. Biochem. Physiol. 73B*, 143

Hirose, K. and Hibija, T. (1968), 4-Chlorotestosterone and rainbow trout. *Bull. Jpn Soc. Sci. Fish. 34*, 473

Hoar, W.S. (1958), Hormones and the metabolism of goldfish. *Can. J. Zool. 36*, 113

Hochachka, P.W. (1962), Thyroidal effects on pathways of carbohydrate metabolism in a teleost. *Gen. Comp. Endocrinol. 2*, 499

Houdebine, L.M., Farmer, S.W. and Prunet, P. (1981), Induction of casein synthesis. *Gen. Comp. Endocrinol. 45*, 61

Huang, M.T. and Lardy, H.A. (1981), Thyroid states and glucose recycling. *Arch. Biochem. Biophys. 209*, 41

Idler, D.R. and Truscott, B. (1966), Testosterone in plasma of skate. *Gen. Comp. Endocrinol. 7*, 375

Idler, D.R. and Truscott, B. (1972), Cortiscosteroids in fish. In: Idler, D.R. (ed.). *Steroids in Non-mammalian Vertebrates*, pp. 127–217. New York, Academic Press

Idler, D.R., O'Halloran, M.J. and Horne, D.A. (1969), Interrenalectomy and glycogen levels of skate. *Gen. Comp. Endocrinol. 13*, 303

Ince, B.W. (1983), Pancreatic control of metabolism. In: Rankin, J.C., Pitcher, T.J. and Duggan, R. (eds). *Control Processes in Fish Physiology*, pp. 89–102. London, Croom Helm

Ince, B.W. and Thorpe, A. (1975), Hormonal effects on plasma free fatty acids in pike, *Esox lucius* L. *Gen. Comp. Endocrinol. 27*, 144

Ince, B.W. and Thorpe, A. (1977a), Plasma insulin in the eel (*Anguilla* L.). *Gen. Comp. Endocrinol. 33*, 453

Ince, B.W. and Thorpe, A. (1977b), Glucose and amino acid-stimulated insulin release. *Gen. Comp. Endocrinol. 31*, 249

Ince, B.W., Lone, K.P. and Matty, A.J. (1982), Anabolic steroids and trout growth. *Br. J. Nutr. 47*, 615

Inui, Y. and Yokote, M. (1977), Effects of glucagon on amino acid metabolism in Japanese eels (*Anguilla japonica*). *Gen. Comp. Endocrinol. 33*, 167

Kochakian, C.D. (1976), Anabolic–androgenic steroids. In: *Handbook of Experimental Pharmacology. 43*, Berlin, Springer

Lam, T.J. (1980), Thyroxine and larval development of tilapia. *Aquaculture 21*, 287

Lam, T.J. (1982), Endocrinology and fish culture. *Can. J. Fish. Aquat. Sci. 39*, 111

Leach, G.J. and Taylor, M.H. (1980), Cortisol and stress in *Fundulus*. *Gen. Comp. Endocrinol. 42*, 219

Leatherland, J.F. (1982), Environmental physiology of teleostean thyroid gland. A review. *Environ. Biol. Fish. 7*, 83

Leatherland, J.F. and Lam, T.J. (1971), Stickleback hypophysis and interrenal. *J. Endocrinol. 51*, 425

Leatherland, J.F. and Sonstegard, R.A. (1980), Seasonal thyroid changes in Coho salmon. *J. Fish Biol. 16*, 539

Leatherland, J.F. and Sonstegard, R.A. (1981), Thyroid and lipids of Coho salmon. *J. Fish Biol. 18*, 634–54

Leatherland, J.F., Coho, C.Y. and Slinger, S.J. (1977), Diet and thyroxine levels of rainbow trout. *J. Fish. Res. Bd Can. 34*, 677–82

Leatherland, J.R. and Nuti, R.N. (1981), Growth hormones in trout. *J. Fish. Biol. 19*, 487

Leibson, L.G., Plisetskaya, E.M. and Mazina, T.I. (1968), Concentration of non-esterifed fatty acids in blood. *Zh. Evol. Biokhim. Fiziol. 4*, 121

Le Ray, C., Bonnet, B., Febvre, A., Vallet, R. and Pic, R. (1970), Quelques activités périphérique des hormones thyroidiennes. *Ann. Endocrinol. Paris. 31*, 567

Le Ray, C., Pic, P., Bonnet, B. and Vallet, F. (1969), Aspects compares de l'activité thyroidienne sur la physiologie de *Mugil auratus*. *Gen. Comp. Endocrinol. 13*, 512

La Roche, G., Johnson, C.L. and Woodall, A.N. (1965), Radiothyroidectomy of rainbow trout. *Gen. Comp. Endocrinol. 5*, 145

Lewander, K., Dare, G., Johansson-Sjobeck, M.L., Larsson, A. and Lidman, U. (1976), Metabolic effects of insulin in the eel. *Gen. Comp. Endocrinol. 29*, 455

Lone, K.P. and Ince, B.W. (1983), Cellular growth responses of trout. *Gen. Comp. Endocrinol. 48*, 32

Lone, K.P. and Matty, A.J. (1980), Effect of methyltestosterone feeding on carp. *Gen. Comp. Endocrinol. 40*, 409

Lone, K.P. and Matty, A.J. (1982), 11-Ketotestosterone and food conversion of carp. *J. Fish Biol. 20*, 93

Lone, K.P. and Matty, A.J (1983), Ethylestrenol and carp tissue. *Aquaculture 32*, 39

Lone, K.P., Ince, B.W. and Matty, A.J. (1982), Dietary protein and ethylestrenol on blood chemistry. *J. Fish Biol. 20*, 597

McBride, J.R. and Overbeeke, V.D.A. (1971), Effects of hormones on sockeye. *J. Fish. Res. Bd Can. 28*, 485

McKeown, B.A., John, T.M. and George, J.C. (1976), Vasotocin effects on plasma of salmon. *Endocrinol Exper. 10*, 45

McKeown, B.A., Leatherland, J.F. and John, T.M. (1975), Growth hormone and prolactin on free fatty acids and glucose in salmon. *Comp. Biochem. Physiol (B). 50*, 425

Majid, A. (1979), Ph.D. Thesis, University of Aston, Birmingham, UK

Markert, J.R., Higgs, D.A., Dye, H.M. and MacQuarrie, D.W. (1977), Growth hormone and growth of coho salmon (*Oncorhynchus kisutch*). *Can. J. Zool 55*, 74

Matty, A.J. (1957), Thyroidectomy of the parrot fish. *J. Endocrinol 15*, 1

Matty, A.J. (1962), Growth hormone effects on *Cottus scorpius* blood. *Nature (Lond.) 195*, 506

Matty, A.J. and Sheltaway, M.J. (1967), Thyroxine and skin purines in trout. *Gen. Comp. Endocrinol. 9*, 473

Matty, A.J., Tsuneki, K., Dickoff, W.W. and Gorbman, A. (1976), Thyroid function in hypophysectomized hagfish. *Gen. Comp. Endocrinol. 30*, 500

Matty, A.J., Chaudry, M.A. and Lone, K.P. (1982), Thyroid hormones and nucleic acids in tilapia. *Gen. Comp. Endocrinol 47*, 497

Mayer, M. and Rosen, F. (1977), Glucocorticoids and skeletal muscle. *Metabolism. 26*, 937

Medda, A.K. and Ray, A.K. (1979), Thyroxine and muscle of Lata fish. *Gen. Comp. Endocrinol 9*, 473

Medda, A.K., Dasmahapatra, A.K. and Ray, A.K. (1980), Effect of sex steroids on Singi fish. *Gen. Comp. Endocrinol. 42*, 427

Morata, P., Vargas, A.M., Pita, M.L. and Sanchey-Medina, F. (1982), Hormonal effects on the liver glucose metabolism in rainbow trout (*Salmo gairdneri*). *Comp. Biochem. Physiol 72B*, 543

Mugiya, Y. and Ichii, T. (1981), Ca uptake in estrogenized rainbow trout. *Comp. Biochem. Physiol 70A*, 97

Mugiya, Y. and Watanabe, E. (1977), Studies on fish scale formation and resorption. *Comp. Biochem. Physiol 57A*, 197

Muller, M.J. and Seitz, H.J. (1982), T_3 induced hepatic gluconeogenesis. *Life Sci 28*, 2243

Muller, R. and Hanke, W. (1974), Adrenocortical hormones and roach. *Gen. Comp. Endocrinol 22*, 381

Murat, J.C. and Serfaty, A. (1971a), L'effet de la thyroxine sur le metabolisme glucidique de la carpe. *J. Physiol Paris 68*, 80

Murat, J.C. and Serfaty, A. (1971b), Thyroxine and carbohydrate in carp. *J. Physiol Paris 68*, 80

Murat, J.C., Plisetskaya, E.M. and Woo, N.Y.S. (1981), Endocrine control of nutrition in cyclostomes and fish. *Comp. Biochem. Physiol 68A*, 149

Narayansingh, T. and Eales, J.G. (1975a), Effects of thyroid hormones. *Comp. Biochem. Physiol 52B*, 399

Narayansingh, T. and Eales, J.G. (1975b), Physiological doses of thyroxine. *Comp. Biochem. Physiol 52B*, 407

Nijkamp, H.J., van Es, A.J.H. and Huisman, A.E. (1974), Energy in chicken and carp. *Eur. Assoc. Anim. Prod. Publ. 14*, 277

Osborn, R.H. and Simpson, T.H. (1978), Thyroid activity in plaice. *J. Fish Biol. 12*, 519

Osborn, R.H., Simpson, T.H. and Youngson, A.F. (1978), Rhythms of thyroid in trout. *J. Fish Biol. 12*, 531

Ottolenghi, C., Puviani, A.C., Baruffaldi, A. and Brighenti, L. (1982), 'In vivo' effects of insulin on carbohydrate metabolism of catfish (*Ictalurus melas*). *Comp. Biochem. Physiol 72A*, 35

Paul, A.K. and Dhar, A. (1980), Thyroxine and glucose metabolism. *Hormone Metabl. Res. 12*, 261

Peter, R.E. (1973), Neuroendocrinology of teleosts. *Amer. Zool. 13*, 743

Peter, R.E. (1979), The brain and feeding behavior. In: Hoar, W.S., Randall, D.J. and Brett, J.R. (eds). *Fish Physiology*, vol. VIII, pp. 121–61. New York, Academic Press

Peter, R.E., Monckton, E.A. and McKeown, B.A. (1976), Gold thioglucose and the brain of goldfish. *Physiol. Behav. 17*, 303

Pickford, G.E., Pang, P.K.T., Weinstein, E., Toretti, J., Hendler, E. and Epstein, F.H. (1970), Cortisol treatment of hypophysectomized *Fundulus*. *Gen. Comp. Endocrinol. 14*, 524

Plack, P.A., Pritchard, D.J. and Frazer, N.W. (1971), Egg proteins in cod serum. *Biochem. J. 121*, 847

Plisetskaya, E.M. (1968), Brain and heart glycogen content and effect of insulin. *Endocrinol. Exper. 2*, 251

Ray, A.K. and Medda, A.K. (1976b), Effect of thyroid hormones and analogues on ammonia and urea excretion in Lata fish (*Ophicephalus punctatus*). *Gen. Comp. Endocrinol. 29*, 190

Sestoft, L. (1980), Calorigenic effect of thyroid hormones. *Clin. Endocrinol. 13*, 489

Simpson, T. (1976), Endocrines and salmonid culture. *Proc. Roy. Soc. Edinburgh (B) 75*, 241

Simpson, T.H. (1978), An interpretation of some endocrine rhythms in fish. In: Thorpe, J.E. (ed.). *Rhythmic Activity of Fishes*, pp. 55–68. London, Academic Press

Smith, M.A.K. and Thorpe, A. (1977), Nitrogen excretion in *Salmo*. *Gen. Comp. Endocrinol. 32*, 400

Storer, J.H. (1967), Effects of cortisol in fish. *Comp. Biochem. Physiol. 20*, 939

Swallow, R.L. and Flemming, W.R. (1966), Liver glycogen levels of *Tilapia mossambica*. *Amer. Zool. 6*, 562

Swift, D.R. (1955), Seasonal thyroid activity of trout. *J. exp. Biol. 32*, 751

Takashima, F., Hibiya, T., Ngan, P.Y. and Aida, K. (1972), Hormones and lipids in rainbow trout. *Bull. Jpn Soc. Sci. Fish. 38*, 43

Tashima, L. and Cahill G.F., Jr (1968), Effects of insulin in toadfish *Opsanus tau*. *Gen. Comp. Endocrinol. 11*, 262

Tata, J.R. (1974), Growth and developmental action of thyroid hormones. In: Greer, M.A. and Salmon, D.H. (eds). *Handbook on Physiology*, vol. III, pp. 469–79. Washington, D.C., American Physiology Society

Tata, J.R. (1980), Growth and developmental hormones. *Biol. Rev. 55*, 285

Thornburn, C.C. and Matty, A.J. (1963), Thyroxine and nitrogen metabolism of goldfish and trout. *Comp. Biochem. Physiol. 8*, 1

Valtonen, T. (1974), Carbohydrate metabolism of liver of whitefish. *Comp. Biochem. Physiol. 47A*, 713

Walton, M.J. and Cowey, C.B. (1979), Gluconeogenesis by isolated hepatocytes from rainbow trout, *Salmo gairdneri*. *Comp. Biochem. Physiol. 62B*, 75

Warman, A.W. and Bottino, N.R. (1978), Lipogenic activity of catfish liver. *Comp. Biochem. Physiol. 59B*, 153

Weigand, M.D. and Peter, R.E. (1980), Plasma F.F.A. in the goldfish. *Comp. Biochem. Physiol. 66A*, 323

White, B.A. and Henderson, N.E. (1977), Annual variations in circulating levels of thyroid hormone in brook trout, *Salvelinus fontinalis*, as measured by radio-immunoassay. *Can. J. Zool. 55*, 475

Whiting, S.J. and Wiggs, A.J. (1978), Effect of nutritional factors and cortisol. *Comp. Biochem. Physiol. 58*, 189

Williams, R.H. (1981), *Text Book of Endocrinology*, Philadelphia, W.B. Saunders

Yaron, Z., Terkalin-Shimony, A., Shahan, Y. and Salzer, H. (1977), Oestradiol in tilapia. *Gen. Comp. Endocrinol. 33*, 45

Young, V.R. and Pluskal, M.G. (1977), Mode of action of anabolic agents. A summary review. *Proc. 2nd Int. Symp. Protein Metab. Nutr.* pp. 17–28

Yu, T.C., Sinnhuber, R.D. and Hendricks, J.D. (1979), Steroid hormone and salmon growth. *Aquaculture 16*, 351

PART THREE: PRODUCTION

8 GROWTH

Malcolm Jobling

8.1 Introduction

The starting point for any discussion of growth processes lies in the simple statement: that which remains in the body must equal that which goes in minus that which comes out, and that which remains is growth. This simple statement regarding growth considers the organism to be an energy depot into which flows energy as food and out of which flows heat energy. It is the basis of the 'balanced energy equation' (Winberg, 1956) and is externally controlled in that growth limitation can occur via factors imposing limits on the rate of food consumption. The inclusion of a feedback loop into the energy equation, in which heat loss (metabolism) is dependent upon food supply, introduces a degree of internal regulation of the growth processes, but overall control remains dependent upon external factors.

An alternative view envisages that the growth of an organism follows a more or less predetermined programme and, therefore, the growth processes are under active internal control, being governed by precise self-regulatory mechanisms. According to this 'feedback-hypothesis' (Weiss, 1952; Weiss & Kavanau, 1957) each system would be regulated via homeostatic mechanisms orientated towards the maintenance of a precise equilibrium, and it is envisaged that growth processes are controlled via numerous unspecified humoral factors (possibly of endocrine origin: Chapter 7). The pursuit of empirical evidence in favour of this hypothesis and attempts to define the humoral growth factors have usually involved the study of regenerative growth of organ systems following surgery (Sidorova, 1978), but, to date, the precise form of the regulatory mechanisms still requires clarification (see Calow (1977) for further discussion).

Workers involved in the study of fish growth, from the standpoint of either management or aquaculture, have favoured the use of models involving some degree of external control over the growth processes and, therefore, the basic bioenergetics of fish growth have been expressed in forms that are usually extensions and developments of either the Pütter-von Bertalanffy anabolism–catabolism model or the balanced energy equation (Elliott, 1979; Ricker, 1979). Although the Pütter–von Bertalanffy model has been widely adopted by fisheries biologists and has been useful in allowing the fitting of curves to growth data, it is not particularly satisfactory when more detailed studies of the factors influencing metabolism and growth are envisaged (Ricker, 1979; Ursin, 1979). Consequently, the balanced energy equation (Winberg, 1956) has been widely applied, in either its basic or a more expanded form. The basic form of the expression is:

$$\text{Rations } (\times \text{ absorption factor}) = \text{metabolism} + \text{growth}$$

but this is often expanded into its many facets by means of an additive series of terms, such that the majority of workers take the IBP energy balance equation (see Chapter 1, and Grodzinski *et al.*, 1975; Bagenal, 1978) as the starting point in empirical studies or as the basis for predictive modelling of fish growth in nature (Kitchell *et al.*, 1977; Kitchell & Breck, 1980). Much of the empirical data have been extensively reviewed elsewhere (Webb, 1978; Brett & Groves, 1979; Elliott, 1979, 1982; Cho *et al.*, 1982) and readers are referred to these discussions of the energy balance equation.

In common with many other studies, the energy balance equation will be taken as the starting point in the present chapter; the components of the equation will be analysed, and the findings and conclusions will be discussed in relation to previous interpretations. The basic form of the energy balance equation is written:

$$pC = R + P$$

and when growth/production (P) is zero, the input in terms of absorbed food energy balances the metabolic output and the status quo is maintained. Under these conditions, metabolism can be broadly divided into energy utilized in fulfilling two functions. First, energy must be expended in activities involved in the maintenance of the internal homeostatic balance. Since the degree of any external disturbance of this balance is likely to be dependent upon the surface area of the organism in contact with the external environment, the energetic costs of overcoming any disturbance are likely to be scaled in proportion to body surface area. As surface area scales in proportion to body weight to the power two-thirds, the energetic costs of maintaining homeostatic balance will increase in proportion to body weight$^{0.67}$ ($W^{0.67}$). The second energy-requiring processes associated with the maintenance of the status quo are those involved in tissue repair and renewal. Since rates of synthesis and turnover of the major body constituents may be expected to be proportional to tissue mass, the metabolism associated with tissue renewal will be assumed to scale in direct proportion to body weight ($W^{1.0}$). Consequently, the balance equation at $P = 0$ can be rewritten:

$$pC = (a_1 W^{0.67} + a_2 W^{1.0})$$

where a_1 and a_2 represent the intensities of 'homeostatic disturbance' and 'tissue renewal' metabolic costs, respectively. Independent assessments of these two metabolic costs are difficult to make and it is more feasible to make an overall estimate of the metabolic costs of maintaining the status quo. This summation will result in maintenance metabolic costs which scale in proportion to body weight to the power 0.67–1.00, the exact value of the scaling coefficient being dependent upon the relative magnitude of an interplay between the various metabolic functions. Very few reliable measurements of maintenance metabolism have been made directly but data exist allowing indirect assessments to be made for a small number of fish species. Estimates of maintenance food intake requirements of fishes can be made by the extrapolation of growth–ration relationships to the point where growth is zero, and since this is the point where energy expenditure is exactly balanced by the energy derived from the food consumed, this gives an indirect assessment of maintenance energy expenditure. The calculation of the scaling coefficient of maintenance metabolism requires that estimates of maintenance requirements of fishes of a wide range of sizes are available, necessitating the construction of growth–ration curves for fishes of different sizes. The amount of practical work involved in data collection is considerable and it comes as no surprise, therefore, that reliable estimates can be made for comparatively few species. The empirical data presented in Table 8.1 provide support for the theoretical expectation of maintenance metabolism scaling coefficients lying within the range 0.67–1.00.

At levels of food intake above those required for maintenance, energy is available for growth/production. Growth/production can be viewed as the accretion of new body tissue, dependent upon biochemical synthetic processes analogous to those involved in the repair

Table 8.1: Weight Scaling Coefficients of Maintenance Requirements for Various Species of Fish Calculated from Growth-ration Relationships.

Species	Scaling coefficient	Source
Cod, *Gadus morhua*	0.88	Jobling, 1982
	0.83	Soofiani, unpublished
Mullet, *Mugil cephalus*	1.04	Flowerdew & Grove, 1980
African catfish, *Clarias lazera*	0.78	Hogendoorn, 1983
Brown trout, *Salmo trutta*	0.72, 0.74	Elliott, 1975
Rainbow trout, *Salmo gairdneri*	0.75	Calculated from Staples & Nomura, 1976

and renewal of body tissues during maintenance. Thus, the division of resources, over and above those required for maintaining the status quo, can be envisaged as being energy *deposited* as growth and the energy *utilized* in growing; that is, the metabolic costs of tissue synthesis and accretion. The expression for the energy balance equation then becomes:

$$pC = \left[(a_1 W^{0.67} + a_2 W^{1.0}) + (rP) \right] + P$$

or

$$pC = \left[\alpha W^{\beta} + rP \right] + P$$

where r represents the metabolic costs per unit growth/production. The form of the energy balance equation presented above implies that there is a fundamental relationship between growth/production and metabolism/respiration. Evidence in favour of this interpretation can be gleaned from studies of production carried out at the

(1) community/population level;
(2) individual, whole animal level;
(3) isolated organ systems and tissues.

In his review of rates of production and energy turnover in terrestrial animal communities, Engelmann (1966) suggested that a linear relationship existed between annual production and annual respiration (kcal m^{-2} year^{-1}). Later, McNeill & Lawton (1970) extended this concept to include aquatic communities, and subsequently, Humphreys (1978, 1979) collected data from 235 published energy budgets and methodically analysed the relationship between production and respiration for a wide diversity of animals. Species were initially separated into one of 14 taxonomic categories (e.g. mice, fish, social insects, molluscs, crustaceans, etc.), and least squares regressions were calculated on the logarithmically transformed production and respiration data for each taxonomic group, treating production as the dependent variable. All the regression lines were significant and none of the slopes differed significantly from 1.0, indicating a fundamental and direct relationship between production and respiration/metabolism. Intercepts of a number of the regression lines differed significantly, with, for example, poikilotherms being significantly different from homeotherms and regression lines for mammals and birds being parallel but separate. This is interpreted as indicating differences in maintenance metabolic costs for the various animal groups.

Early in the present century a number of workers investigated the changes in metabolism which occurred during animal development. These studies were evaluated by Brody (1942) in the light of his own work on growth and metabolism of young rats, and he was led to propose the concept of an 'organizational energy cost': the more rapid the growth rate, the greater the oxygen consumption or rate of heat production. This concept then appears to have been neglected for a quarter of a century before being restated by Krieger (1966) and Ashworth (1969), the latter attempting to establish a link between the 'organizational energy cost' and post-feeding metabolic rates. It was found that the metabolic rates of previously malnourished infants were substantially elevated during the period of 'recovery growth', leading to the suggestion that this increase in metabolic rate was related to undefined 'energy costs of growth' (Krieger, 1966; Ashworth, 1969). Later studies demonstrated that the increase in metabolic rate associated with 'recovery growth' was not a special case, but that the link between metabolism and growth rates also applied to normally growing individuals (Freeman, 1964; Krieger & Whitten, 1969; Brooke & Ashworth, 1972; Alvear & Brooke, 1978; Krieger, 1978). The most important conclusions of these studies can be summarized as follows:

(1) There is a significant positive correlation between rates of weight gain and metabolism.
(2) Different patterns of metabolism exist between subjects of similar nutritional status (given similar meals) but growing at different rates.

Table 8.2: Relationship Between Dry Weight Gain and Oxygen Consumption of African Catfish, *Clarias lazera*, of Different Sizes Fed at Various Ration Levels (calculated from Hogendoorn, 1983). Oxygen consumption (ml) = $a + b$ dry weight gain (mg); a = maintenance oxygen consumption; b = oxygen consumption per unit dry weight deposition.

Initial weight of fish, (g) mean (range)	n	a	$b \pm$ S.E.
1.57 (1.50–1.65)	4	162.6	0.554 ± 0.017
7.28 (6.84–7.99)	4	1069.9	0.359 ± 0.011
43.58 (40.2–50)	4	3974.4	0.390 ± 0.099
91.9 (82.9–96.9)	4	9119.9	0.469 ± 0.142

(3) Subjects that are growing slowly, whether because of nutritional restriction or age-related effects, show little or no increase in metabolic rate following food intake.

Thus, the evidence from work with homeotherms (rats, chickens and man) suggests that rates of growth and metabolism are linked. Evidence for such relationships in poikilotherms is scarce, but some information is available. Jobling (1981) re-examined published data (Brett and co-workers, 1969, 1976, 1979) relating to growth and respiratory metabolism of sockeye salmon, *Oncorhynchus nerka*, held under various rearing conditions. A close relationship was found between growth rate and metabolic rate. Oxygen consumption of the starfish, *Asterias rubens*, has been studied during the period of recovery growth following long-term starvation. Analysis of the results of this investigation showed that there was a close relationship between the magnitude of the increase in metabolic expenditure following feeding and the growth displayed by the experimental animals. It was suggested that this increased metabolic expenditure may reflect the energetic costs of growth (Vahl, 1983). Recently, Hogendoorn (1983) studied the effects of different feeding levels on growth, body composition and oxygen consumption of the African catfish, *Clarias lazera*, of different sizes. Results of this detailed investigation can be used to examine growth–metabolism (oxygen consumption) relationships, and data for fishes of different sizes, at different levels of feeding, are summarized in Table 8.2. Analysis of these data showed that the maintenance oxygen consumption increased in proportion to body weight to the power 0.816 and that

Figure 8.1: The Relationship Between Dry Weight Gain and Oxygen Consumption in the African Catfish, *Clarias lazera* (calculated from Hogendoorn, 1983). (Note that the figure axes are given as grams dry weight gain and litres oxygen consumed, as opposed to milligrams weight gain and millilitres oxygen consumed in the regression equation presented on page 219.)

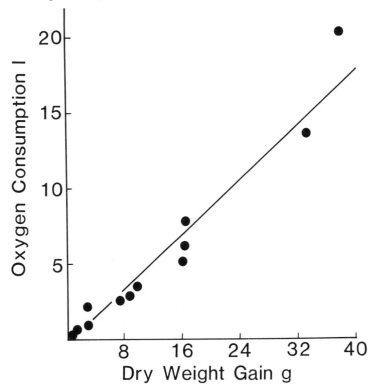

there were no significant differences between the slopes of the growth–metabolism regression lines, leading to the conclusion that the 'metabolic costs' per unit dry weight gain were independent of fish size. The 'oxygen consumed for growth purposes' was then calculated for each group of fish by subtracting the estimated maintenance consumption from the total amount of oxygen consumed throughout the entire experimental period. The data for fishes of different sizes, fed at various ration levels, are plotted in Figure 8.1, and the results show that there is a highly significant relationship ($r^2 = 0.994$; $P > 0.001$) between consumption and weight gain:

$$Y = 0.463X - 269.3$$

where Y is oxygen consumption in millilitres and X is weight gain in milligrams dry weight. This confirms that there is a fundamental relationship between metabolism (respiration) and growth in the intact animal.

A small number of studies have been carried out to investigate differences in heat production/respiration between various organ systems and to relate these differences to metabolic function. Despite the practical problems involved in making such measurements, attempts have been made to calculate the contribution made by various organs to total metabolic heat production, and the results of such investigations suggest a close association between rates of metabolism and rates of protein synthesis (Webster, 1981). The links between metabolism and protein synthesis are also apparent at the level of the whole animal (Millward *et al.*, 1979; Reeds & Harris, 1981; Webster, 1981), there being a tendency for metabolic rate and the rate of protein synthesis to change in the same direction. However, so few attempts have been made to measure both rates simultaneously that there is insufficient data for the presentation of a quantitative relationship between the two.

How much of the metabolic expenditure of growing fishes can be attributed to the costs of protein deposition? In terms of energetics, protein growth is undoubtedly more expensive than the accumulation of fat; there are, however, a number of problems involved in the estimation of the costs of protein synthesis. Energy expenditure for the formation of the peptide bond can be calculated, but costs such as those required for RNA synthesis and template formation are less certain. Published estimates for the costs of protein synthesis usually lie within the range 70–100 mmol ATP g^{-1} protein, depending upon the assumptions made whereas the costs of lipid and glycogen synthesis lie within the ranges 15–25 mmol ATP g^{-1} and 10–12 mmol ATP g^{-1}, respectively (Grisolia & Kennedy, 1966; Buttery & Annison, 1973; Coulson *et al.*, 1978; Hommes, 1980).

Hogendoorn's (1983) growth–metabolism data for *Clarias lazera* have been examined in the light of changes in body composition occurring during the growth of the fish, and the amount of protein gain has been calculated. The protein gain–oxygen consumption relationship is shown in Figure 8.2, and the regression line ($r^2 = 0.955$; $P > 0.001$) can be described by the equation:

$$Y = 0.758X - 367.5$$

Figure 8.2: The Relationship Between Protein Growth and Oxygen Consumption in the African Catfish, *Clarias lazera*. (Note that the figure axes are given as grams protein gain and litres of oxygen consumed, whereas the regression equation on page 220 is in terms of milligrams protein gain and millilitres oxygen consumed.)

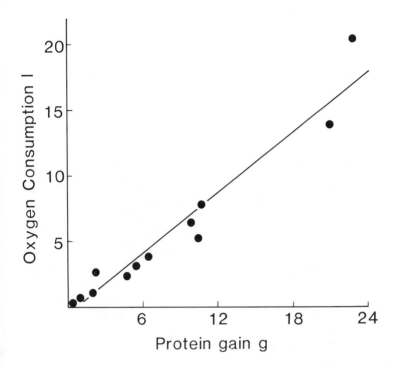

where Y is oxygen consumption in millilitres and X is protein gain in milligrams. Mean values of protein gain and oxygen consumption were 6234 mg protein and 4360 ml oxygen, respectively. This gives an estimate of approximately 700 ml oxygen consumed per gram protein gain. Assuming that this oxygen is used directly for glucose oxidation, this gives an ATP yield and utilization of approximately 185 mmol ATP g^{-1} per gram protein deposited. This is somewhat higher than the theoretical estimates of the costs of protein synthesis, but these theoretical calculations refer to costs of *synthesis* rather than costs of protein *deposition*. Agreement between evaluations of these two costs is to be expected only if an increase of 1 g in protein synthesis results in the accretion/deposition of 1 g of new protein tissue. Protein growth in skeletal muscle requires levels of protein synthesis in excess of the amount deposited (Millward *et al.*, 1975, 1981; Laurent *et al.*, 1978), and studies with intact animals suggest that protein deposition requires a disproportionate (1.5–2.2 times) increase in protein synthesis (Reeds *et al.*, 1980; Reeds & Harris, 1981) leading to a theoretical estimate of 105–220 mmol ATP per gram protein deposited. Assuming that protein deposition in fishes leads to the same order of increase in protein synthesis as it does in mammals, one is led to conclude that a substantial portion of the metabolic expenditure in growing fishes may be attributable to the costs of protein growth.

If, as has been suggested, much of the metabolic expenditure in growing individuals is related to the costs of growing, then variations in growth rate would be expected to be mirrored in variations in metabolic rates and vice versa. Thus, any growth process which affects only certain size groups within an animal population may be expected to lead to alterations in size–metabolism relationships, with size-dependent patterns of somatic or reproductive growth being reflected in changes in metabolic weight exponents. Similarly, the metabolic patterns of poikilotherms would be expected to reflect seasonal changes in somatic and reproductive growth rates. For example, rates of protein synthesis in isolated hepatocytes of *Trematomus hansoni* have been shown to be influenced by the reproductive status of the fish, with synthetic rates of cells from gravid females being 50% greater than those of hepatocytes from immature females and males (Haschemeyer & Mathews, 1983), and such differences may be expected to lead to detectable differences in metabolic energy utilization.

This is an attractive hypothesis, but convincing evidence is scarce since metabolic rates can be affected by a large variety of factors and, in poikilotherms, it is difficult to study metabolic changes in isolation from the seasonal temperature cycle. Thus, although there is much information available regarding seasonal variations in the metabolic rates of poikilotherms, most studies have been made with regard to examining the effects of temperature adaptations. Temperature-independent seasonal variations in metabolic rates have been observed in a number of species (Berg *et al.*, 1958; Berg & Ockelmann, 1959; Barnes *et al.*, 1963; Phillipson & Watson, 1965; Calow, 1975; Vahl, 1978), with the peak of the annual metabolic cycle being observed during the reproductive period, and recently Vahl & Sundet (1983) reported that metabolic rates of maturing Icelandic scallops, *Chlamys islandica*, were greater than those of immatures. It is tempting, therefore, to suggest that these high rates of metabolism are associated with the costs of gametogenesis and vitellogenesis, but such a claim would not be completely justified since the observed increases in metabolic rate could have been caused by factors other than growth-related phenomena. Thus, clear links between reproductive growth costs and metabolic rates remain to be established and the challenge of demonstrating the presence of such links offers interesting opportunities for future research.

8.2 Interpretations and Applications of Growth–Metabolism Relationships

In the traditional presentation of the energy balance equation

$$pC = (R_S + R_A + R_F) + P$$

metabolism (R) is considered to consist of several distinct components which can be partitioned into resting metabolism (R_S), activity metabolism (R_A) and 'specific dynamic action' (R_F). Resting metabolism, which may also be encountered in the guise of 'standard', 'basal' or 'low routine' metabolism (see Brett, 1972, for further discussion and definition), is generally considered to reflect the energy use of an unfed and resting animal in a 'normal physiological state'. Usually, little indication is given as to what is meant by the expression 'normal physiological state', and a lack of standardization of the procedures employed in attempts to measure resting metabolism has certainly led to experiments being carried out on animals of different physiological status. For example, the length of time animals are deprived of food markedly affects their metabolic rate, with increased periods of food deprivation leading to reductions in metabolism (Beamish, 1964; Jobling, 1980; Woo & Cheung, 1980), possibly due to the reductions in rates of protein synthesis and turnover that occur with increased length of time of food deprivation (Jackim & La Roche, 1973; Smith & Haschemeyer, 1980; Smith, 1981; Lied *et al.*, 1982). Thus, although the concept of resting (basal) metabolism is an attractive one, it is difficult to make representative measurements, and, irrespective of whether 'resting metabolism' is determined directly or predicted from measurements made on active animals, such estimates must be considered as being, at best, rough approximations.

The development of tunnel respirometers (Brett, 1964; Smith &

Newcomb, 1970) has allowed measurement of oxygen consumption associated with different levels of forced, steady-state swimming activity under controlled laboratory conditions. The results of such investigations indicate that swimming at maximum speeds may lead to a 15-fold increase in metabolic rate and the energetic costs of swimming are considerable (reviewed by Beamish, 1978). A major problem arises, however, in relating these laboratory measurements to levels of activity shown by free-swimming fishes in the wild. Results of telemetric studies have shown that a number of fish species are sedentary (Young *et al.*, 1972; Hawkins *et al.*, 1974, 1979; Diana, 1980), and it has been suggested that free-swimming fishes use relatively little energy in locomotory activity (Holliday *et al.*, 1974; Priede & Young, 1977), but measurements based solely on monitoring fish movements are likely to be unreliable. Recent advances in telemetric methods for monitoring physiological activities (ECG and EMG) of fishes have led to the possibility of making more accurate assessments of energy use of fishes in the wild (Priede, 1978; Weatherley *et al.*, 1982), but in free-swimming fishes, heart rate is likely to be a better indicator of total metabolism than of energy expenditure involved in locomotor activity alone. On the other hand, EMG recordings are an index of locomotory activity, and EMG monitoring should allow energetic costs of activity to be estimated for fishes under natural conditions.

The specific dynamic action (R_F) is usually defined as the energy used during the course of digestion, assimilation and storage of products of digestion, in the deamination of amino acids and for the synthesis of nitrogenous excretory products. Specific dynamic action has also been called the calorigenic or thermic effect and, among ruminant nutritionists, heat increment of feeding (HI) has been the favoured term. The recent revival of interest, among human nutritionists, in the effects of feeding on metabolic heat production has led to the coining of the phrase 'dietary induced thermogenesis' (DIT), but all of these terms are, presumably, used to describe the same series of physiological/biochemical processes occurring in different animal groups. Recognizing that there was a lack of knowledge as to the processes contributing to 'SDA', Ware (1975) suggested a broader definition: '... SDA represents an entropic tax paid during food conversion'. Work with mammals, during the first 30 years of the current century, suggested that 'SDA' varied with the type of food, and the accepted view, presented in textbooks on nutrition and bioenergetics can be summarized as follows. If each food component (protein, fat and carbohydrate) is fed separately, the respective heat production in homeothermic animals would be equivalent to 30% of the caloric content of protein, 13% for lipid and 5% for carbohydrate. Animals eating mixed diets experience SDAs dependent upon the respective proportions of the three basic food components, and hence SDA may be expressed in terms of the energy content of the food, e.g. kilojoules per gram dry matter (Davidson & Passmore, 1969; Ganong, 1969; Harper, 1971; McDonald *et al.*, 1973).

This interpretation of 'SDA' as being a fundamental property of the diet has also been applied in work with fish species and it is assumed that the magnitude of the SDA depends upon the quality and quantity of food consumed such that:

$$R_F = mC$$

where C is the quantity of food consumed and m is a proportionality constant dependent upon the composition of the diet and reaching a maximum of 0.3 when pure protein is fed (Weatherley, 1976; Jobling, 1981).

Recent work (summarized in Jobling, 1981, 1983) has questioned the validity of this traditional view and suggests that SDA may represent an energy expenditure associated with growth, such that:

$$R_F = rP$$

where P is growth/production and r is the energy cost per unit growth. An examination of the two interpretations of SDA, within the context of the energy balance equation, reveals fundamental differences. The traditional interpretation of SDA suggests that SDA and growth are directly *competitive*, that is, a diet inducing a high SDA will reduce the amount of food energy available for growth. The alternative interpretation implies that SDA and growth are *interactive*, and that high rates of growth will be reflected in high rates of metabolism and high SDA. Thus, the traditional approach to SDA implies that best growth will be given by diets which induce lowest SDA per energy intake, whereas the alternative view is that diets promoting good growth should also induce high SDA. The way in which results from 'SDA investigations' are interpreted has obvious implications for nutritionists and aquaculturalists intending to formulate diets promoting optimum rates of growth, and hence it is important that increased effort be directed towards solving the practical problems involved in the collection of reliable empirical data related to the SDA phenomenon.

A number of studies carried out under laboratory conditions have shown that there is a relationship between feeding rates and metabolic expenditure, leading to a close relationship between metabolism and growth (Brett, 1976; Staples & Nomura, 1976; Vivekanandan & Pandian, 1977; Soofiani & Hawkins, 1982; Hogendoorn, 1983). This suggests that if reliable data for these three primary variables (feeding rate/ration, metabolism and growth) are available, then it should be possible to develop predictive equations that could be applied to natural populations of fishes. For example, if data for metabolic rates of wild fishes were available (estimated by ECG techniques), it should be possible to estimate rates of growth and food consumption. This type of holistic approach is not new and has previously applied to the analysis of growth relationships of cod, *Gadus morhua*, and haddock, *Melanogrammus aeglefinus* (Kerr, 1982). Kerr (1982) suggested that food intake and growth could be estimated from levels of activity (swimming) metabolism, a conclusion that appears to be based upon the adoption of the traditional concept of the SDA component of metabolism. The model developed by Kerr (1982) predicts that there is a general tendency for activity metabolism to increase with increasing food intake. With slight modification, this model could be written to incorporate both SDA and activity metabolism, such that metabolic expenditure would be predicted with increased ration level, partly due to the increased SDA and partly due to the increased foraging activity required for prey capture. The attractions of this approach are both simplicity and generality, but if holistic models are to have any value it is essential that precise empirical descriptions of growth–metabolism relationships be made available so that the formulation of reliable predictive equations is made possible.

8.3 Acknowledgements

I should like to express my thanks to H. Hogendoorn, O. Vahl and N.M. Soofiani for allowing me access to their unpublished results and, at the same time, take this opportunity to apologize to them if they feel that my use of their data has been an abuse.

References

Alvear, J. and Brooke, O.G. (1978), Specific dynamic action in infants of low birth weight. *J. Physiol. 275*, 54P

Ashworth, A. (1969), Metabolic rates during recovery from protein-calorie malnutrition: the need for a new concept of Specific Dynamic Action. *Nature (Lond.), 223*, 407

Bagenal, T. (ed.) (1978), *IBP Handbook No. 3: Methods for Assessment of Fish Production in Fresh Waters.* Oxford, Blackwell

Barnes, H., Barnes, M. and Finlayson, D.M. (1963), The seasonal changes in body weight, biochemical composition and oxygen uptake of two common Boreo-Arctic cirripedes, *Balanus balanoides* and *B. balanus. J. mar. biol. Ass. UK, 43*,185

Beamish, F.W.H. (1964), Influence of starvation on standard and routine oxygen consumption. *Trans. Am. Fish. Soc. 93*, 103

Beamish, F.W.H. (1978), Swimming capacity. In: Hoar, W.S. and Randall, D.J. (eds). *Fish Physiology*, vol. VII, pp. 101–87. New York, Academic Press

Berg, K. and Ockelmann, K.W. (1959), The respiration of freshwater snails. *J. exp. Biol. 36*, 690

Berg, K., Lumbye, E. and Ockelmann, K.W. (1958), Seasonal and experimental variations of the oxygen consumption of the limpet *Ancylus fluviatilis. J. exp. Biol. 35*, 43

Brett, J.R. (1964), The respiratory metabolism and swimming performance of young sockeye salmon. *J. Fish. Res. Bd Can. 21*, 1183

Brett, J.R. (1972), The metabolic demand for oxygen in fish, particularly salmonids, and a comparison with other vertebrates. *Resp. Physiol. 14*, 151

Brett, J.R. (1976), Feeding metabolic rates of young sockeye salmon, *Oncorhynchus nerka*, in relation to ration level and temperature. *Fish. Mar. Serv. Res. Dev. Tech. Rep. 675*, 43 pp

Brett, J.R. (1979), Environmental factors and growth. In: Hoar, W.S., Randall, D.J. and Brett, J.R. (eds). *Fish Physiology, vol. VIII*, pp. 599–675. New York, Academic Press

Brett, J.R. and Groves, T.D.D. (1979), Physiological energetics. In: Hoar, W.S., Randall, D.J. and Brett, J.R. (eds). *Fish Physiology, vol. VIII*, pp. 279–352. New York, Academic Press

Brett, J.R., Shelbourn, J.E. and Shoop, C.T. (1969), Growth rate and body composition of fingerling sockeye salmon, *Oncorhynchus nerka*, in relation to temperature and ration size. *J. Fish. Res. Bd Can. 26*, 2363

Brody, E. (1942), Litter size, growth rate and heat production of suckling rats. *Am. J. Physiol. 138*, 180

Brooke, O.G. and Ashworth, A. (1972), The influence of malnutrition on the postprandial metabolic rate and respiratory quotient. *Brit. J. Nutr. 27*, 407

Buttery, P.J. and Annison, E.F. (1973), Considerations of the efficiency of amino acid and protein metabolism in animals. In: Jones, J.G.W. (ed.). *The Biological Efficiency of Protein Production*, pp. 141–71. Cambridge, Cambridge University Press

Calow, P. (1975), The respiratory strategies of two species of freshwater gastropods (*Ancylus fluviatilis* Müll and *Planorbis contortus* Linn) in relation to temperature, oxygen concentration, body size and season. *Physiol. Zool. 48*, 114

Calow, P. (1977), Ecology, evolution and energetics: a study in metabolic adaptation. *Adv. Ecol. Res. 10*, 1

Cho, C.Y., Slinger, S.J. and Bayley, H.S. (1982), Bioenergetics of salmonid fishes: energy intake, expenditure and productivity. *Comp. Biochem. Physiol. 73B*, 25

Coulson, R.A., Herbert, J.D. and Hernandez, T. (1978), Energy for amino acid absorption, transport and protein synthesis *in vivo. Comp. Biochem.*

228 Fish Energetics

Physiol. 60A, 13

Davidson, S. and Passmore, R. (1969), *Human Nutrition and Dietetics*, 4th ed. Edinburgh, Livingstone

Diana, J.S. (1980), Diel activity pattern and swimming speeds of northern pike (*Esox lucius*) in Lac Ste Anne, Alberta. *Can. J. Fish Aquat. Sci. 37*, 1454

Elliott, J.M. (1975), The growth rate of brown trout, *Salmo trutta* L. fed on reduced rations. *J. Anim. Ecol. 44*, 823

Elliott, J.M. (1979), Energetics of freshwater teleosts. In: Miller, P.J. (ed.) *Fish phenology: Anabolic Adaptiveness in Teleosts*, pp. 29–61. New York, Academic Press

Elliott, J.M. (1982), The effects of temperature and ration size on the growth and energetics of salmonids in captivity. *Comp. Biochem. Physiol. 73B*, 81

Engelmann, M.D. (1966), Energetics, terrestrial field studies and animal productivity. *Adv. Ecol. Res. 3*, 73

Flowerdew, M.W. and Grove, D.J. (1980), An energy budget for juvenile thick-lipped mullet, *Crenimugil labrosus* (Risso). *J. Fish Biol. 17*, 395

Freeman, B.M. (1964), The effect of diet and breed upon the oxygen requirements of the domestic fowl during the first fortnight of post-embryonic life. *Brit. Poult. Sci. 5*, 263

Ganong, W.F. (1969), *Review of Medical Physiology*, 4th edn. Los Altos, Calif., Lange Medical Publications

Grisolia, S. and Kennedy, J. (1966), On specific dynamic action, turnover and protein synthesis. *Perspect. Biol. Med. 9*, 578

Grodzinski, W., Klekowski, R.Z. and Duncan, A. (eds). (1975), *IBP Handbook No. 24: Methods for Ecological Bioenergetics*. Oxford, Blackwell

Harper, H.A. (1971), *Review of Physiological Chemistry*, 13th edn. Los Altos, Calif., Lange Medical Publications

Haschemeyer, A.E.V. and Mathews, R.W. (1983), Temperature dependency of protein synthesis in isolated hepatocytes of Antarctic fish. *Physiol. Zool. 56*, 78

Hawkins, A.D., MacLennan, D.N., Urquhart, G.G. and Robb, C. (1974), Tracking cod, *Gadus morhua* L., in a Scottish sea loch. *J. Fish. Biol. 6*, 225

Hawkins, A.D., Urquhart, G.G. and Smith, G.W. (1979), Ultrasonic tracking of juvenile cod by means of a large spaced hydrophone array. In: Amlaner Jr, C.J. and MacDonald, D.W. (eds). *A Handbook on Biotelemetry and Radio Tracking*, pp. 461–70. Oxford, Pergamon

Hogendoorn, H. (1983), Growth and production of the African catfish, *Clarias lazera* (C & V). III. Bioenergetic relations of body weight and feeding level. *Aquaculture 35*, 1

Holliday, F.G.T., Tytler, P. and Young, A.H. (1974), Activity levels of trout (*Salmo trutta*) in Airthrey Loch, Stirling and Loch Leven, Kinross. *Proc. R. Soc. Edinb. 74B*, 315

Hommes, F.A. (1980), The energy requirement for growth: a re-evaluation. *Nutr. Metab. 24*, 110

Humphreys, W.F. (1978), Ecological energetics of *Geolycosa godeffroyi* (Araneae: Lycosidae) with an appraisal of production efficiency in ectothermic animals. *J. Anim. Ecol. 47*, 627

Humphreys, W.F. (1979), Production and respiration in animal populations. *J. Anim. Ecol. 48*, 427

Jackim, E. and La Roche, G. (1973), Protein synthesis in *Fundulus heteroclitus* muscle. *Comp. Biochem. Physiol. 44A*, 851

Jobling, M. (1980), Effects of starvation on proximate chemical composition and energy utilization of plaice, *Pleuronectes platessa* L. *J. Fish Biol. 17*, 325

Jobling, M. (1981), The influences of feeding on the metabolic rate of fishes: a short review. *J. Fish Biol. 18*, 385

Jobling, M. (1982), Food and growth relationships of the cod, *Gadus morhua* L., with special reference to Balsfjorden, north Norway. *J. Fish Biol. 21*, 357

Jobling, M. (1983), Towards an explanation of specific dynamic action (SDA). *J. Fish Biol. 23*, 549

Kerr, S.R. (1982), Estimating the energy budgets of actively predatory fishes. *Can. J. Fish Aquat. Sci. 39*, 371

Kitchell, J.F. and Breck, J.E. (1980), Bioenergetics model and foraging hypothesis for sea lamprey (*Petromyzon marinus*). *Can. J. Fish Aquat. Sci. 37*, 2159

Kitchell, J.F., Stewart, D.J. and Weininger, D. (1977), Applications of a bioenergetics model to yellow perch (*Perca flavescens*) and walleye (*Stizostedion vitreum vitreum*). *J. Fish. Res. Bd Can. 34*, 1922

Krieger, I. (1966), The energy metabolism in infants with growth failure due to maternal deprivation, under-nutrition or causes unknown. 1. Metabolic rate calculated from the insensible loss of weight. *Pediatrics 38*, 63

Krieger, I. (1978), Relation of specific dynamic action of food to growth in rats. *Am. J. Clin. Nutr. 31*, 764

Krieger, I. and Whitten, C.F. (1969), Energy metabolism in infants with growth failure due to maternal deprivation under-nutrition or causes unknown. 2. Relationship between nitrogen balance, weight gain and post-prandial excess heat production. *J. Pediat. 75*, 374

Laurent, G.L., Sparrow, M.P., Bates, P.C. and Millward, D.J. (1978), Turnover of muscle protein in the fowl. Collagen content and turnover in cardiac and skeletal muscles of the adult fowl and the changes during stretch-induced growth. *Biochem. J. 176*, 419

Lied, E., Lund, B. and von de Decken, A. (1982), Protein synthesis *in vitro* by expaxial muscle polyribosomes from cod, *Gadus morhua. Comp. Biochem. Physiol. 72B*, 187

McDonald, P., Edwards, R.A. and Greenhalgh, J.F.D. (1973), *Animal Nutrition*, 2nd edn. London, Longman

McNeill, S. and Lawton, J.H. (1970), Annual production and respiration in animal populations. *Nature (Lond.) 225*, 472

Millward, D.J., Garlick, P.J., Stewart, R.J.C., Nnanyelugo, D.O. and Waterlow, J.C. (1975), Skeletal-muscle growth and protein turnover. *Biochem. J. 150*, 235

Millward, D.J., Holliday, M.A., Bates, P.C., Dalal, S., Cox, M. and Heard, C.R.C. (1979), The relationship between dietary state, thyroid hormones, oxygen consumption and muscle protein turnover. *Proc. Nutr. Soc. 38*, 33A

Millward, D.J., Bates, P.C., Brown, J.G., Cox, M. and Rennie, M.J. (1981), Protein turnover and the regulation of growth. In: Waterlow, J.C. and Stephen, J.M.L. (eds). *Nitrogen Metabolism in Man*, pp. 409–18. London, Applied Science

Phillipson, J. and Watson, J. (1965), Respiratory metabolism of the terrestrial isopod *Oniscus asellus* L. *Oikos 16*, 78

Priede, I.G. (1978), Behavioural and physiological rhythms of fish in their natural environment, as indicated by ultrasonic telemetry of heart rate. In: Thorpe, J.E. (ed.). *Rhythmic Activity of Fishes*, pp. 153–68. New York, Academic Press

Priede, I.G. and Young, A.H. (1977), The ultrasonic telemetry of cardiac rhythms of wild brown trout (*Salmo trutta* L.) as an indicator of bio-energetics and behaviour. *J. Fish Biol. 10*, 299

Reeds, P.J. and Harris, C.I. (1981), Protein turnover in animals: man in his context. In: Waterlow, J.C. and Stephen, J.M.L. (eds). *Nitrogen Metabolism in Man*, pp. 391–408. London, Applied Science

Reeds, P.J., Cadenhead, A., Fuller, M.F., Lobley, G.E. and McDonald, J.D. (1980), Protein turnover in growing pigs. Effects of age and food intake. *Brit. J. Nutr. 43*, 445

Ricker, W.E. (1979), Growth rates and models. In: Hoar, W.S., Randall, D.J.

and Brett, J.R. (eds). *Fish Physiology*, vol. VIII, pp. 677–743. New York, Academic Press

Sidorova, V.F. (1978), *The Postnatal Growth and Restoration of Internal Organs in Vertebrates* (transl. B.M. Carlson). Littleton, Mass., PSG Publishing

Smith, L.S. and Newcomb, T.W. (1970), A modified version of the Blazka respirometer and exercise chamber for large fish. *J. Fish. Res. Bd Can. 27*, 1321

Smith, M.A.K. (1981), Estimation of growth potential by measurement of tissue protein synthetic rates in feeding and fasting rainbow trout, *Salmo gairdneri* Richardson. *J. Fish Biol. 19*, 213

Smith, M.A.K. and Haschemeyer, A.E.V. (1980), Protein metabolism and cold adaptation in Antarctic fishes. *Physiol. Zool. 53*, 373

Soofiani, N.M. and Hawkins, A.D. (1982), Energetic costs at different levels of feeding in juvenile cod, *Gadus morhua* L. *J. Fish Biol. 21*, 577

Staples, D.J. and Nomura, M. (1976), Influence of body size and food ration on the energy budget of rainbow trout, *Salmo gairdneri* Richardson. *J. Fish Biol. 9*, 29

Ursin, E. (1979), Principles of growth in fishes. In: Miller, P.J. (ed.). *Fish Phenology: Anabolic Adaptiveness in Teleosts*, pp. 63–87. New York, Academic Press

Vahl, O. (1978), Seasonal changes in oxygen consumption of Iceland scallop (*Chlamys islandica* (O.F. Müller)) from 70°N. *Ophelia 17*, 143

Vahl, O. (1983), The relationship between specific dynamic action (SDA) and growth in the sea star, *Asterias rubens* L. *Oecologia 61*, 122

Vahl, O. and Sundet, J.H. (1983), Is sperm really so cheap? *18th European Marine Biology Symposium, Oslo*

Vivekanandan, E. and Pandian, T.J. (1977), Surfacing activity and food utilization in a tropical air-breathing fish exposed to different temperatures. *Hydrobiologia 54*, 145

Ware, D.M. (1975), Growth, metabolism and optimal swimming speed of a pelagic fish. *J. Fish. Res. Bd Can. 32*, 33

Weatherley, A.H. (1976), Factors affecting maximization of fish growth. *J. Fish. Res. Bd Can. 33*, 1046

Weatherley, A.H., Rogers, S.C., Pincock, D.G. and Patch, J.R. (1982), Oxygen consumption of active rainbow trout, *Salmo gairdneri*, derived from electromyograms obtained by radiotelemetry. *J. Fish Biol. 20*, 479

Webb, P.W. (1978), Partitioning of energy into metabolism and growth. In: Gerking, S.D. (ed.). *Ecology of Freshwater Fish Production*, pp. 184–214. Oxford, Blackwell

Webster, A.J.F. (1981), The energetic efficiency of metabolism. *Proc. Nutr. Soc. 40*, 121

Weiss, P. (1952), Self-regulation of organ growth by its own products. *Science 115*, 487

Weiss, P. and Kavanau, J.L. (1957), A model of growth and growth control in mathematical terms. *J. Gen. Physiol. 41*, 1

Winberg, G.G. (1956), Rate of metabolism and food requirements of fish. *Fish. Res. Bd Can. Transl. Ser. 194*, 1960

Woo, N.Y.S. and Cheung, S.I. (1980), Metabolic effects of starvation in the snakehead, *Ophiocephalus maculatus*. *Comp. Biochem. Physiol. 67A*, 623

Young, A.H., Tytler, P., Holliday, F.G.T. and MacFarlane, A. (1972), A small sonic tag for measurement of locomotory behaviour in fish. *J. Fish Biol. 4*, 57

9 ENERGETICS OF REPRODUCTION

R.J. Wootton

9.1 Introduction

A fish can be viewed as a system that takes an input, food, providing energy and materials, which it 'maps' into an output, the progeny (Pianka, 1976). Its biological success is the number of progeny produced in its lifetime that reach sexual maturity relative to the number produced by other fish in the population. This 'mapping' depends on the somatic body of the fish which provides the framework and machinery that converts food into offspring. Reproduction usually requires more than just the production of gametes (Miller, 1979). It may involve the development of secondary sexual characters such as breeding colours and morphological features; pheromones may be released, and other secretions can include mucus for attaching the eggs to the substrate, or for nest building. All these will require the expenditure of energy in addition to the energy spent on the production of gametes. Reproduction frequently depends on complex behaviour which can include migration to the spawning grounds, territorial defence of spawning areas, complex courtship sequences and parental care. These behavioural activities also demand an expenditure of energy. Thus the energy costs of reproduction are threefold: first, those of the primary sex products, the eggs or sperm; secondly, those of the secondary sexual characteristics; and thirdly, those of reproductive behaviour.

Reproductive effort is the proportion of the energy income of the fish that is devoted to reproduction, so that it is a measure of the efficiency

231

with which energy is diverted into reproduction (Gadgil & Bossert, 1970). An increase in reproductive effort would be expected to increase the number of progeny produced per unit of food consumed, but this would be at the cost of reducing the energy available for the maintenance and growth of the somatic body (Chapter 1). Thus two questions are posed. For a given population, how does reproductive effort change as environmental variables, particularly the availability of food and population density, change? How does reproductive effort vary between populations and species?

The three-spined stickleback, *Gasterosteus aculeatus*, is a convenient species to use in an intensive analysis of the energetics of reproduction (Wootton, 1976). It is easy to breed in the laboratory so that experimental manipulations can be carried out at each stage of its reproductive cycle. It is widely distributed and shows significant inter-population variation in its reproductive characteristics; for example some populations are anadromous, so that comparative studies are possible. Male and female sticklebacks have contrasting reproductive strategies. The male typically shows well developed secondary sexual characteristics including a distinctive breeding coloration, and a major modification of the kidneys into an organ that produces mucus for gluing a nest. He has a complex repertoire of reproductive behaviour including territorial defence, nest-building, courtship and parental care of the eggs and newly hatched fry. Sperm production is low, the testes form about 2% of the body weight in a sexually mature male, and a male will perform relatively few fertilizations in a breeding season. Reproductive investment by the female is primarily cytoplasmic. Immediately before a spawning, the ovaries form up to 30% of the total body weight, and the spawned eggs may represent about 20% of the body weight. A female is capable of spawning many times in a breeding season at intervals of a few days, so that she has the potential of producing two or three times her own body weight of eggs in a breeding season. The reproductive energetics of the stickleback, especially the female, will be described in detail, and where possible comparisons will be made with other species.

9.2 Energetics of Gamete Production

9.2.1 Energy Content of Gonads and Gametes

Ovaries and Eggs. The mean value of the energy content of newly spawned eggs or ripe ovaries for 50 teleost species was 23.48 kJ g^{-1} dry wt (Wootton, 1979). Stickleback eggs had a mean energy content of 22.6 kJ g$_{-1}$, with a similar value for ripe ovaries (Wootton, 1974; Wootton & Evans, 1976), although a higher value for the energy content of ripe ovaries of 26.15 kJ g^{-1} was found for a different population (Meakins, 1976). In the medaka, *Oryzias latipes*, there was a significant difference between the energy content of eggs, 23.60 kJ g^{-1}, and the ovaries at 34.27 kJ g^{-1} (Hirshfield, 1980), and in the northern anchovy, *Engraulis mordax*, the energy content of the ovaries, 23.89 kJ g^{-1}, was higher than that of the eggs, 22.80 kJ g^{-1} (Hunter & Leong, 1981).

Testes. Both the small size of the testes and the small quantity of sperm produced have prevented the measurement of their energy contents in the stickleback. Measurements on other species suggest that the energy content of the testes is similar or lower than that of ovaries in the same species. In the pike, *Esox lucius*, just prior to spawning, the energy content of the testes was 22.60 kJ g^{-1} dry wt, and that of the ovaries was 24.69 kJ g^{-1} (Diana & MacKay, 1979).

9.2.2 Energetics of Gonad Maturation

Maturation of the Ovaries. In many populations of the stickleback, sexual maturity is reached at 1 year old, with few fish surviving to reach a second breeding season (Wootton, 1976). Breeding takes place in late spring and summer, and by September the young fish can usually be sexed by the macroscopic appearance of the gonads. The rates of food consumption and of somatic and ovarian growth in a population of sticklebacks living in a backwater of the River Rheidol in Mid-Wales were estimated between mid-September and the start of the breeding season in mid-May (Wootton *et al.*, 1978, 1980). The rate of food consumption was low during the late autumn and winter, but then increased sharply from February onwards. Throughout the period of low food consumption, the ovaries showed a slow but steady growth, although there were periods when the soma showed a significant drop in its total energy content. In March and April, the ovaries and the soma grew rapidly. During the period of low food consumption, the ovaries seemed to be insulated from the effects of an inadequate energy income whereas the soma suffered a significant depletion. This depletion was probably largely of high-energy storage components including fat bodies around the viscera, because dry weight was relatively stable. Over the winter period, the ovaries grew at a rate of about 0.2–0.3 J day^1, which represented only about 1% of the daily energy income. But during this period, the energy income was barely sufficient to cover the energy expenditures associated with the basic maintenance of the female. In the 8-month period from mid-September to mid-May, the estimated total energy income was 17.15 kJ while the ovaries grew by 0.51 kJ (2.9% of income) and the soma by 0.66 kJ (3.8% of income).

The suggestion that, during this period of ovarian maturation, the ovaries were relatively insulated from variations in energy income was supported by an experimental study in which females were collected at monthly intervals throughout the year, then fed constant rations which varied from starvation to satiation for 21 days. There were no significant effects of ration level on the size of the ovaries throughout most of the non-breeding period, but the soma and especially the liver were sensitive to the ration level. The results suggested that the liver acted as a buffer between the ovaries and the soma so that ovarian maturation could be maintained even at low rates of energy input (Allen & Wootton, 1982a). This contrasts with results that were obtained for the winter flounder, *Pseudopleuronectes americanus*, which suggested that, at low rations, the soma was favoured over the ovaries (Tyler & Dunn, 1976). The winter flounder is a relatively long-lived species, so females will typically have several breeding opportunities.

A comparison with the stickleback is also provided by energetics of ovarian maturation in large, freshwater predatory fish. In a population of pike (*Esox lucius*) in Lac Ste Anne, Alberta, spawning took place in April and early May. Ovarian growth started in August but 81% of the total increase was achieved between October and March. This ovarian growth during the winter was not at the expense of a depletion of the soma, but was supported by winter feeding, although the daily ration was estimated at only 3.8 kJ kg^{-1} between October and March compared with 72.8 kJ kg^{-1} between May and September (Diana, 1979; Diana & MacKay, 1979). The largemouth bass, *Micropterus salmoides*, is a piscivore with a more southerly distribution than the pike or stickleback. Spawning occurs in early summer, and, like the stickleback, the male constructs and guards a nest but the female plays no part in parental care. Energy intake by female largemouth bass in Watts Bar Reservoir, Tennessee, was high between April and December when water temperatures were above 10°C, but low in late winter. Growth of the ovaries took place in two phases, the first in autumn and the second in early spring prior to spawning. In the period of low food consumption between mid-December and mid-March, the rate of ovarian growth was estimated at 1.84 J g^{-1} day^{-1}, but the rate of energy intake was insufficient to fully meet the costs of ovarian growth and somatic maintenance so the balance was met by the depletion of stored body energy (Adams *et al.*, 1982). A similar depletion of the soma to support ovarian growth during the winter occurred in the perch, *Perca fluviatilis*, in Lake Windermere (Craig, 1977).

These studies on the energetics of ovarian maturation illustrate the ability of the fish to maintain ovarian growth even during periods of low energy intake by adopting either an extremely low routine metabolism (pike) or by transferring resources from the soma to the ovaries (stickleback, largemouth bass and perch). This regulation of the partitioning of energy ensures that the ovaries are ripe at the time of the year most suitable for the production of young, even when this requires ovarian growth in a period unfavourable for feeding.

Maturation of the Testes. In the stickleback, spermatogenesis results in the testes containing functional sperm by the end of autumn, at which time the testes form only 1–2% of body weight, a size that is not exceeded even during the breeding season. Although it has been suggested that testicular tissue may be energetically more costly to produce than ovarian tissue (Ursin, 1979), any differences are unlikely to be important in the stickleback, which seems to produce only relatively small amounts of sperm. The energy costs associated with the maturation of the testes are probably negligible compared with the costs of ovarian maturation. In the pike in Lac Ste Anne, testicular growth occurred entirely in August. The energy for this growth may have come from the liver, which showed a loss in energy content almost equivalent to the testicular gain in the same period. Just prior to spawning, the energy content of the testes of a 2-year-old pike was 74.9 kJ compared with 1122 kJ for the ovaries. After spawning the energy content of the testes was 23.0 kJ and that of the ovaries 26.7 kJ (Diana & MacKay, 1979). Female rations were 1.5 to 2.3 times the ration of males, but gonadal production was 15 times higher (Diana, 1979). In the largemouth bass, the energy required for ovarian growth was also much greater than for the testes (Adams *et al.*, 1982). This pattern of a smaller energy cost to testicular maturation is not universal. In the arctic cod, *Boreogadus saida*, the testes of mature fish form 10–27% of the body weight, and the ovaries form a similar proportion (Craig *et al.*, 1982).

In the stickleback and largemouth bass, the male defends a nest and assumes the parental role, so that although the energy cost of gonadal maturation is smaller than in the female, the energy costs of behaviour associated with reproduction are probably much higher (see below).

9.2.3 Energetics of Egg Production

Once breeding has started, the female stickleback is capable of spawning several times at intervals of a few days; typically, the interval between spawnings is 4–7 days at temperatures between 15 and 20°C. The number of eggs produced per spawning is a function of body size, and the relationship between fecundity per spawning (n) and length (L) takes the form:

$$n = aL^b$$

where a and b are constants. The values of a and b may vary both between populations and between polymorphic forms within a stickleback population, though b is typically between 2 and 3 (Wootton, 1973a). This form of the relationship between fecundity and body size is usual for teleost fishes (Wootton, 1979) and shows that if reproduction involves the diversion of resources away from growth, the future fecundity of the fish will be reduced because of the reduction in growth.

The final maturation of the ovaries in the stickleback involves a rapid increase in their weight: from about 8% of the total body weight to about 20–30% immediately before spawning. This rapid increase in weight involves an increase in both the weight of dry matter and the water content in the ovaries (Wootton, 1974). The final stages of oocyte maturation which immediately precede ovulation probably involve a hydration of the ovaries (Wallace & Selman, 1979). With the expulsion of the eggs at spawning, the weight of the ovaries collapses back down to about 8% of total body weight.

Components of the energy budget over the interval between successive spawnings were estimated in two experiments in which females received high rations of two different foods (Wootton, 1974; Wootton & Evans, 1976). The average values of the estimated components are shown in Table 9.1.

The energy content per unit weight of the soma declined significantly just prior to spawning, but recovered rapidly after spawning suggesting that at high food levels spawning does not require a net depletion of energy from the soma (Wootton, 1974). Although food was always available to the females, their voluntary consumption varied signifi-

Table 9.1: Average Values for Some Components of the Energy Budget of Female *G. aculeatus* for the Interval Between Successive Spawnings. (Data from Wootton, 1974, and Wootton & Evans, 1976.)

Variable	Food	
	Minced beef	*Tubifex*
Length of inter-spawning interval (h)	87.5	96
Mean fresh weight of female (g)	1.122	1.331
Total food consumption (J)	Not estimated	3520
Egg production (J)	730	850
Somatic growth (J)	90	110
Total production (J)	820	960
Net gain (J day^{-1})	220	240
Reproductive effort[a]	Not estimated	24

[a] Reproductive effort defined as: (energy content of eggs/energy content of food consumed during interspawning interval) × 100.

cantly, with the final 24 h before spawning marked by a significant reduction in the amount of food eaten (Wootton & Evans, 1976). This variation in energy content and voluntary food intake emphasizes that the female can exert control over the timing of her energy intake and regulate the pattern of transfer of energy from the soma to the ovaries within the inter-spawning interval. Fish do not act as mere passive recipients of an energy input as some models of energy partitioning unintentionally imply.

Effect of Food Level on the Energetics of Egg Production. Egg production by female sticklebacks is sensitive to the availability of food. Because of the significant correlation between body size and fecundity, a high rate of food consumption before the breeding season which resulted in the female starting breeding at a high body size would ensure that her fecundity per spawning was high (Wootton, 1973b). Once the breeding season has started, food supply might affect both the fecundity per spawning and the number of spawnings.

In the stickleback, the fecundity per spawning is primarily determined by the weight of the female at the start of the inter-spawning interval (Wootton, 1977), although the average ration consumed per day of the interval also has a slight effect so that females on high rations are slightly more fecund than those on a low ration. In an experiment in which females were fed enchytraeid worms, the weight of the female accounted for just over 50% of the variance in fecundity, and ration accounted for a further 3% (Table 9.2) (Fletcher & Wootton, unpublished). This strong relationship between fecundity and body weight means that once a female becomes committed to a spawning, irrespective of her energy input, she has only slight regulatory powers over the energy that will be exported in the form of eggs. Females provided with an ad lib supply of tubificid or enchytraeid worms consumed sufficient food both to produce eggs and to show some somatic production. One female, with an initial weight of 1.102 g, spawned 10 times in the 40-day experimental period, producing a total of 9400 J of eggs, and also showed 2850 J of somatic production. Thus her total production represented a daily 306 J. But females fed a restricted ration of worms tended to lose weight. It was estimated that, for a female weighing 1.2 g at the start of the experiment, a daily ration of 780 J was required if the female was to maintain a constant net weight. This represents the consumption of worms weighing about 8% of the body weight of a female per day (Wootton, 1977). For a non-spawning female, a daily ration equivalent to about 2% of her body weight was sufficient to maintain her at constant body weight (Allen & Wootton, 1982b).

Table 9.2: Multiple Regression Analysis of the Relationship Between Fecundity per Spawning, n, Daily Ration, C (grams fresh weight), and Post-spawning Weight of Female, W (grams fresh weight) for *G. aculeatus* Fed Enchytraeid Worms.

$$\text{Model: } \log_e n = a + b_1 \log_e C + b_2 \log_e W$$

Predictor variable	Regression coefficient ±1 S.E.	Significance level, P	Change in r^2
$\log_e W$	0.47 ±0.063	$<10^{-4}$	0.52
$\log_e C$	0.13 ±0.035	0.0003	0.03
Intercept	4.62 ±0.099		

NB r^2 measures the proportion of the variance in the dependent variable that is accounted for by a predictor variable.

A female that was committed to a spawning, but was not supplied with sufficient food during the inter-spawning interval to meet the energy cost of the eggs, subsidized her egg production from her body. This meant that reproductive effort over an inter-spawning interval, defined as the ratio of the energy content of the eggs to the energy content of food consumed during the interval, was inversely related to ration. At a given ration, reproductive effort was positively correlated with the weight of the female (Table 9.3) (Fletcher &Wootton, unpublished). Thus the largest females on the lowest rations showed the highest reproductive efforts.

Table 9.3: Multiple Regression Analysis of the Relationship Between Reproductive Effort, RE[a], Daily Ration, C, (grams fresh weight) and Female Post-spawning Weight, W (grams fresh weight), for *G. aculeatus* Fed Enchytraeid Worms.

Model: $RE = a + b_1 \log_e C + b_2 \log_e W$

Predictor variable	Regression coefficient ±1 S.E.	Significance level, P	Change in r^2
$\log_e R$	−22.5 ±3.01	$<10^{-4}$	0.32
$\log_e W$	16.4 ±5.47	0.0034	0.05
Intercept	−20.92		

[a] RE defined as: (Energy content of eggs/energy content of food consumed per inter-spawning interval) × 100.

For a given ration and weight of female, there was a significant positive relationship between reproductive effort and fecundity per spawning. This effect of reproductive effort was relatively weak; a 20% increase in reproductive effort led to about a 10% increase in fecundity.

Growth rate was measured as the daily specific growth rate calculated from: $(\log_e W_{i+1} - \log_e W_i)/\text{ISI}$, where W_{i+1} and W_i were the weights immediately after the $i+1$ and ith spawnings, and ISI was the length of the inter-spawning interval in days. As would be expected from other studies on the relationship between growth, body weight and ration (Brett, 1979; Elliott, 1979), growth rate was positively related to ration and inversely to body weight. There was also a weak inverse relationship between growth rate and reproductive effort, although this relationship was strongest for those fish receiving the lowest ration.

Although fecundity per spawning was largely determined by the weight of the female at the start of the inter-spawning interval, food supply was the most important factor determining the number of spawnings over a breeding season and hence the total fecundity per breeding season. Females on low rations ceased spawning once their weight had declined to 70–80% of their initial weight (Wootton, 1977, 1979). This lower fecundity of the females on low rations still required a higher reproductive effort over the breeding season than that shown by fish on a high ration, though reproductive effort was much more variable among the females at the low rations (Table 9.4).

Partial correlation analysis showed that over the whole breeding season at a given ration and initial body weight, there was a significant positive correlation between fecundity and reproductive effort ($r = 0.47$) and a significant negative correlation between specific growth rate and reproductive effort ($r = -0.46$).

Table 9.4: Effect of Ration on Mean Fecundity, Daily Specific Growth Rate and Reproductive Effort Measured Over a Breeding Season for Female *Gasterosteus aculeatus* Fed Enchytraeid Worms.

Ration as per cent initial body weight	Total fecundity ±1 S.E.	Specific growth rate ±1 S.E.	Reproductive effort[a] ±1 S.E.
2 (N=7)	208 ±48	−0.0057 ±0.0010	39.9 ±11.30
4 (N=10)	331 ±109	−0.0045 ±0.0018	27.0 ±5.72
8 (N=11)	555 ±172	0.0030 ±0.0011	16.2 ±1.77
16 (N=8)	973 ±153	0.0041 ±0.0007	15.4 ±1.59

[a] Reproductive effort defined as: (Energy content of eggs/energy content of food consumed over a breeding season) × 100.

Thus, in the female stickleback, body weight largely determines the number of eggs produced at a spawning, but food supply determines the total number of spawnings during a breeding season, and so the breeding season fecundity. Females on low rations show high reproductive efforts and subsidize the cost of egg production by a reduction in their body size. At high rations, females do not merely increase their fecundity, but also grow in size. This growth at high rations in breeding females may indicate that such females are producing eggs at the maximum rate that is physiologically or morphologically possible, so that no more energy can be partitioned into reproduction but energy can be partitioned into growth. Gravid females are extremely distended, which suggests that the physical limits set by the size of the abdominal cavity are reached. The distended belly is a releaser for male courtship. When presented with a choice between two gravid females the male may prefer the more distended (and more fecund) one (Rowland, 1982). Choice by males may act as a selection pressure on females to produce as many eggs as possible at each spawning, even, as for females on low rations, at the cost of fewer spawnings. The growth seen at high rations in the breeding season may improve the chances of the female surviving to spawn again the following year. In the natural populations from which the females were taken, most fish reached sexual maturity after 1 year but few survived into a second winter (Allen & Wootton, 1982c; Wootton, 1984). However, there were a few fish that probably did survive to a second breeding season. The causes for the failure of most fish to survive to a second breeding season are not known. There was no significant relationship between reproductive effort and mortality in the experiments, but the number of females used was small so the mortality difference would have had to be large to be detected. Sticklebacks kept in the laboratory after the completion of their first breeding season usually survive reasonably well. They do not show the irreversible degenerative changes that characterize obligate, semelparous fish, in which all the adults die during or soon after their first breeding season.

Egg Production in the Medaka. The medaka, *Oryzias latipes*, is a small cyprinodont fish which will spawn daily during the breeding season, producing up to 50 eggs per day for as long as 4 months (Hirshfield, 1980). Like the stickleback, it is a short-lived fish. Hirshfield (1977, 1980) kept females at three different temperatures, 25, 27 and 29°C and fed them different rations. Both growth and daily egg production were affected by food ration, although temperature and ration interacted. At all temperatures, there was a positive relationship between growth and ration. At 25 and 29°C, there was also a positive relationship between daily egg production and ration, but at 27°C there was no significant relationship. Total production, measured as the sum of energy content of eggs produced plus the energy gained as growth, was positively related to ration, but not to temperature. The estimated average daily total production was 24.7 J at 29°C, 30.1 J at 27° and 29.7 at 25°C. These values are about one-tenth of the daily production achieved by a female stickleback.

At the two higher temperatures, there was an inverse relationship between reproductive effort and ration, with reproductive effort defined as the ratio of egg energy to food energy over a 25-day experimental period. At the lowest temperature, the relationship was positive. The rate of food consumption at 25°C was low, so that the highest ration was similar to the lowest ration at 27 and 29°C. The mean reproductive effort for the 25-day experimental period was significantly higher at 29°C than at 27°C for a given ration and fish weight. The means were 14% at 20°C and 11% at 27°C. At 25°C reproductive efforts were lower than at the higher temperatures, but a direct comparison was not possible because of the different relationship between reproductive effort and ration.

As in the stickleback, reproductive effort was inversely related to growth, although the relationship was not significant at 25°C. There was also a strong correlation between reproductive effort and the number of moribund or dead fish. Hirshfield (1980) suggested that females with a higher reproductive effort lost more weight and this loss led to a higher mortality. Mortality was also high at the highest temperature. Development of the eggs and young would be faster at the higher temperature, reducing their period of vulnerability, and this, associated with the higher mortality of adults, would favour higher reproductive efforts.

The medaka fed low rations sacrificed growth rather than reproduction, but at high food levels reproductive effort was not kept high and some growth took place. It is possible that at high rations, the physiological and morphological limits to egg production had been reached, so that females could not channel any more resources to reproduction and so allocated the surplus to growth. Hirshfield (1980) argued that, at high rations, the increased fecundity that would result from a high reproductive effort would simply be outweighed by the corresponding increased costs to survivorship and growth. In both the stickleback and medaka, fecundity is a function of body size, so that growth will yield a higher future fecundity if the female can survive to enjoy it.

A field study of two medaka populations in Japan suggested that at the height of the breeding season the reproductive effort was 10–18% (Hirshfield, 1977, 1980).

Egg Production in the Northern Anchovy. A field study of the marine clupeid, *Engraulis mordax*, yielded estimates of the energy cost of egg production (Hunter & Leong, 1981). From histological studies of the ovaries and the proportion of recently spawned fish that were captured throughout the year, it was estimated that an average female spawned at intervals of 6 to 10 days, and achieved 15 spawnings between February and September and five more between October and January. Previous estimates of spawning frequency for this species had been much lower. The spawning frequency implied that in a reproductively active female, batches of eggs were successively undergoing vitellogenesis, so that the high frequency of spawning depended on the reservoir of unyolked, small oocytes in the ovaries. Within the population, the fecundity per spawning was a function of body size, but showed little seasonal variation, a situation similar to that noted in the stickleback. There was some evidence that fecundity per spawning may vary between populations. The energy lost from the ovaries at each spawning was equivalent to about 50% of the pre-spawning total, so that the high frequency of spawnings could only be achieved by the transfer of resources into the ovaries. During the annual bloom of zooplankton between April and July, the anchovy rapidly accumulated body fat, but this fat declined to a minimum in the following spring, the time of maximum spawning. This temporal pattern suggests that there is a time lag of about a year in the effect of the zooplankton bloom on reproduction. It was estimated that a female weighing 16.4 g would store 57 kJ in the form of fat, accumulated at a rate of 0.837 kJ day^{-1}. The total stored was equivalent to 13 spawnings and the rate of accumulation equivalent to a spawning every 5 days. From laboratory studies on growth and the field data, it was estimated that on average between ages 1 and 2, food equivalent to 532 kJ was consumed, growth was 26 kJ, and 42 kJ of eggs were produced. This is a mean annual reproductive effort of $(42/523) \times 100$, i.e. 8%. These estimates of energy input and output imply a ration of 1.46 kJ daily, equivalent to a consumption of copepods weighing about 4% of the body weight of the anchovy. Reproductive effort increased with age so that for fish aged 3–4 years it was 11%. These reproductive efforts are much higher than the 1% estimated for another Pacific clupeid, *Sardinops caerula* (Lasker, 1970). The difference emphasizes the importance of obtaining accurate estimates of the average number of spawnings in serially spawning fish (Wootton, 1979).

A decline in the Pacific sardine population was accompanied by an increase in the length of the anchovy spawning period, which, prior to the decline, had been restricted to the winter quarter. This suggests that the anchovy has increased its average number of spawnings per year, which could be a consequence of an improved food supply. This would be comparable to the stickleback, in which food supply affects primarily the number of spawnings rather than the fecundity per spawning. There is a further similarity. It was estimated that egg cannibalism by the anchovy could account for 17% of the daily egg production during the peak spawning period. In the stickleback, egg cannibalism by both males and females is frequent (Wootton, 1976).

Other Estimates of Reproductive Effort. Constantz (1979) compared the energy costs of reproduction in two populations of the live-bearing cyprinodont *Poeciliopsis occidentalis*, the Gila topminnow. One population lived in a spring, which provided a nearly constant abiotic environment, and the other lived in a wash prone to flash-flooding. In both the spring and the wash, fish became sexually mature about a year after birth, but in the spring, some fish born in February grew rapidly and matured within 5 months of birth but had disappeared by September. The average annual allocation of energy to reproduction was 2.83 kJ by long-lived spring fish, 0.79 kJ by short-lived spring fish and 2.41 kJ by long-lived wash fish. The annual reproductive efforts for the three groups were 2.1, 3.5 and 4.8%, respectively, where reproductive effort was defined as the proportion of the total energy budget allocated to reproduction. Age-specific reproductive effort for the long-lived spring fish varied between 3.1 and 6.5%, but for the short-lived spring fish and the wash fish, reproductive effort increased with age, peaking at 5.2% and 9.8%, respectively.

In the wash females, there was an inverse relationship between egg size and fecundity, but this trade-off between size and fecundity was not present in the spring fish. At a given length, the average wash female had a higher reproductive weight than the spring fish. Constantz (1979) interpreted these two relationships as indicating that reproductive output by the spring fish was more limited by food supply, so that the reproductive mass occupied less of the abdominal cavity than was available, whereas the wash fish were limited by abdominal volume from further increases in reproductive mass. Experimental studies on another live-bearing cyprinodont, *Poecilia reticulata*, had shown that a restriction in food supply both reduced the number of young born and the recruitment of mature oocytes (Hester, 1964).

Diana (1983) estimated that for the pike in Lac Ste Anne the typical annual reproductive effort of males aged between 1 and 3 years was 7–10%, and for the females it was 14–16 %. In the first year of life, the pike allocated about 42% of their energy income to growth, but after sexual maturation less than 10% was so allocated.

Two earlier studies on long-lived marine fishes had provided estimates on the annual reproductive effort. In the American plaice, *Hippoglossoides platessoides*, MacKinnon (1972, 1973) estimated that the reproductive effort of a mature female was 7.5%, but that effort tended to increase with the age and size of the female. For North Sea cod, *Gadus morhua*, Daan (1975) estimated the average annual reproductive effort as 1.3% in a 2-year-old female, increasing to 11–12% for an 8- to 25-year-old fish.

Variations in the Energy Content of Eggs. Although both the anchovy and Gila topminnow showed significant seasonal changes in the size and total energy content of the eggs, there was no evidence that these changes were due to alterations in either the quantity or quality of food available. In the stickleback, little change in either the size or chemical composition of the eggs was found over a wide range of rations (Fletcher & Wootton, unpublished). It is possible that in the face of adverse feeding conditions, fish will tend to maintain the quality of the eggs, although this may be at the cost of a reduction in fecundity either through a reduction in fecundity per spawning or the number of spawnings (Chapter 1). Too few studies on the relationship between food supply and egg quality are available to test this hypothesis.

9.3 Energetics of Secondary Sexual Characteristics

Little is known about the energy costs of the development of secondary sexual characteristics such as breeding coloration or morphological characteristics. For most characteristics, the costs may be small compared with that of gamete production and reproductive behaviour, but in the male stickleback there is some indirect evidence that one secondary sexual characteristic is energetically costly. The kidneys of the male become modified during the breeding season to produce the mucus which is used to glue the nest together. A lower proportion of males kept on a low ration (2% body weight of enchytraeid worm) built nests than males kept on higher rations, although the males were kept in individual tanks and so did not have to defend a nest site. Post-mortem analysis showed that when males of a similar length were compared, the males on the low ration had significantly smaller kidneys although their testes were not significantly smaller (Stanley, 1983). This suggests that the development of the kidneys into effective mucus-producing organs may have been inhibited at low rations.

9.4 Energetics of Reproductive Behaviour

9.4.1 Energy Costs of Spawning Migration

Although some populations of sticklebacks do show an anadromous, spawning migration from the sea into fresh water, the energetics of this migration have yet to be studied. The American shad, *Alosa sapidissima*, also shows a spawning migration into fresh water and while in fresh water does not feed, so the energy costs of maintenance and migration must be met by the energy stored in the body on entry into the river. On the eastern seaboard of North America, shad spawn in rivers from Florida to the St. Lawrence estuary. Whereas the northern populations are iteroparous, the southern populations are semelparous. This cline in post-spawning survival generated the hypothesis that the cline was regulated in part by the energetics of migration and spawning (Glebe & Leggett, 1981a).

The energy costs of migration and spawning were estimated for shad spawning in a northern river, the Connecticut, and a southern river, St. Johns in Florida. The estimates were made from changes in the quantities of lipid and protein in the bodies of male and female shad during their time in fesh water (Glebe & Leggett, 1981a,b). In the northern river, there was significant post-spawning survival, but survival of fish making a run late in the season was poorer than for fish at the peak of the run. The late run used about 60% of their total somatic reserves of energy, whereas early and peak-run fish used 45–50%. Early and peak-run males used about 30% of their reserves in the upstream migration to the spawning areas, whereas the females used about 26%. The return to the sea expended a further 20% of the reserves. During the upstream migration, there was no evidence that material was transferred from the soma to the gonads in either males or females. In contrast, both males and females in the St Johns River population showed a significant increase in gonad weight during the upstream migration, so that in a large female the energy content of the ovaries increased by 65%. The St. Johns River fish used a much higher percentage of their somatic energy reserves in migration and spawning, with males using about 65% and females up to 80%. The energy required for an average St Johns River female to migrate up stream and spawn, 9200 kJ, was similar to the energy expenditure by an average Connecticut River female completing the entire migration back to the sea, 8600 kJ. St Johns River fish allocated 16% of their body reserves to reproductive products compared with the 7% allocated by the Connecticut River fish.

The percentage loss of somatic energy reserves by the semelparous St. Johns River shad was comparable to the losses that occurred during the upstream migration and spawning of the semelparous Pacific salmon, *Oncorhynchus*. The losses by the iteroparous Connecticut shad were comparable to the losses shown by the anadromous, iteroparous Atlantic salmon, *Salmo salar*. These comparisons suggest that expenditure of more than about 60% of the somatic reserves on migration and spawning precludes post-spawning survival. One interpretation of these results is that the inter-population differences within the shad and the inter-specific differences within the salmonids may be caused by the different energy costs of migration and spawning. However, Glebe & Leggett (1981b) argued that semelparous populations of the shad are found in short southern rivers where the energy costs of migration ought to permit a return migration to the sea. They also noted that many semelparous populations of *Oncorhynchus* species have short spawning migrations. They suggest that energy constraints are not the main determinants of the differences in life-history patterns, but that differences in energy allocation provide the mechanism by which the different life-history patterns are achieved. The particular life-history patterns are selected by the environmental and demographic characteristics the populations experience. This interpretation places emphasis on the physiological mechanisms by which the energy partitioning is regulated.

In the iteroparous shad in the Connecticut River, the late-run fish suffered a significantly higher mortality, having expended a higher percentage of their energy reserves. This suggests that even where significant post-breeding survival is adaptive, the energy allocation to migration is close to the minimum required to achieve this survival. Such allocation would ensure that within the constraints imposed by environment and demography, the energy allocation to reproduction is maximized (Glebe & Leggett, 1981b).

9.4.2 Energetics of Territorial Behaviour

In many species, including the stickleback, reproductive behaviour includes territoriality. The energy costs of holding a reproductive territory are of two forms. First, there is the cost of the locomotion involved in the defence of the territory. Secondly, there is the energy loss associated with the restricted foraging provided within the territory. Although no accurate estimates of the energy costs of territorial behaviour are available for the male stickleback, two lines of evidence suggest that the costs may be significant. At a given density of males, a low ration led to fewer males holding a territory. Additionally, males that were in territorial groups lost weight when supplied with a ration that was at least a maintenance ration for solitary males (Stanley, 1983).

The energy costs of territorial behaviour for male pupfish, *Cyprinodon*, were calculated by combining estimates of the energy costs of locomotion from respirometry studies with behavioural observations on the amount of locomotion shown by reproductively active males (Feldmeth, 1983). In both an experimental situation and in the field, territorial males spent a high proportion of their time actively swimming in defence of their territories during the day. It was estimated that an active male would expend 0.32 kJ during a 16 h day, whereas a male merely holding position would expend only 0.16 kJ. However, it was found that non-territorial males that tried to sneak fertilizations also had high energy expenditures, so that the dominant, territorial males were not paying a high energy cost relative to other social groups in the population. An exception to this was probably provided by mature females, which formed slow-moving schools in areas away from the breeding territories and probably had relatively low locomotory costs. These females had to meet the high metabolic cost of egg production. This study emphasizes the importance of measuring the energy costs of reproduction relative to some control groups that are not reproducing, but which may still be having to meet high energy costs for other reasons.

9.4.3 *Energetics of Parental Behaviour*

The energy costs of parental behaviour result partly from the direct energy costs of the behaviour and partly from the restrictions placed on the foraging behaviour of the fish by the requirements of parenthood. A parental male stickleback fans the eggs in the nest, spending high proportions of the available time in this activity and close to the nest. Although no direct estimates of the energy costs of fanning have been made, parental males that were fed a low food ration showed significantly less fanning than better fed males (Stanley, 1983). In the convict cichlid, *Cichlasoma nigrofasciatum*, both sexes show parental care, but most of the fanning of eggs is done by the female. When parental pairs were fed reduced rations, the females showed a significant reduction in the time spent fanning, whereas there was little change in the males (Townshend & Wootton, in preparation). In both the stickleback and the convict cichlid, the parental fish restrict almost all their activities to an area around the spawning site, so that they are relying on a relatively small area to provide their food requirements.

9.5 Significance of the Energy Costs of Reproduction

The studies described show that it is possible to obtain estimates of the energy costs of reproduction in fish and to explore the effects of environmental changes on the proportion of energy devoted to reproduction. What significant biological problems can be answered by such studies?

Life-history theory attempts to predict what life-history patterns will be adaptive in particular environments. A central assumption of the theory is that there is a trade-off between present reproduction and future expected reproduction; that is, the decision to reproduce here and now will have an adverse effect on the animal's reproductive output in the future (Williams, 1966) (Chapter 1). This effect may result from a reduction in the chances of the animal surviving to reproduce again, or from an effect on the number of progeny that will be produced at a future reproductive attempt. This last possibility is relevant for fish in which growth is indeterminate so that an increase in body size occurs after sexual maturity is reached. In female fish, fecundity is a function of body size, so that if reproduction reduces growth, this reduces future fecundity. Although the correlations between reproduction, post-breeding survivorship and growth can be estimated without any consideration of the mechanisms causing the correlations, a more satisfactory analysis would identify the causal relationships between these life-history traits. The concept of reproductive effort was introduced in this context. It is assumed that, at a given age, an increase in reproductive effort will increase fecundity but will reduce future expected reproduction by reducing either the post-spawning survivorship or post-spawning growth. Energy spent on a current reproductive attempt is not available to be spent on activities related to survivorship or on somatic growth. These considerations suggest that reproductive effort should vary in a predictable way between life histories in which the patterns of age-specific mortality and fecundity vary.

So far, estimates of the annual reproductive effort of fishes have clustered around or below 10%, irrespective of the life-history strategy. In a long-lived fish, the pike, a high proportion of the energy input was allocated to growth during the juvenile period. With the onset of sexual maturation, the proportion allocated to growth declined sharply, but the proportion allocated to maintenance increased from about 60% to 80% (Diana, 1983). In a short-lived fish, the stickleback, both somatic growth and sexual maturation were achieved in the first year of life. A high allocation of energy to reproduction during the breeding season was mirrored by a high allocation to somatic growth during the first 3 or 4 months of life (Allen & Wootton, 1982c). The life-history strategies differ less in the proportion of energy annually allocated to reproduction than in the temporal patterning of that allocation relative to the allocation to building the somatic framework.

The experimental studies on the stickleback and medaka have shown that, at a given ration and size of fish, there was a positive relationship between reproductive effort and fecundity but a negative relationship with growth. In the medaka, there was also a negative correlation between reproductive effort and survivorship. These studies provide experimental support for the central assumption of life-history theory and suggest that at least one component of the trade-off between present and future reproduction reflects the pattern of energy partitioning between the gonad and the soma. They do not demonstrate that such phenotypic co-variations reflect parallel genetic co-variations such that similar trade-offs would emerge under selection for life-history traits (Reznick, 1983). The comparisons of the energy costs of the iteroparous and semelparous populations of shad do provide evidence that post-breeding survivorship is related to the pattern of energy partitioning in natural proportions that probably differ genetically. The comparisons also support the prediction of life-history theory that the reproductive effort in a semelparous life history is higher than in an iteroparous life history, given that the comparisons are for reasonably similar organisms.

These studies further suggest that fish have significant regulatory powers over energy partitioning between the gonad and soma and exert that regulatory ability to achieve particular reproductive outputs. This regulatory aspect should be emphasized because there is a tendency to model the energy partitioning of a fish as though the fish is passive and helpless in the face of fluctuations in the food supply. The models often assume that there are certain fixed energy expenditures which the fish has to meet before investing energy in reproduction. Only energy surplus to these fixed requirements is available for reproduction. The studies on the stickleback and medaka showed that, at low rations, some fish would allocate less energy to maintenance than was required to maintain a constant body weight even though this caused a significant reduction in size and, in the case of the medaka, a significant increase in mortality.

A cautionary note must be added. In the experimental studies, a control group of non-reproductive fish was not maintained at similar ration levels, and it is possible that if a sufficiently long experimental time period is used, the growth and survival of a reproductively active group do not differ significantly from a non-reproductive group. Differences in appetite, absorption efficiency and conversion efficiency between reproductive and non-reproductive groups could partly or wholly compensate for any short-term effects of reproduction.

Reznick (1983) compared the allocation of energy between reproductive and somatic tissues in female guppies, *Poecilia reticulata*, which were closely related genetically. Females that reproduced were not significantly smaller at the end of the experiment than females that were prevented from reproducing by being mated with impotent males. The non-reproducing females did store more body fat, but the energy represented by this extra fat was significantly less than expected if all the energy not allocated to reproduction was allocated to the soma. In these guppies there was almost no phenotypic trade-off between growth and reproduction. Energy intended for reproduction was not efficiently shunted into growth. In complete contrast, when guppies from different populations were compared, there was a significant trade-off between the energy allocation to growth and reproduction.

Iles (1974) noted that in North Sea herring, *Clupea harengus*, the onset of sexual maturation was not signalled by a sudden change in growth pattern which would indicate that growth was disrupted or disturbed by the allocation of resources to the gonads. He argued that the neuro-endocrine control of feeding, growth, storage and reproduction regulated the interrelationships between these processes so that effects of environmental variability were minimized. In the herring, these processes showed clear seasonal patterns with gonad maturation taking place during a period of reduced food consumption, but linked to the period of high food consumption by a cycle of storage and transfer (Iles, 1974; Wootton, 1979). This control effectively 'masks' any cost, in terms of somatic growth, of reproduction.

A classic statement on the partitioning problem in life-history theory is due to Fisher (1958), who suggested it would be important to know both the physiological mechanisms by which resources are made available to the reproductive organs and the environmental conditions which would favour a higher or lower allocation to reproduction. Recently, most attention has been paid to the second half of the problem, but the two are closely interconnected. It is now important that mechanisms by which partitioning is controlled are analysed in the context of the life histories of species.

Acknowledgements

I wish to thank Mr D. Fletcher, who, as an NERC postgraduate student, collected much of the data on the effects of ration on *G. aculeatus*.

References

Adams, S.M., McLean, R.B. and Parrotta, J.A. (1982), Energy partitioning in largemouth bass under conditions of seasonally fluctuating prey availability. *Trans. Am. Fish. Soc. 111*, 549

Allen, J.R.M. and Wootton, R.J. (1982a), Effect of food on the growth of carcass, liver and ovary in female *Gasterosteus aculeatus* L. *J. Fish Biol. 21*, 537

Allen, J.R.M. and Wootton, R.J. (1982b), The effect of ration and temperature on the growth of the three-spined stickleback, *Gasterosteus aculeatus* L. *J. Fish Biol. 20*, 409

Allen, J.R.M. and Wootton, R.J. (1982c), Age, growth and rate of food consumption in an upland population of the three-spined stickleback, *Gasterosteus aculeatus* L. *J. Fish Biol. 21*, 95

Brett, J.R. (1979), Environmental factors and growth. In: Hoar, W.S., Randall, D.J. and Brett, J.R. (eds). *Fish Physiology, vol. VIII*, pp. 599–675. New York, Academic Press

Constantz, G.D. (1979), Life history patterns of a livebearing fish in contrasting environments. *Oecologia 40*, 189

Craig, J.F. (1977), The body composition of adult perch, *Perca fluviatilis*, in Windermere, with reference to seasonal changes and reproduction. *J. Anim. Ecol. 46*, 617

Craig, P.C., Griffiths, W.B., Haldorson, L. and McElderry, H. (1982), Ecological studies of arctic cod (*Boreogadus saida*) in Beaufort Sea coastal waters, Alaska. *Can. J. Fish Aquat. Sci. 39*, 395

Daan, N. (1975), Consumption and production in North Sea cod, *Gadus morhua*: an assessment of the ecological status of the stock. *Neth. J. Sea Res. 9*, 24

Diana, J.S. (1979), The feeding pattern and daily ration of a top carnivore, the northern pike (*Esox lucius*). *Can. J. Zool. 57*, 2121

Diana, J.S. (1983), An energy budget for northern pike (*Esox lucius*). *Can. J. Zool. 61*, 1968

Diana, J.S. and MacKay, W.C. (1979), Timing and magnitude of energy deposition and loss in the body, liver, and gonads of northern pike (*Esox lucius*). *J. Fish. Res. Bd Can. 36*, 481

Elliott, J.M. (1979), Energetics of freshwater teleosts. *Symp. zool. Soc. Lond. 44*, 29

Feldmeth, C.R. (1983), Cost of aggression in trout and pupfish. In: Aspey, W.P. and Lustick, S.I. (eds). *Behaviour Energetics: The Cost of Survival in Vertebrates*, pp. 117–38. Columbus, Ohio State University Press

Fisher, R.A. (1958), *The Genetical Theory of Natural Selection*. New York, Dover

Gadgil, M. and Bossert, W.H. (1970), Life historical sequences of natural selection. *Amer. Nat. 104*, 1

Glebe, B.D. and Leggett, W.C. (1981a), Temporal, intra-population differences in energy allocation and use by American shad (*Alosa sapidissima*) during spawning migration. *Can. J. Fish Aquat. Sci. 38*, 795

Glebe, B.D. and Leggett, W.C. (1981b), Latitudinal differences in energy allocation and use during the freshwater migrations of American shad (*Alosa sapidissima*) and their life history consequences. *Can. J. Fish Aquat. Sci. 38*, 806

Hester, F.J. (1964), Effects of food supply on fecundity in the female guppy *Lebistes reticulatus* (Peters). *J. Fish. Res. Bd Can. 21*, 757

Hirshfield, M.F. (1977), The reproductive ecology and energetics of the Japanese medaka *Oryzias latipes*. Ph.D. Thesis, University of Michigan, Ann Arbor

Hirshfield, M.F. (1980), An experimental analysis of reproductive effort and cost in the Japanese medaka, *Oryzias latipes*. *Ecology 61*, 282

Hunter, J.R. and Leong, R. (1981), The spawning energetics of female northern anchovy, *Engraulis mordax*. *Fish. Bull. 79*, 215

Iles, T.D. (1974), The tactics and strategy of growth in fishes. In: Harden Jones, F.R. (ed.). *Sea Fisheries Research*, pp. 331–45. London, Elek Scientific

Lasker, R. (1970), Utilization of zooplankton energy by a Pacific sardine population in the California current. In: Steele, J.H. (ed.). *Marine Food Chains*, pp. 265–84. Edinburgh, Oliver & Boyd

MacKinnon, J.C. (1972), Summer storage of energy and its use for winter metabolism and gonad maturation in American plaice (*Hippoglossoides platessoides*). *J. Fish. Res. Bd Can. 29*, 1749

MacKinnon, J.C. (1973), Analysis of energy flow and production in an unexploited marine flatfish population. *J. Fish. Res. Bd Can. 30*, 1717

Meakins, R.H. (1976), Variations in the energy content of freshwater fish. *J. Fish Biol. 8*, 221

Miller, P.J. (1979), A concept of fish phenology. *Symp. zool. Soc. Lond. 44*, 1

Pianka, E.R. (1976), Natural selection of optimal reproductive tactics. *Amer. Zool. 16*, 775

Reznick, D. (1983), The structure of guppy life histories: the trade-off between growth and reproduction. *Ecology 64*, 862

Rowland, W.J. (1982), Mate choice by male sticklebacks, *Gasterosteus aculeatus*. *Anim. Behav. 30*, 1093

Stanley, B.V. (1983), Effect of food supply on reproductive behaviour of male *Gasterosteus aculeatus*. Ph.D. Thesis, University of Wales

Tyler, A.V. and Dunn, R.S. (1976), Ration, growth and measures of somatic and organ condition in relation to meal frequency in winter flounder, *Pseudopleuronectes americanus*, with hypothesis regarding population homeostasis. *J. Fish. Res. Bd Can. 33*, 63

Ursin, E. (1979), Principles of growth in fishes. *Symp. zool. Soc. Lond. 44*, 63

Wallace, R.A. and Selman, K. (1979), Physiological aspects of oogenesis in two species of sticklebacks, *Gasterosteus aculeatus* L. and *Apeltes quadracus* (Mitchill). *J. Fish. Res. Bd Can. 14*, 551

Williams, G.C. (1966), *Adaptation and Natural Selection*. Princeton, Princeton University Press

Wootton, R.J. (1973a), Fecundity of the three-spined stickleback, *Gasterosteus aculeatus* L. *J. Fish Biol. 5*, 683

Wootton, R.J. (1973b), The effect of size of food ration on egg production in the female three-spined stickleback, *Gasterosteus aculeatus* L. *J. Fish Biol.* 5, 89

Wootton, R.J. (1974). The inter-spawning interval of the female three-spined stickleback, *Gasterosteus aculeatus. J. Zool. Lond. 172*, 331

Wootton, R.J. (1976), *The Biology of the Stickleback*. London, Academic Press

Wootton, R.J. (1977). Effect of food limitation during the breeding season on the size, body components and egg production of female sticklebacks (*Gasterosteus aculeatus*). *J. Anim. Ecol. 46*, 823

Wootton, R.J. (1979). Energy costs of egg production and environmental determinants of fecundity in teleost fishes. *Symp. zool. Soc. Lond. 44*, 133

Wootton, R.J. (1984), *A Functional Biology of Sticklebacks*. London, Croom Helm

Wootton, R.J. and Evans, G.W. (1976). Cost of egg production in the three-spined stickleback (*Gasterosteus aculeatus*). *J. Fish Biol. 8*, 385

Wootton, R.J., Allen, J.R.M. and Cole, S.J. (1980). Energetics of the annual reproductive cycle in female sticklebacks, *Gasterosteus aculeatus* L. *J. Fish Biol. 17*, 387

Wootton, R.J., Evans, G.W. and Mills, L. (1978). Annual cycle in female three-spined sticklebacks (*Gasterosteus aculeatus* L.) from an upland and lowland population. *J. Fish Biol. 12*, 331

PART FOUR: ENERGY BUDGETS

10 LABORATORY STUDIES OF ENERGY BUDGETS

A.E. Brafield

10.1 Introduction

An energy budget is a balance sheet of energy income set against energy expenditure. A fish in a laboratory experiment is an open thermodynamic system exchanging energy with its surroundings in three ways: heat, work, and the potential energy of biochemical compounds. It can be shown that

$$C = P + R + U + F$$

where C (for consumption) is the energy content of the food eaten, P the energy in growth materials (production), R the net loss of energy as heat (R standing for respiration), U (urinal loss) the energy lost in nitrogenous excretory products, and F the energy lost in the faeces. It is assumed in this equation that work done by the fish on its surroundings, and vice versa, is small enough to be ignored (Wiegert, 1968; Brafield & Llewellyn, 1982).

The compilation of energy budgets for fish has a fairly short but very active history, with the result that the literature on the subject is considerable. No historical survey is attempted here, and as far as possible only recent references will be quoted. Instead, the intention is to indicate where the major difficulties lie when erecting energy budgets, and where useful developments in techniques and concepts are appearing. A short review of some of the more important literature is desirable by way of introduction, however, in the context of fish energetics in general and energy budgets for various fish species in particular.

257

In 1977, Kitchell *et al*. elaborated and subdivided the basic energy budget equation, and developed a bioenergetics model which they applied to yellow perch (*Perca flavescens*) and walleye (*Stizostedion vitreum*). Then, in a vintage year, Braaten (1979), Brett & Groves (1979) and Fischer (1979) published comprehensive reviews of the problems and difficulties implicit in evaluating and relating the various energy budget components in fish studies. At the same time, Elliott (1979) surveyed the energetics of freshwater teleosts, mainly trout, and erected some growth models.

There have been numerous attempts to compile energy budgets for particular fish species, or at least to relate quantitatively and reliably some of the factors in the energy budget equation. Among the more recent of these, in chronological order, are the studies on perch (*Perca fluviatilis*) by Solomon & Brafield (1972), on brown trout (*Salmo trutta*) by Elliott (1976b), on tilapia (*Tilapia mossambica*, now *Sarotherodon mossambicus*) by Mironova (1976), on rainbow trout (*Salmo gairdneri*) by Staples & Nomura (1976), on thick-lipped mullet (*Crenimugil labrosus*) by Flowerdew & Grove (1980), on stickleback (*Gasterosteus aculeatus*) by Wootton *et al*. (1980), on mullet (mainly *Rhinomugil corsula*) by Kutty (1981), and on sockeye salmon (*Oncorhynchus nerka*) by Brett (1983). These are necessarily only a selection of the more recent and familiar budgets published, but they include discussions of other and earlier work.

No attempt is made here to compile a table comparing the results of such studies. The variability in technique, nature of food, feeding regime, respiratory rate, temperature, fish size and age, and a host of other factors is so complex that even if compiling a comprehensive table of energy budgets were feasible it might not reveal much of value. Nevertheless, some attempts have been made to tabulate energy budgets for various species (e.g. Brett & Groves, 1979; Fischer, 1979). Brett & Groves (1979) were able to arrive at the following generalized budgets for young fish feeding well (figures are percentages of food energy, C):

Carnivores: $100\,C = 29\,P + 44\,R + 7\,U + 20\,F$

Herbivores: $100\,C = 20\,P + 37\,R + 2\,U + 41\,F$

These equations show some familiar features of fish energy budgets, such as the lower assimilation efficiency (higher percentage for F) in herbivores, the relatively small amounts of energy lost in nitrogenous waste (U), and the higher value for heat loss (R) than for the energy in growth materials (P) in both equations. Comparisons between P and R can be illuminating, for these two channels represent the energy used in some way by the fish, and they can be seen as competing with one another for the resource C, for energy is either 'banked in a deposit account' (P) or 'spent' in activity (R). In this context the recent work by Koch & Wieser (1983) with roach (*Rutilus rutilus*) is interesting, for they found that swimming is significantly reduced (saving 1485 kJ kg^{-1}) during the period of gonadal growth (costing 364 kJ kg^{-1}). They claim this to be the first published evidence for a poikilotherm that reduction of locomotor activity (part of R) can compensate for the cost of producing gonadal tissue (part of P). These matters are discussed in detail in Chapter 1.

Returning to the basic energy budget equation ($C = P + R + U + F$), it can be seen that if faecal energy (F) is subtracted from food energy (C), one has the energy absorbed and assimilated by the fish ($P + R + U$). Of this, the energy of nitrogenous waste (U) is excreted, leaving $P + R$ as an indicator of the metabolizable energy. The energy lost as heat (R) can be subdivided (e.g. Chapter 1, and Brett & Groves, 1979, Figure 18; Elliott, 1979, Figure 1) but this is not necessary in the present context; nor is it useful here to separate P into somatic and gonadal growth. Ideally, all five factors in the equation should be measured independently but simultaneously, over a fairly long period, in attempts to erect energy budgets, but this is not always practicable. There are numerous difficulties, both technical and conceptual, in evaluating each of the five channels. It is the aim here to discuss some of these problems and indicate where improvements have been and are being made.

10.2 Food Energy (C) and Faecal Energy (F)

Energetics aspects of fish nutrition are considered in some earlier chapters, and in addition there are recent reviews by Cowey & Sargent (1979), Jauncey (1982a) and Jobling (1983b). This section is not, therefore, concerned with the various aspects of nutrition as a process (ration size, feeding regime, protein-sparing effects of carbohydrate and fat, digestibility and the like) but with calculating food energy and faecal energy in laboratory experiments directed at determining overall energy balance. Making estimations of C and F is not difficult, for one only needs to know the energy content of representative samples of the food and the faeces and the total amounts of food actually consumed and of the faeces produced.

Natural foods are often given in energy budget experiments and no problems arise if the food is consistent in size and energy content, such as the *Gammarus* fed to perch (*Perca fluviatilis*) by Solomon & Brafield (1972) and to brown trout (*Salmo trutta*) by Elliott (1976b). Alternatively, food can be given in the form of pellets, a convenient method because the experimenter can vary their size and energy content and the relative proportions of carbohydrate, fat and protein that they contain. There is danger that some nutrients may be leached out of the pellets into the water, causing overestimation of C. This can be avoided by minimizing the interval between introducing the food and its being eaten by the fish, and by promptly removing any food left uneaten. Pellets are often bound with an alginate or similar binding agent to reduce the danger of them disintegrating. In this laboratory we use the alginate Glucol E/RH2 (Alginate Industry Ltd), which is soluble in cold water but when treated with appropriate alkali is chemically modified and becomes insoluble on drying. The manufacturers recommend making a 2% solution of the glucol in distilled water (by mechanical stirring), adding $1.cm^3$ of a 10% solution of Na_2CO_3 to $100\,cm^3$ of the glucol, and stirring into this weighed quantities of whichever dried food ingredients are desired. We roll out the resulting paste between foil-covered boards, store it frozen, and then cut it into pellets. The pellets are dried at $70°C$ and weighed before feeding to fish.

The energy content of food material is usually measured by burning weighed samples in a bomb calorimeter, which can be calibrated using known weights of thermochemical standard benzoic acid (26.45 kJ g^{-1}). An alternative but less satisfactory method of measuring food energy is to analyse the food material and multiply the proportions of carbohydrate, protein and fat by suitable energy conversion factors (weight-specific heats of combustion). Mean values generally accepted as reliable for complete combustion of carbohydrate, protein and fat are 17.2, 23.6 and 39.5 kJ g^{-1}, respectively (Brafield & Llewellyn, 1982; Jobling, 1983b), although Brett & Groves (1979) point out that whereas the latter value is appropriate for the saturated fats of a mammal, a more appropriate figure for the highly unsaturated fats associated with fish may be 36.2 kJ g^{-1}. The energy content of pellets comprising a mixture of the three basic nutrients is usually in the range 20–25 kJ g^{-1}.

Removal of faeces as soon as possible after they are produced minimizes the underestimation of faecal energy (F) which could arise from material dissolving into the surrounding water. Little can be done in energy budget experiments about faecal material actually voided from the anus in solution. Fortunately, any error arising from ignoring this is likely to be small, for Elliott (1976a) found that dissolved organic material contributed only 4% or less to the total faeces of brown trout (*Salmo trutta*). Measuring the energy content of dried and weighed faeces by standard bomb calorimetry can be difficult if the amount of available material is small. If a miniature bomb calorimeter is not available, faecal matter can be mixed with a known weight of benzoic acid to obtain sufficient material to bomb, the energy value of the acid then being deducted from the recorded energy content to arrive at that of the faecal sample. If faeces are plentiful, numerous samples can be bombed, of course, in order to monitor any variations in faecal energy content.

If C and F are accurately measured in a long-term energy budget experiment, it can be confidently claimed that the difference between the two is the energy absorbed and assimilated by the fish. Thus the usual index of assimilation efficiency, perhaps better called absorption efficiency, is $(C-F)/C$. Expressed as a percentage this indicates the proportion of food energy actually taken into the fish. It is not appropriate here to consider these efficiencies in detail; suffice it to say that the general equations prepared by Brett & Groves (1979) and quoted above suggest typical values of 80 and 59% for carnivorous and herbivorous fish, respectively. The lower value for the herbivores reflects the greater proportion of indigestible matter in their diet. In this respect fish are typical of carnivores and herbivores in general, which on average have assimilation efficiencies of 80 and 57%, respectively (Brafield & Llewellyn, 1982).

10.3   Energy in Growth Materials (P)

Some of the energy in the food consumed by a growing fish is retained in growth materials. The complex interrelationships between feeding rate, growth rate and metabolic rate are discussed in Chapter 8. Here we are concerned with how best to estimate the component P in an energy budget: the energy retained in growth. Gain in wet weight usually provides the basic information and so energy-budget experiments often last several weeks, long enough for significant weight change to have occurred. Weighing the fish during the course of an experiment is generally avoided as the process causes considerable stress; initial and final wet weights are adequate. The initial dry weight can then be estimated from knowledge of the gain in wet weight, the final dry weight, and the ratio of wet weight to dry weight. Error can arise here, as no account is taken of any changes in the wet weight to dry weight ratio over the course of the experiment, and it is well known that this ratio varies in response to several influences. Homogenization of the fish carcass allows bomb calorimetry of representative samples to arrive at an energy value (kilojoules per gram dry weight) which can be multiplied by the calculated dry weight gain to produce an estimate of P. It is best to express the energy content of the fish in terms of ash-free dry weight, for ash (skeletal and other incombustible material) is often responsible for 10–15% of the dry weight. Jauncey (1982b) found that some of his poorly fed or starving tilapia (*Sarotheroden mossambicus*) had ash contents as high as 20% or more of the dry weight.

This method assumes that the wet weight to dry weight ratio, and the energy value of the fish tissue per gram dry weight, were the same at the beginning of the experiment, maybe a month before, as at the end. (Workers with rats, which seem more consistent in energy content than fish, often take the initial energy content as that found for other rats of the same weight and condition; but data for fish are usually too sparse and variable to allow this.) The assumption that the energy value is unchanged during the course of an experiment seems unavoidable but is most unlikely to be valid. The energy value of fish varies with body size, feeding level, dietary quality, condition and the like. This must be largely due to varying proportions of fat and protein in the tissues, the former having a higher energy value per gram (section 10.2). If stored fat rather than protein is being respired by a poorly fed or starving fish, there will be a larger fall in energy content per unit weight than would appear to be the case from considering the fall in weight itself. When energy intake is below maintenance (when the fish is losing weight), the amount of protein respired will depend upon the protein-sparing effect of carbohydrate or fat (Elliott, 1979). Solomon & Brafield (1972), Spillett (1978) and Musisi (1984) in this laboratory have approached the problem by calculating the extent to which the weight-specific energy value of their fish might have altered during the course of an experiment if such changes were responsible for the imbalances found. The estimated differences between initial and final energy values necessary to generate exactly 100% balances are often quite small.

The problem of inaccuracies arising from unmonitored changes in energy value during the course of an energy budget experiment is really quite acute. Weatherley & Gill (1983) have found that the energy content of rainbow trout (*Salmo gairdneri*) rises as the fish grows, from about 24.7 kJ g^{-1} dry weight for a 10 g dry weight fish to about 28.5 kJ g^{-1} dry weight for a 75 g dry weight fish, this increase being due mainly to a relative increase in lipid content.

Several workers have found a negative correlation between lipid content and water content of fish (e.g. Flowerdew & Grove, 1980). Poor feeding or starvation tends to cause a lower lipid level and a higher water content (Elliott, 1979). Caulton (1977, 1982) has made calculations relating measured weight loss in starving *Tilapia rendalli* with losses estimated from metabolic rate and substrate depletion. Over a 10-day period at 18°C, for example, the mean fall in wet weight was 2.02 g per fish. To maintain routine metabolism, 11.7 kJ was calculated to have been released, 5.1 kJ from the respiration of lipid and 6.6 kJ from the respiration of protein. This was estimated to have entailed a loss of 0.13 g lipid and 0.28 g protein, together with an associated loss of 1.76 g water. The total estimated loss of weight, 2.17 g, satisfactorily matches the 2.02 g actually measured, and Caulton found similarly close agreements between estimated and measured weight losses with groups of fish at five other temperatures.

A fish on sub-maintenance rations is, by definition, losing weight, and therefore suffering a net loss of energy in its tissues. In such a case the energy budget equation could be written $C + P_1 = R + U + F$, where P_1 represents the energy taken from materials in the body. Energy budget balances for fish on sub-maintenance rations, or for starving fish, are sometimes poor. Inaccurate estimates of P may well be responsible, resulting from invalid assumptions about the energy value of the fish.

An interesting and rather novel alternative to monitoring growth by weight increase is concerned with tissue concentrations of the nucleic acids. Measurement of the RNA : DNA ratio affords the most satisfactory index, as estimations of either RNA or DNA alone can be misleading. The concentration of DNA can rise, for example, if cellular metabolites such as protein or fat are being depleted; and it can fall if cellular enlargement rather than cellular multiplication is occurring (Villareal, 1983). Similar difficulties can arise if the concentration of RNA alone is estimated, although Kayes (1979) has shown that the RNA concentration in the skeletal muscle of black bullhead (*Ictalurus melas*) can be a more precise growth rate indicator that the RNA : DNA ratio.

It has been clearly demonstrated over the last 10 years or so that the RNA : DNA ratio is a useful indicator both of recent growth rate and of long-term growth of fish (see Villareal, 1983, for a summary of the literature). A recent example of such studies is the work of Wilder & Stanley (1983) with brook trout (*Salvelinus fontinalis*) and Atlantic salmon (*Salmo salar*). They extracted RNA and DNA from white muscle, analysed them by ultraviolet absorbance, and found that RNA : DNA ratios were significantly correlated with growth rates estimated as weight increases.

Changes in the RNA : DNA ratio reflect changes in protein synthesis, for an increase in the RNA concentration of muscle is followed by a rise in protein synthesis. These variations are controlled by external factors such as the season (Bulow *et al.*, 1981) and by internal factors such as hormones (Lone & Matty, 1982; Chapter 7). Ideally, perhaps, one should monitor the RNA : DNA ratio as well as assessing growth by changes in weight, if it is felt desirable to distinguish between mere accumulation of metabolites (such as fat) and the role played in growth by cell multiplication.

10.4 Energy Lost as Heat (*R*)

Measurement of the heat lost by a fish has caused more difficulty than that of any other component in the energy budget equation. Here an attempt is made to discuss and evaluate three distinct approaches to estimating *R*: indirect calorimetry based on measurement of oxygen consumption; indirect calorimetry from oxygen, carbon dioxide and ammonia measurements; and direct calorimetry.

10.4.1 *Indirect Calorimetry from Oxygen Consumption*

The most widely used method for calculating the heat lost by a fish is a form of indirect calorimetry in which the rate of oxygen consumption is measured. The total amount of oxygen consumed over the period under consideration is multiplied by an energy equivalent (see below) to produce an estimate of the energy lost as heat.

The measurement of oxygen uptake rates has gained greatly from the development of polarographic oxygen electrodes. As long as care is taken over such considerations as acclimatization to the conditions in the respirometer, the respiratory rate of a fish can now be measured easily and accurately. Respiratory rates are extremely variable, of course, and the terms standard, routine and active are commonly used. These three levels of metabolism have been succinctly defined by Brett (1962) and Kutty (1981) and are discussed in Chapter 1. Standard metabolism is as near the basic rate as normal experimental techniques allow, with the fish quiescent and maintaining the minimum energy turnover for survival. The standard rate is not often measured, for laboratory fish are rarely in this state. More commonly, the fish under study is being fed at intervals and can swim freely in a fairly large respirometer. Such routine metabolism represents 'normal' activity, in which the fish can swim at will but is protected from disturbing stimuli. The term active rate was originally used to reflect the maximum activity of which a fish is capable, but is now often used to describe any level of forced activity. The tunnel respirometers used by Brett (1964, 1973) measure various levels of active metabolism.

Respirometers of the type devised by Solomon & Brafield (1972), in which routine levels of metabolism can be monitored, are useful for energy budget experiments where continuous measurement of oxygen uptake over several weeks is desirable. The vessel containing the fish has a food inlet tube at the top and a faeces outlet tube at the bottom. A pump maintains a regular and continuous flow of water through this vessel and then past an oxygen electrode connected to an oxygen meter and a strip-chart recorder. A fish in such a respirometer appears to be free of stress and, compared to the natural situation, is probably leading rather a quiet life. It may therefore be showing a generally lower level of activity than would be the case where food must be searched for and where predators must be evaded. Consequently, the respirometer fish may spend a smaller proportion of available energy in activity (R), allowing more scope for growth (P). Since respirometers measuring routine metabolic levels can operate for long periods and allow the fish to be fed, they are particularly suitable for the study of increased metabolic rate following a meal (specific dynamic action). This is not an appropriate place to contemplate the mysteries of SDA, but it is discussed in Chapter 8 and has recently been considered by Soofiani & Hawkins (1982) and Jobling (1983a).

Once the total oxygen consumed by the fish over the experimental period has been calculated, it is multiplied by a suitable coefficient (here referred to as Q_{ox}) to arrive at the energy lost as heat (R). This coefficient is variously referred to as the oxycalorific coefficient, the oxycaloric or oxycalorific equivalent, or the energy equivalent. Although application of a Q_{ox} results in a quantity now normally expressed in joules rather than calories, the use of the word oxycalorific is still valid, even desirable, as it reflects the use of oxygen to arrive at an estimate of *heat* (Latin 'calor'). R is the only component of the energy budget where energy is manifest as heat, the others (C, P, U and F) reflecting potential energy in chemical bonds.

Values for Q_{ox} vary with the respiratory substrate. In the case of carbohydrate, 2833 kJ are released as heat in the complete combustion of 1 mol of glucose by 6 mol of oxygen (192 g). So the Q_{ox} for carbohydrate is 14.76 kJ released as heat per gram of oxygen consumed (14.76 J mg^{-1} oxygen). Thus if 50 mg of oxygen, for example, were consumed in respiring carbohydrate, 738 J would be released as heat, as $O_2 \times Q_{ox} = R$. Fats vary in their composition in such a way that their Q_{ox} values also vary, but an average in general use is 13.72 J mg^{-1} oxygen. The case of protein is more complex, as the Q_{ox} varies with the nitrogenous end-product. The following equation (Brafield & Llewellyn, 1982) represents the production of ammonia by the respiration of 100 g protein:

$$(4.42\ C,\ 7.00\ H,\ 1.44\ O,\ 1.14\ N) + 4.6\ O_2 \longrightarrow$$
$$1.14\ NH_3 + 4.42\ CO_2 + 1.79\ H_2O$$

The sulphur in the protein has been ignored both because it is only about 1% and because its excretory form is variable and often unknown. The energy lost as heat is 1967 kJ, as it is the energy value of the 100 g protein (2364 kJ) minus that of the 1.14 mol of ammonia (397 kJ). When this is divided by the amount of oxygen involved (147.2 g), the Q_{ox} is seen to be 13.36 J mg^{-1} oxygen. Ammonia is the most common excretory product in teleosts (section 10.5), but sometimes appreciable amounts of urea are produced. Similar calculations to those above for ammonia production indicate a Q_{ox} of 13.60 J mg^{-1} oxygen for protein respiration when urea is the end-product. These and some similar values for Q_{ox} from the literature are shown in Table 10.1.

The Q_{ox} of 13.36 J mg^{-1} oxygen derived above refers to the respiration of protein when ammonia is the nitrogenous end-product. But suppose the nitrogenous waste leaves the fish as NH_4^+. Gnaiger (1983) has pointed out that the enthalpy of protonation of NH_3 to

Table 10.1: Some Oxycalorific Coefficients (Q_{OX}). Figures are for joules lost as heat per milligram oxygen consumed.

	Brafield & Llewellyn (1982)	Elliott & Davison (1975)	Gnaiger[a] (1983)
Carbohydrate	14.76	14.77	14.72
Fat	13.72	13.72	13.75
Protein to ammonia	13.36	13.39	13.97
Protein to urea	13.60	13.60	13.69

[a] Converted from kilojoules per mol oxygen.

NH_4^+ is -52 kJ mol^{-1}, a substantial amount. He claims that neglecting to take this into account when applying a Q_{ox} 'entails a significant error'. He is making an important point, but it is not clear whether the excretory product of fish is only, or even predominantly, NH_4^+ rather than NH_3. Hard evidence on this point is patchy and sometimes apparently conflicting. Evans *et al.* (1982) have accumulated a considerable body of evidence supporting their model of an active Na^+/NH_4^+ exchange mechanism at the fish gill, and they suggest that passive outward diffusion of NH_4^+ will supplement this in marine fish. On the other hand, Kormanik & Cameron (1981), while conceding that the Na^+/NH_4^+ exchange mechanism is a significant excretory pathway in some species, state that it 'appears to be unimportant or lacking in others'. They suggest that the diffusive loss across the gills probably occurs mostly as NH_3 rather than NH_4^+. Cameron & Heisler (1983) write that under normal conditions (e.g. low external NH_3) 'diffusive movement of NH_3 appears to account adequately for ammonia excretion' in rainbow trout (*Salmo gairdneri*). Factors affecting the NH_3/NH_4^+ equilibrium are subtle, and until the relative importance of the three possible excretory pathways (active Na^+/NH_4^+ exchange and passive diffusion of NH_4^+ or of NH_3) is known for the fish under investigation, a Q_{ox} value based on ammonia production is probably adequate. In any case, as far as complete energy budgets are concerned, misunderstanding of the NH_3/NH_4^+ situation may cause errors in Q_{ox} and therefore in R, but these will be offset by contrary errors in estimating the energy loss in nitrogenous waste (U).

Since fish are generally respiring a mixture of substrates, a composite Q_{ox} will usually need to be calculated. If fat ($Q_{ox} = 13.72$) and protein ($Q_{ox} = 13.36$ if ammonia is produced) are being respired in the ratio 7 : 3, for example, the appropriate Q_{ox} will be 13.61 J mg^{-1} oxygen. If detailed information on substrates is not available, some generally acceptable Q_{ox} can be used. For starving fish, for example, Brett & Groves (1979) suggest 4.63 kcal dm^{-3} oxygen, equivalent to 13.56 J mg^{-1} oxygen.

It is unfortunate if the composite Q_{ox} used in an energy budget experiment has to be based on the assumption that substrates are being respired in the proportions in which they occur in the food, as this is most unlikely to be the case. If assimilation efficiencies of carbohydrate, fat and protein vary, the proportions of these three actually absorbed will differ from their proportions in the food. For this reason faeces should be analysed whenever possible to determine the proportions of the three basic nutrients that they contain. For some years in this laboratory (e.g. Spillett, 1978; Musisi, 1984) we have been feeding various fish species three artificial pelleted diets: one high in protein, one largely of protein and carbohydrate and one largely of protein and fat. Faecal levels of these three nutrients are usually very small, but occasionally the amount of fat in the faeces has been quite high, requiring a downward adjustment in the composite Q_{ox} used. But even if the proportions of protein, carbohydrate and fat in the food and in the faeces are known (and therefore the proportions actually absorbed and assimilated), it is still dangerous to assume that one knows the proportions in which they are actually respired. Interconversions between assimilated protein, carbohydrate and fat will occur, and the three will be contributing differentially to growth and therefore to respiration. This problem is overcome by the other indirect calorimetry method, shortly to be described.

There is a complication in estimating R for a fish on sub-maintenance rations (one feeding but losing weight) in that part of the total heat produced is due to respiration of dietary substrates and part to respiration of the fish's own tissues. We follow the method devised here by Spillett (1978). The weights of protein, carbohydrate and fat absorbed by the fish during the course of the experiment (from the proportions in food and faeces) are calculated and converted to energy lost as heat by multiplying each of them by the appropriate heat of combustion: 19.67 kJ g^{-1} for protein to ammonia (see above), 17.2 kJ g^{-1} for carbohydrate and 39.5 kJ g^{-1} for fat (section 10.2). The three resulting values for heat losses are divided by the appropriate Q_{ox} values (see above) to arrive at estimates of the oxygen consumed in respiring the dietary protein, carbohydrate and fat, respectively. The sum of these three indicates the oxygen consumed in respiring dietary nutrients. If this is subtracted from the total oxygen consumption of the fish as measured by respirometry, one arrives at the amount of oxygen consumed by the fish in respiring its own tissues. When this is multiplied by the composite Q_{ox} appropriate for the respiration by a poorly fed fish of its own tissues, one has the heat loss resulting from such respiration. Adding this to the heat loss previously calculated as arising from respiration of dietary substrates produces a reliable estimate of the total heat loss (R).

10.4.2 Indirect Calorimetry from Oxygen, Carbon Dioxide and Ammonia Measurements

For many years, workers on poultry, rats, cattle or man himself have applied equations that accurately predict R from measurements of oxygen consumption, carbon dioxide production and nitrogenous excretion (Brafield & Llewellyn, 1982). With the recent improvement in techniques for measuring carbon dioxide and ammonia production by aquatic animals, the application of a similar equation to fish has become feasible. The following should be applicable to energy budget work with fish:

$$R = 11.18\,A + 2.61\,B - 9.55\,N$$

where R is the total heat loss (in joules) and A, B and N are the oxygen consumed, carbon dioxide produced and ammonia produced, respectively (all three in milligrams). This equation differs very slightly from the first version (Brafield & Llewellyn, 1982) owing to more precise basic data and calculation. The derivation, a fairly lengthy process, is based on the method developed by Weir (1949) for use with mammals. The Q_{ox} values used are taken from the first column of Table 10.1. Respiratory quotient (RQ) values were taken as 1.0, 0.71 and 0.96 for carbohydrate, fat and protein to ammonia respectively (the 0.96 deriving from the equation for the respiration of 100 g protein given earlier, which indicates that 4.42 mol carbon dioxide correspond to 4.6 mol oxygen), which on conversion to RQ values by weight give 1.375, 0.976 and 1.32. It was assumed that 7.6 mg oxygen are used in respiring protein to produce 1 mg ammonia (again from the equation given earlier, which indicates that 147.2 g oxygen are involved in producing 19.4 g ammonia).

Application of this equation to the calculation of R depends on accurate measurements of oxygen consumption (see section 10.4.1), ammonia production (see section 10.5) and carbon dioxide production.

Kutty (1968) has discussed the difficulties associated with accurately estimating carbon dioxide dissolved in water, and has reviewed the early literature. The main problem is that the considerable amount of bound carbon dioxide normally present, especially in hard water, makes the carbon dioxide released by a fish a relatively small increase. The Maros–Schulek titrimetric method has been widely used (e.g. Thillart & Kesbeke, 1978) and has been modified and improved over the years (see Kutty, 1981). Burggren (1979) took water samples with gas-tight syringes and measured their carbon dioxide content with a Radiometer CO_2 electrode. He found the electrodes to vary in their performance and took care to use only those with high stability and responsiveness. Thillart *et al.* (1983) have also used Radiometer CO_2 electrodes. It seems likely that electrode systems will replace other methods of measuring carbon dioxide concentration, just as they have in the case of oxygen measurement. The availability of a reliable method is particularly welcome as it will enourage expansion in important areas of fish physiology such as the metabolic implications of different RQ values (Kutty, 1981) and the various anaerobic metabolic pathways that occur in fish (Thillart & Kesbeke, 1978).

Musisi (1984), in the experiments about to be described, measured carbon dioxide production with a Radiometer CO_2 electrode (type E 5036) connected to a PHM 71 Mk2 Acid–Base Analyzer. (The principle and practice of the method is explained in the Radiometer electrode operating instructions.) The system was calibrated with water through which various percentages of carbon dioxide had been passed. From a table comparing these percentages with the corresponding carbon dioxide concentrations (in ppm) a calibration graph was constructed linking meter readings with carbon dioxide in ppm. Thus meter readings from the CO_2 electrode in the energy budget experiments could be converted into carbon dioxide concentrations (ppm) in the water leaving the continuous-flow respirator.

Numerous energy budgets for tilapia (*Sarotherodon mossambicus*) have been compiled by Musisi (1984) in this laboratory, but only six of these involved simultaneous measurements of oxygen consumption, carbon dioxide production and ammonia production, for fish of about the same size fed super-maintenance rations. The results of these, in which R could be calculated from the equation given above as well as from oxygen consumption multiplied by a Q_{ox}, are shown in Table 10.2. The apparatus used was based on that of Solomon & Brafield (1972), with the addition of the CO_2 electrode in a second

Table 10.2: Results of Six 30-day Energy Budgets For *Sarotherodon mossambicus* (Musisi, 1984) in Which Energy Lost as Heat was Estimated by Oxygen Consumption Multiplied By a Q_{ox} (R_1) and Also From the Equation Involving Oxygen, Carbon Dioxide and Ammonia (R_2). All figures are in kilojoules except for the balances (B_1 and B_2), which are percentages. See text for details.

Experiment number	C	P	R_1	R_2	U	F	B_1 $(P + R_1 + U + F)/C$	B_2 $(P + R_2 + U + F)/C$
1	427.9	56.8	294.5	284.1	35.6	50.7	102.3	99.8
2	585.9	34.2	444.0	420.4	40.1	109.3	107.1	103.1
3	553.4	71.1	247.3	220.2	52.1	103.2	85.6	80.7
4	633.5	60.8	218.5	212.1	36.0	152.5	73.8	72.8
5	654.1	82.1	388.0	362.2	34.1	178.6	104.4	100.4
6	540.7	38.5	393.4	376.8	30.4	122.1	108.1	105.0

electrode chamber in series with the oxygen electrode chamber. Each of these six experiments ran continuously for 30 days at 27°C. The means of the initial and final wet weights ranged from 30.0 to 51.4 g. Three meals a day were given, each 4% of body weight. The diets used were dried *Tubifex* pellets (experiment 1) and artificial pellets high in protein (experiment 2), high in both protein and carbohydrate (3 and 4) or high in both protein and fat (5 and 6). Column U in Table 10.2 shows estimates of energy lost in ammonia derived from the modified Bower & Holm-Hansen (1980) method described in section 10.5.

Table 10.2 shows values for R derived from two methods: R_1 from oxygen consumption multiplied by a composite Q_{ox}, and R_2 from application of the equation shown above involving oxygen, carbon dioxide and ammonia measurements. Estimates of R by the two methods correspond fairly closely and produce comparable balances, though the values for R_1 are all slightly higher than those for R_2. This comparison of the two indirect methods of measuring R is discussed in section 10.6.

10.4.3   *Direct Calorimetry*

This method measures the heat produced by an animal by monitoring the rise in temperature of its surroundings. The basic technique has been in use with mammals for 200 years, but has been applied to aquatic animals only recently (Brafield & Llewellyn, 1982). Work on fish is particularly sparse: Davies (1966) attempted to measure the heat production of goldfish (*Carassius auratus*) directly, with a differential micro-calorimeter; and Brett & Groves (1979) cite experiments by Smith (1976) on four species of salmonids. Smith used a modified adiabatic calorimeter but found it less sensitive to metabolic fluctuations than an indirect method involving measurement of oxygen consumption.

The direct calorimeter built by Lowe (1978) in this laboratory was primarily intended for use with a polychaete, but to test the versatility of the basic design he enlarged the chamber containing the experimental animal and measured heat loss from the stickleback *Gasterosteus aculeatus* (and also from a decapod crustacean). The fish was enclosed in a vacuum-jacketed glass tube 9 cm long and 3.75 cm in diameter (100 cm^3). Air-saturated water at 15°C was pumped through the vessel continuously, at a rate of 4.30 cm^3 min^{-1}. Two thermistors, one just before the entrance to the vessel and one just past the exit, monitored the temperature of the water passing

Table 10.3: Heat Loss From a Stickleback (of 1.80 g wet weight) Measured Simultaneously by Direct (R_d) and by Indirect (R_i) Calorimetry. Figures are in joules per hour (from Lowe, 1978).

	R_d	R_i
Lowest rate	5.12	5.23
Highest rate	8.05	8.01
Mean of 49 hourly readings	6.28	6.07

them. They were incorporated into a Wheatstone bridge circuit in such a way that the temperature difference between them, due to heat loss by the fish, was displayed on a strip-chart recorder. A current was passed through a wire coil (of known heat production) in a twin vessel, for calibration purposes. After passing across the thermistor near the exit of the fish vessel, water flowed through a small chamber carrying an oxygen electrode connected to an oxygen meter and a second strip-chart recorder. (The whole apparatus, in the form with the smaller tubes for work with polychaetes, is shown in Brafield & Llewellyn, 1982, Figure 5.1.) In this way heat loss was measured by direct and indirect calorimetry simultaneously. (The advantage of this, in the experiments with polychaetes for which the apparatus was originally designed, is that anaerobic respiration can be identified; for heat loss from anaerobic metabolism contributes to the estimation of R in the case of the direct method but not in the case of the indirect one. In theory at least, the same principle could be applied to work on anaerobic respiration in fish.) The composite Q_{ox} chosen for use in the indirect method (13.55 J mg^{-1} oxygen) was based on the assumption that the fish was respiring its own tissues, as it had been starved for over a day.

After the fish had been in the apparatus for about a day, for acclimatization and heat equilibration, continuous strip-chart records were made for 49 h. From these records the mean rates of heat loss for each hour were calculated for both the direct and indirect methods. These hourly means varied considerably over the 49 h of the experiment, due to the fluctuating activity of the fish, but values from the two methods for any given hour matched each other very closely. The lowest and highest hourly rates of heat loss are shown in Table 10.3, along with the mean values of all 49 h. Although the means for hourly heat loss from the direct and the indirect methods are similar (equivalent to 3.49 and 3.37 J g^{-1} h^{-1}, respectively), the former is decidedly higher than the latter. This comparison between measurements of heat loss by direct and indirect calorimetry are discussed in section 10.6.

10.5 Energy Lost in Nitrogenous Excretory Products (U)

Ammonia is the chief excretory product of teleosts and is released mainly from the gills. (The question of whether it is actually released as NH_3 or as NH_4^+ has been briefly discussed in section 10.4.1.) Under some circumstances, however, significant amounts of urea may be produced, and Elliott (1976a) has summarized the methods available for measuring it. Other nitrogenous substances such as amino acids, uric acid and creatine may be excreted but are almost always in extremely small quantities.

The rate at which a fish excretes ammonia is affected by many factors, such as dietary quantity and quality (Elliott, 1976a; Brett & Groves, 1979). Savitz *et al.* (1977), working with largemouth bass (*Micropterus salmoides*), found that nitrogen excretion increased linearly as nitrogen absorption increased, and that the relationship between nitrogen retention and nitrogen absorption could also be described by a linear equation. Nitrogen in the diet of a fish will either appear in the faeces or be absorbed, and in the latter case will either be stored in growth material or be excreted. It is becoming fashionable to erect nitrogen budgets, which relate nitrogen gains to losses. A further interesting aspect of nitrogen metabolism concerns a temporary rise in ammonia excretion following a meal – a manifestation of specific dynamic action. The concern here, however, is with the energy lost in nitrogenous excretion (U) as a component in the energy budget. Early workers tended to discount this channel, believing the energy content of ammonia to be negligible, but 347.9 kJ mol^{-1} (about 20.5 J mg^{-1}) is an appropriate figure. Nevertheless, U is generally the smallest component in an energy budget. In the generalized budget for carnivorous fish presented by Brett & Groves (1979), given in section 10.1, U is only 7% of C (the energy of the food), and the mean value for U as a percentage of C for the six budgets shown in Table 10.2 is 6.8%. The amount of energy lost in nitrogenous excretion is therefore modest but significant, and account must be taken of it when compiling energy budgets. This requires an accurate method of measuring ammonia.

There are various methods available for measuring total ammonia (i.e. both NH_3 and NH_4^+) in aqueous solution. Thillart & Kesbeke (1978) and Cameron & Heisler (1983) have made enzymatic assays, and the latter authors have also used an ammonia electrode which senses the partial pressure of ammonia in solution. Over the years, however, the most widely used method has probably been the one associated with Barthelot, and later with Solórzano, in which the ammonia reacts with phenol and hypochlorite in the presence of nitroprusside to form a measurable colour, indophenol blue. Bower & Holm-Hansen (1980) have summarized the evolution of the technique. Their own improvement is to substitute sodium salicylate for phenol, thereby avoiding the production of highly volatile chlorophenol which can be toxic to both fish and experimenter. Bower & Holm-Hansen (1980) used the commercial bleach 'Clorox' as the source of sodium hypochlorite, but Musisi (1984) has found 'Voxsan' (used at swimming baths) to be more stable. She used this method, calibrated with a range of ammonium sulphate concentrations, to determine the amount of ammonia excreted by tilapia (*Sarotherodon mossambicus*) in the long-term energy budgets shown in Table 10.2. Samples of water were taken for analysis at frequent intervals and at various times relative to the giving of a meal. The mean hourly ammonia production was calculated and scaled up to estimate total ammonia production for the 30 days of each experiment. Multiplying this total by the energy value of ammonia (see above) gave a value for U. Urea was also measured, by the Conway micro-diffusion method, but only insignificant traces were found.

The rate of ammonia production was so small in Musisi's (1984) experiments that the flow through the respirometer had to be stopped for several hours to allow the ammonia to build up to levels that could be reliably measured. Although this is a common practice, it entails two hazards. First, oxygen deficiency may occur if the water is not aerated, yet if it is aerated some ammonia may escape (Weiler, 1979), resulting in underestimation of its rate of production. Secondly, the ammonia may build up to levels that harm the fish. Elliott (1976a), for example, was aware of this danger and checked that such levels were not reached in his experiments. It is the un-ionised molecule (NH_3) which is toxic, not the ammonium ion (NH_4^+), however, and Trussell (1972) has shown that the former is a very small percentage of the total ammonia at the pH levels and temperatures generally involved (though it rises with increasing pH and temperature).

As an alternative to ammonia measurement, U can be calculated from the oxygen consumption, as shown by Brafield & Solomon (1972). They used an erroneous figure for the energy value of ammonia, as Elliott & Davison (1975) pointed out, but the principle is sound. From the equation given in section 10.4.1 for the respiration of protein with ammonia production, it can be seen that 397 kJ will be lost in ammonia (1.14 mol) for every 147.2 g oxygen involved. Thus 2.70 J are lost in ammonia for each milligram of oxygen consumed in respiring protein. So, if the proportion of the total measured oxygen consumption which is involved in respiring protein can be estimated, multiplying it (in milligrams) by 2.70 will produce an estimate of the energy (in joules) lost in ammonia (U). Similar calculations yield a value of 2.46 J lost in urea per milligram oxygen consumed in respiring protein when urea is the nitrogenous end-product (Brafield & Llewellyn, 1982). Elliott & Davison (1975) produced very similar values to these: 2.59 J mg^{-1} oxygen for the case of ammonia production and 2.41 for urea.

10.6 Concluding Remarks

Much has been achieved in the study of fish energetics since the classical and seminal work of Winberg (1956) nearly 30 years ago. The establishment and development of the concept of the energy budget, now framed in the familiar equation, has stimulated much research and provoked much thought. The light that such studies shed on the ways in which fish partition the energy resources at their disposal is considerable. Many energy budgets have been compiled, by many workers, on many species of fish. The quantity and scope of the information in the literature is bewildering. The variety of methods used and the many different ways of processing and presenting energy budget data make comparisons between budgets, and hence the establishment of reliable generalizations and principles, extremely difficult. This situation, of increasingly detailed information lending itself to decreasingly effective synthesis, seems likely to continue. On the other hand, the basic techniques used to estimate the various components of the energy budget have stood the test of time fairly well, and when an energy budget in the region of 100% is achieved, it seems likely that all channels have been measured with an acceptable degree of accuracy.

Numerous potential sources of error when erecting energy budgets have been identified, and in most cases improved methods have helped to covercome the problems. For example, the use of food pellets bound with an alginate has reduced overestimation of the energy in the food consumed (C), while allowing artificial diets to be given to fish of whatever nutrient composition the experimenter chooses. Other difficulties are more intractable, such as those involved with assessing the energy of growth materials (P) merely from wet weight increase, wet weight to dry weight ratio and final energy content. Several of these potential sources of error, and the ways in which they can be minimized, have been discussed here.

Estimation of three of the factors in the energy budget equation $(C, P$ and $F)$ rely in principle on a single tool which is accurate and simple to use, the bomb calorimeter. Chemical methods for estimating the amounts of nitrogenous excretory products produced by a fish (in order to calculate U) are continually being refined; and generally the alternative method, of estimating U from oxygen consumption, is available. The energy lost as heat (R) is still the channel presenting the most significant practical problems, though here, too, recent improvements in technique are making estimation of R more reliable (as long as care and thought are devoted not only to the design and use of the apparatus but also to the nature of the subsequent calculations). So why does one still get the odd rogue budget which does not balance? There seems to be no simple or single answer to this, but if each bad budget is inspected, it is usually possible to identify unusual features that may have been responsible for the imbalance. (It is not, of course, scientific to suggest that it is simply Nature getting her own back on the experimenter rudely probing her secrets.)

Three approaches to the estimation of heat loss (R) have been described here. Two are indirect methods: one involving oxygen consumption and a Q_{ox}, the other an equation incorporating oxygen consumption and carbon dioxide and ammonia production. The third is direct calorimetry. The two indirect methods have been applied simultaneously by Musisi (1984) and the correspondence between them (R_1 and R_2 in Table 10.2) is close. Simultaneous direct and indirect calorimetry (the version involving oxygen consumption and Q_{ox}) has been achieved by Lowe (1978) and again the correspondence is close (Table 10.3). However, it would be premature to claim on the basis of this comparative evidence alone (I know of no other) that the three methods are equally effective, or that as a result it does not matter very much which of the three is used. (Certainly, in the case of energy studies with man, significant differences have sometimes been found between the results of direct and indirect calorimetry (Webb, 1980; Webb *et al.*, 1980), which remain unexplained in spite of the most careful consideration of the possible causes.) In any case, each of these three methods of estimating heat loss (R) has its advantages and disadvantages. Indirect calorimetry involving oxygen consumption and a Q_{ox} is widely used, easy to carry out, and in many cases seems to produce adequate results. The indirect method involving the equation which incorporates oxygen consumption and carbon dioxide and ammonia production has distinct advantages. One is that it removes the need to assume that substrates are respired in the proportions in which they occur in the food (or in the proportions in which they are absorbed). Another is that it sheds light on anaerobic pathways and on carbohydrate–fat interconversions, as it will pick up unusually high respiratory quotients. It could be applied with advantage to studies of fish energetics where plentiful data are available on oxygen consumption, carbon dioxide production and ammonia excretion (e.g. Kutty, 1981). Direct calorimetry is technically the most difficult of the three methods of measuring R, but it does bring the satisfaction of knowing that heat loss is being measured by doing just that — measuring *heat* loss rather than measuring something else from which heat loss is deduced.

What road will laboratory studies of fish energy budgets follow now? New techniques are available which should increase accuracy in the measurement of several of the energy budget components, particularly heat loss. Two such techniques new to fish physiology — direct calorimetry and the application of equations involving gaseous exchange and nitrogenous excretion rates — have been in use by workers on the energetics of man and other mammals for many years. Hopefully, much closer notice will be taken by the fish physiologist and the mammal physiologist of one another's technical and theoretical advances. The rigorous experimental and conceptual analysis shown, for example, in the papers by Webb (1980) and Webb *et al.* (1980), mentioned above, is very stimulating. Finally, it is hoped that other bridges will be built, particularly between laboratory studies and field studies of fish energy budgets.

References

Bower, C.E. and Holm-Hansen, T. (1980), A salicylate–hypochlorite method for determining ammonia in seawater. *Can. J. Fish. Aquat. Sci. 37*, 794

Braaten, B.R. (1979), Bioenergetics — a review on methodology. In: Halver, J.E. and Tiews, K. (eds). *Proc. World Symp. on Finfish Nutrition and Fishfeed Technology*, vol. 2, pp. 461-504. Berlin, Heenemann

Brafield, A.E. and Llewellyn, M.J. (1982), *Animal Energetics*, Glasgow, Blackie

Brafield, A.E. and Solomon, D.J. (1972), Oxy-calorific coefficients for animals respiring nitrogenous substrates. *Comp. Biochem. Physiol. 43A*, 837

Brett, J.R. (1962), Some considerations in the study of respiratory metabolism in fish, particularly salmon. *J. Fish. Res. Bd Can. 19*, 1025

Brett, J.R. (1964), The respiratory metabolism and swimming performance of young sockeye salmon. *J. Fish. Res. Bd Can. 21*, 1183

Brett, J.R. (1973), Energy expenditure of sockeye salmon, *Oncorhynchus nerka*, during sustained performance. *J. Fish. Res. Bd Can. 30*, 1799

Brett, J.R. (1983), Life energetics of sockeye salmon, *Oncorhynchus nerka*. In: Aspey, W.P. and Lustick, S.I. (eds). *Behavioural Energetics: The Cost of Survival in Vertebrates*, pp. 29–63. Columbus, Ohio State University Press

Brett, J.R. and Groves, T.D.D. (1979), Physiological energetics. In: Hoar, W.S., Randall, D.J. and Brett, J.R. (eds). *Fish Physiology, vol. VIII*, pp. 279–352. New York, Academic Press

Bulow, F.J., Zeman, M.E., Winningham, J.R, and Hudson, W.F. (1981), Seasonal variations in RNA–DNA ratios and in indicators of feeding, reproduction, energy storage, and condition in a populaton of bluegill, *Lepomis macrochirus* Rafinesque. *J. Fish Biol. 18*, 237

Burggren, W.W. (1979), Bimodal gas exchange during variation in environmental oxygen and carbon dioxide in the air-breathing fish *Trichogaster trichopterus*. *J. exp. Biol. 82*, 197

Cameron, J.N. and Heisler, N. (1983), Studies of ammonia in the rainbow trout: physico-chemical parameters, acid–base behaviour and respiratory clearance. *J. exp. Biol. 105*, 107

Caulton, M.S. (1977), The effect of temperature on routine metabolism in *Tilapia rendalli* Boulenger. *J. Fish Biol. 11*, 549

Caulton, M.S. (1982), Feeding, metabolism and growth of tilapias: some quantitative considerations. In: Pullin, R.S.V. and Lowe-McConnell, R.H. (eds). *The Biology and Culture of Tilapias. ICLARM Conference Proceedings 7*, pp. 157–80. Manila, Philippines, International Center for Living Aquatic Resources Management

Cowey, C.B. and Sargent, J.R. (1979), Nutrition. In: Hoar, W.S., Randall, D.J. and Brett, J.R. (eds). *Fish Physiology, vol. VIII*, pp. 1–69. New York, Academic Press

Davies, P.M.C. (1966), The energy relations of *Carcassius auratus* L. II. The effect of food, crowding and darkness on heat production. *Comp. Biochem. Physiol. 17*, 983

Elliott, J.M. (1976a), Energy losses in the waste products of brown trout (*Salmo trutta* L.). *J. Anim. Ecol. 45*, 561

Elliott, J.M. (1976b), The energetics of feeding, metabolism and growth of brown trout (*Salmo trutta* L.) in relation to body weight, water temperature and ration size. *J. Anim. Ecol. 45*, 923

Elliott, J.M. (979), Energetics of freshwater teleosts. In: Miller, P.J. (ed.) *Fish Phenology: Anabolic Adaptiveness in Teleosts. Symp. Zool. Soc. Lond. 44*, pp. 29–61. London, Academic Press

Elliott, J.M. and Davison, W. (1975), Energy equivalents of oxygen consumption in animal energetics. *Oecologia 19*, 195

Evans, D.H., Claiborne, J.B., Farmer, L., Mallery, C. and Krasny, E.J. (1982), Fish gill ionic transport: methods and models. *Biol. Bull. 163*, 108

Fischer, Z. (1979), Selected problems of fish bioenergetics. In: Halver, J.E. and Tiews, K. (eds). *Proc. World Symp. on Finfish Nutrition and Fishfeed Technology*, vol. I, pp. 17–44. Berlin, Heenemann

Flowerdew, M.W. and Grove, D.J. (1980), An energy budget for juvenile thick-lipped mullet, *Crenimugil labrosus* (Risso). *J. Fish Biol. 17*, 395

Gnaiger, E. (1983), Calculation of energetic and biochemical equivalents of respiratory oxygen consumption. In: Gnaiger, E. and Forstner, H. (eds). *Polarographic Oxygen Sensors. Aquatic and Physiological Applications*, pp. 337–45. Berlin, Springer

Jauncey, K. (1982a), Carp (*Cyprinus carpio* L.) nutrition — a review. In: Muir, J.F. and Roberts, R.J. (eds). *Recent Advances in Aquaculture*, pp. 215–63. London, Croom Helm

Jauncey, K. (1982b), The effects of varying dietary protein level on the growth, food conversion, protein utilization and body composition of juvenile tilapias (*Sarotherodon mossambicus*). *Aquaculture 27*, 43

Jobling, M. (1983a), Towards an explanation of specific dynamic action (SDA). *J. Fish Biol. 23*, 549

Jobling, M. (1983b), A short review and critique of methodology used in fish growth and nutrition studies. *J. Fish Biol. 23*, 685

Kayes, T. (1979), Effect of hypophysectomy and beef growth hormone replacement therapy, over time and at various dosages, on body weight and total RNA–DNA levels in the black bullhead (*Ictalurus melas*). *Gen. Comp. Endocr. 37*, 321

Kitchell, J.F., Stewart, D.J. and Weininger, D. (1977), Applications of a bioenergetics model to yellow perch (*Perca flavescens*) and walleye (*Stizostedion vitreum vitreum*). *J. Fish. Res. Bd Can. 34*, 1922

Koch, F. and Wieser, W. (1983), Partitioning of energy in fish: can reduction of swimming activity compensate for the cost of production? *J. exp. Biol. 107*, 141

Kormanik, G.A. and Cameron, J.N. (1981), Ammonia excretion in animals that breathe water: a review. *Mar. Biol. Letters 2*, 11

Kutty, M.N. (1968), Respiratory quotients in goldfish and rainbow trout. *J. Fish. Res. Bd Can. 25*, 1689

Kutty, M.N. (1981), Energy metabolism of mullets. In: Oren, O.H. (ed.). *Aquaculture of Grey Mullet*, pp. 219–64. Cambridge, Cambridge University Press

Lone, K.P. and Matty, A.J. (1982), Cellular effects of adrenosterone feeding to juvenile carp, *Cyprinus carpio* L., effect on liver, kidney, brain and muscle protein and nucleic acids. *J. Fish Biol. 21*, 33

Lowe, G.D. (1978), The measurement by direct calorimetry of the energy lost as heat by a polychaete, *Neanthes* (= *Nereis*) *virens* (Sars). Ph.D. thesis, University of London

Mironova, N.V. (1976), Changes in the energy balance of *Tilapia mossambica* in relation to temperature and ration size. *J. Ichthyol. 16*, 120

Musisi, L. (1984), The nutrition, growth and energetics of tilapia, *Sarotherodon mossambicus*. Ph.D. thesis, University of London

Savitz, J., Albanese, E., Evinger, M.J. and Kolasinski, P. (1977), Effect of ration level on nitrogen excretion, nitrogen retention and efficiency of nitrogen utilization for growth in largemouth bass (*Micropterus salmoides*). *J. Fish Biol. 11*, 185

Smith, R.R. (1976), Studies on the energy metabolism of cultured fish. Ph.D. thesis, Cornell University

Solomon, D.J. and Brafield, A.E. (1972), The energetics of feeding, metabolism and growth of perch (*Perca fluviatilis* L.). *J. Anim. Ecol. 41*, 699

Soofiani, N.M. and Hawkins, A.D. (1982), Energetic costs at different levels of feeding in juvenile cod, *Gadus morhua* L. *J. Fish Biol. 21*, 577

Spillett, P.B. (1978), The nutrition and energy relations of the fish *Perca fluviatilis* L. and *Carassius auratus* L. Ph.D. thesis, University of London

Staples, D.J. and Nomura, M. (1976), Influence of body size and food ration on the energy budget of rainbow trout *Salmo gairdneri* Richardson. *J. Fish Biol. 9*, 29

Thillart, G.V.D. and Kesbeke, F. (1978), Anaerobic production of carbon dioxide and ammonia by goldfish *Carassius auratus* L. *Comp. Biochem. Physiol. 59A*, 393

Thillart, G.V.D., Randall, D. and Hoa-Ren, L. (1983), CO_2 and H^+ excretion by swimming coho salmon, *Oncorhynchus kisutch. J. exp. Biol. 107*, 169

Trussell, R.P. (1972), The percent un-ionized ammonia in aqueous ammonia solutions at different pH levels and temperatures. *J. Fish. Res. Bd Can. 29*, 1505

Villareal Gonzales, C.A. (1983), The role of light and endocrine factors in the development of bimodality of growth in the juvenile Atlantic salmon (*Salmo salar* L.). Ph.D. thesis, University of Stirling

Weatherley, A.H. and Gill, H.S. (1983), Protein, lipid, water and caloric contents of immature rainbow trout, *Salmo gairdneri* Richardson, growing at different rates. *J. Fish Biol. 23*, 653

Webb, P. (1980), The measurement of energy exchange in man: an analysis. *Amer. J. Clin. Nutr. 33*, 1299

Webb, P., Annis, J.F. and Troutman, S.J. (1980), Energy balance in man measured by direct and indirect calorimetry. *Amer. J. Clin. Nutr. 33*, 1287

Weiler, R.R. (1979), Rate of loss of ammonia from water to the atmosphere. *J. Fish. Res. Bd Can. 36*, 685

Weir, J.B. de V. (1949), New methods for calculating metabolic rate with special reference to protein metabolism. *J. Physiol. 109*, 1

Wiegert, R.G. (1968), Thermodynamic considerations in animal nutrition. *Amer. Zool. 8*, 71

Wilder, I.B. and Stanley, J.G. (1983), RNA–DNA ratio as an index to growth in salmonid fishes in the laboratory and in streams contaminated by carbaryl. *J. Fish Biol. 22*, 165

Winberg, G.G. (1956), Rate of metabolism and food requirements of fishes. *Beloruss. State University, Minsk* (Fish. Res. Bd Can. Transl. Ser. No. 194, 1960)

Wootton, R.J., Allen, J.R.M. and Cole, S.J. (1980), Energetics of the annual reproductive cycle in female sticklebacks, *Gasterosteus aculeatus* L. *J. Fish Biol. 17*, 387

11 FIELD STUDIES OF ENERGY BUDGETS

N.M. Soofiani and A.D. Hawkins

11.1 Introduction

An energy budget provides a powerful framework for identifying the most important aspects of the life of any fish. It enables the various inputs and outputs to the animal to be examined in common, transferable units, and permits some distinction to be made between the different ways in which energy is expended in the wild. The main components of the budget, the energy taken in as food (C), the energy lost as nitrogenous excretory products (U) and as faeces (F), the energy dissipated in metabolism and devoted to different activities (R), and the surplus energy available for growth of the somatic and reproductive tissues (P), are all relevant to an understanding of the survival, growth and overall production of fish in the wild. The food consumed by fish, and its relationship to growth, are of particular importance, since they are crucial for the understanding and modelling of marine ecosystems (Steele, 1974). The food consumption itself can be applied directly in studies of the transfer of energy within food chains and of the influence of a species at one trophic level upon another. Moreover, a knowledge of the energy transfer from one level to another is essential for determining the natural mortality of prey species, a parameter which is not readily estimated by any other method. Such data are especially relevant to the management of multi-species fisheries, where one commercially valuable species feeds upon another.

In principle, each of the primary variables within an energy budget is observable, but in practice the individual components can seldom be measured with great precision. This is particularly so for a population of fish in the field, where the animals cannot be placed within a calorimeter or restrained for experimental study. It is often necessary to obtain estimates by several independent, indirect methods. A particular advantage of any energy budget, however, is that it provides a balanced equation relating these various components to one another. If, for any reason, one of the quantities cannot be measured, then it can subsequently be estimated by difference from the others.

The preparation of an energy budget in the field requires a wide knowledge of the animal's ecology, including estimates of feeding rates, growth, and activity. In addition, detailed ancillary information is required on the physiological energetics of the fish which can only be obtained in the laboratory. The conduct of these parallel and complementary studies can rarely be completed on a short time scale. Any review of energy budgets in the field must inevitably concentrate on these methodological problems, and we shall therefore begin by considering the various ways in which information on the energy inputs and outputs can be gained for fish in the wild.

11.2 The Main Components of the Energy Budget

The main elements in an energy budget for fish have recently been considered by Kitchell *et al.* (1977), Braaten (1979), Elliott (1979) and Fischer (1979). The various terms in current use have been set out by Brafield & Llewellyn (1982) and have already been defined in Chapters 1 and 10. We recognize that there are difficulties in establishing the precise relationships between the different elements but we will not dwell upon these problems here.

11.2.1 Measurement of the Energy Input (C)

The main energy input to the fish — the income to the budget — is the food consumed. However, it occasionally happens that fish do not feed, or are feeding poorly, and in this case the main energy resources are provided internally, the fish utilizing its hoarded capital. The 'income' in this case shows up as negative growth, or a loss in weight of the animal's tissues, accompanied by a distinctly different chemical composition of the tissues remaining after the period of starvation.

In most cases, however, fish are consuming food. In laboratory studies, or where fish are being cultured, the energy intake is relatively easy to determine. Live food or artificial diets are provided in measured quantities, and have a known energy content. Indeed, there is some measure of control over the composition of the food, which can be varied to suit the requirements of the experiment. The various problems encountered in the selection and presentation of a particular food, and in determining its composition are examined in Chapters 3 and 5. By contrast, in the field, there is no control over the quantities of food taken by the fish, or the composition of the diet, and the food and energy intake must be measured directly or estimated by some means.

The nature and quality of food consumed by fish in the wild can easily be determined by examination of the stomach contents, but estimation of the actual energy intake per unit time is more difficult. Of course, it is possible to avoid the problem altogether, by determining the total outgoings of the fish and assuming that this is equal to the income. Winberg (1956) suggested that the metabolizable part of the income $(P+R)$, after deduction of F and U, was about 80% of the total, the metabolic rate in the wild (R) being about twice the routine rate measured in the laboratory. On this basis an estimate of the energy intake can be made simply by measuring growth, and estimating or measuring the routine rate of metabolism. The metabolic rate can be corrected for fish of differing size by applying an exponent of 0.8 to the weight. However, more recent studies have shown that the metabolism of active fish is rarely a constant multiple of the routine rate (Ware, 1978), and the weight exponent may vary with the degree of activity (Brett, 1965). Nevertheless, Winberg's equation has been applied in numerous studies of the energy budgets of fish, and with appropriate adjustments to the various assumptions can still provide a useful model, along with the more sophisticated versions developed in recent years (e.g. those of Kerr, 1971; and Ursin, 1979).

As an alternative to these indirect methods of estimation, which rely on manipulation of the energy budget equation, food intake can also be measured directly from the quantities of food contained within the stomachs of fish (Elliott & Persson, 1978; Mann, 1978; Windell, 1978). Bajkov (1935) first pointed out that a knowledge of the quantities of food digested by fish in a given time offered a means for estimating the food intake. Essentially, the rate of elimination or evacuation of food from the stomach equals, on average, the rate of gastric evacuation. Having measured the rate of gastric evacuation under differing conditions in the laboratory, and having observed the change in the stomach contents of wild fish over a given interval of time, it is possible to calculate the rate of food ingestion.

Laboratory experiments to measure gastric evacuation rates have generally employed standard diets, or have been confined to measurements on a few standard food items. Recent studies, reviewed by Windell (1978), have shown that the rate of evacuation is exponential in most species. There is a voluminous literature on this topic and for a critical review the reader should refer to Chapter 5. It is particularly important that these studies should take account of the various factors that can influence evacuation. These include the degree of food deprivation of the fish (see Chapter 5), the composition of the food (Elliott, 1979) and the water temperatures (Jones, 1974; Elliott, 1979), and perhaps also the size of fish (Pandian, 1967a) and the size of the food item. Application of the results of these experiments usually involves several assumptions, and especially that feeding proceeds at the same rate throughout the day. Moreover, sampling of the stomach presents a number of technical difficulties for fish in the field (Hyslop, 1980). One particular problem is that some species may regurgitate food material on being captured. Another is that predatory fish consuming large items relatively infrequently are difficult to sample representatively. There may be a bias towards the capture of hungry fish, with empty stomachs. Despite these difficulties, however, there have been several refined applications of this method. Following a series of gastric evacuation experiments on brown trout *Salmo trutta* (Elliott, 1972, 1975a), two elegant methods for estimating food consumption were outlined by Elliott & Persson (1978). Both methods relied on a series of stomach samples being taken at intervals over 24 h. The first method assumed that the feeding rate was constant, and the second assumed that this rate decreased with time within the period between samples. The latter assumption is more applicable to large fish, feeding close to the satiation limit. Both methods represent a great improvement over many of those previously applied, but they do rely on stomach sampling at short intervals, and they are really only applicable to species feeding on large numbers of small prey items, and showing a rapid elimination of food from the stomach. The method is not easily applied to the larger carnivores, which eat relatively large items infrequently, and which often have a slow elimination rate. Diana (1979) has recently considered this problem for a top carnivore, the northern pike *Esox lucius*, and has adopted a rather different method of analysis based on the assessment of meal size and frequency from field collections. Daan (1973), Jones (1974) and Hawkins *et al.* (1985) have estimated food consumption rates for the cod *Gadus morhua*, another predatory fish consuming large items, from the mean weight of food in the stomachs of a large sample of fish. In all these examples, ancillary information was required on the rate of evacuation, and the value taken for this rate had a great influence upon the estimates obtained.

Other methods which have been adopted for estimating the food requirements of fish have depended heavily upon the results of laboratory experiments. They have been reviewed by Ricker (1971) and Braaten (1979). Though they may involve assumptions which can be questioned, they are nevertheless valuable in providing an independent estimate of food intake. Of these methods, perhaps the most common is to estimate food consumption from the growth of the fish (Brocksen *et al.*, 1968; Davis & Warren, 1971; Carline & Hall, 1973; Elliott, 1975b). If the growth rate of fish is known for fish in the wild, then the food intake to produce that growth rate can be estimated from the results of experiments in the laboratory, where the growth rates of fish fed on different rations are observed. Ricker (1971), in reviewing this method, drew attention to several key precautions to be taken in applying it, and these have since been elaborated by other workers (Mann, 1978; Braaten, 1979). It is important that the laboratory feeding experiments should be conducted at the appropriate time of the year, under environmental conditions as close as possible to those in the wild. The laboratory animals should be in similar condition to fish from the wild at that time, and should show similar levels of metabolic activity; they should be neither more nor less active than their wild counterparts. The food consumed should be similar to that making up the natural diet, and should preferably not be treated in any way which would make it more readily assimilated, or different in composition or energy content. Growth in the laboratory should ideally be measured in units of energy, as well as weight, since fish fed on differing rations can vary greatly in their energy content, increasing in fat and decreasing in moisture content at the highest ration levels (Staples & Nomura, 1976; Hawkins *et al.*, 1985). It is usually necessary to feed the fish over a very wide range of ration levels, including starvation, maintenance (yielding zero growth), and to excess.

A great many growth/ration experiments have been performed for fish in the laboratory, and have been reviewed by Elliott (1979), Condrey (1982), and others. With a wide range of ration levels the growth can be either positive or negative around the maintenance level, and is usually expressed as specific growth rate (Braaten,

1979), to take account of size differences between the experimental fish. At a given temperature, for fish of a given size range, the relationship between specific growth rate and ration expressed as a percentage of body weight is linear (Condrey, 1982), as in the example shown in Figure 11.1.

Figure 11.1: The Relationship Between Specific Growth Rate in Weight and Ration Level for Juvenile Cod in Laboratory Feeding Experiments. (a) 7°C; (b) 10°C; (c) 15°C; (d) 18°C. Data from Soofiani (1983).

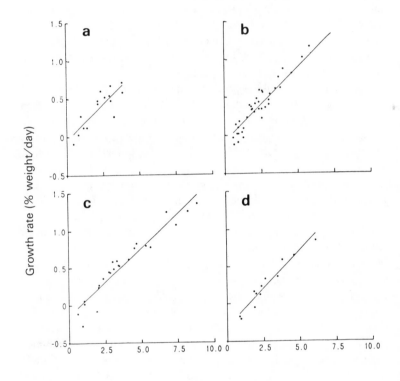

Ration (% weight/day)

The other information which is essential before food intake can be estimated by this method is a knowledge of the growth of wild fish. The measurement of growth rate will be discussed later, but it should be mentioned here that it is usually desirable to measure growth rates by two independent methods, to ensure their validity. Ideally, the comparison of growth rates in the laboratory and in the field should be done in energy terms though this is not always practicable.

Ricker (1971) has stressed the value of this method of estimating food consumption, which requires a minimum of experimental facilities. Perhaps the most difficult condition to fulfil is the matching of activity levels in the laboratory with those in the field, which are rarely known. Fish kept under confined conditions and fed individually cannot be expected to behave in the same way as fish in the wild, particularly predatory fish which must search out and attack their prey, and which may incur heavy metabolic costs in doing so. Despite these difficulties, Carline & Hall (1973) reported that the growth efficiencies of coho salmon (*Oncorhyncus kisutch*) held in artificial streams and in aquaria were remarkably similar, and suggested that this method could provide reliable estimates of food consumption in the wild. Moreover, Elliott (1975b) has shown that consumption–growth relationships for salmonids kept in aquaria do not differ greatly from those obtained from fish in model streams.

Food intake can also be estimated by constructing a nitrogen budget for the fish (Ricker, 1971; Braaten, 1979). The rate of nitrogen intake is estimated by summing estimates of the rate of nitrogen accumulated in growth, and the rate of nitrogen loss through defaecation and excretion (Birkett, 1969). This method has mainly been applied by Soviet workers.

A particular problem encountered in estimating the energy intake of fish in the field is the conversion of food intake, often derived simply in terms of weight, to units of energy. Where the estimate is based on stomach contents analysis, it may be necessary to capture and measure the energy content of particular food organisms, or alternatively to analyse the proportion of carbohydrate, fat and protein, and multiply by appropriate energy conversion factors (see Chapter 10). It is necessary to take account of seasonal changes in the energy content of the food, which can be pronounced.

Earlier, we mentioned non-feeding fish where the energy resources are derived mainly from breakdown of the body tissues. Here, the energy utilized can be measured by monitoring depletion of the body reserves. A good example of this approach is provided by Idler & Clemens (1959), in their studies of the changes in body composition of non-feeding salmonids migrating upstream. It should be stressed that there are problems in calculating the energy released, since a particular efficiency must be assumed for the energy conversion process.

11.2.2 *Measurement of the Energy Accumulated as Growth (P)*

A proportion of the energy accumulated by the fish is allocated to growth of the somatic and reproductive tissues. A comprehensive review of the various methods for studying growth in the field is beyond the scope of this chapter. There is an extensive literature on the growth of fish, though much of it is concerned with modelling growth in wild fish populations with the intention of producing maximum yield to man through fishing.

It is now commonplace for fisheries scientists to sample fish populations for the estimation of growth. The principal method is to measure mean lengths and mean live weights for fish at different ages. This method often assumes that fish of different ages and lengths are equally susceptible to capture. If there is any sampling bias during capture, say towards larger, older fish, then the mean length of fish of a given age tends to be an overestimate. Furthermore, this method relies on an effective method of age determination. The latter is most commonly carried out by reading the growth zones on otoliths, scales or other skeletal parts by methods which have been found satisfactory for many species of fish, but which are essentially based on the subjective interpretation of visible patterns of winter and summer growth. Difficulties in 'reading' otoliths and scales may introduce major errors into ageing, especially for tropical species where separation between the different zones may not be obvious. Even for temperate fishes, environmental changes and activities like spawning may alter the structure of these hard parts. In these circumstances it is advisable to employ an alternative means of determining the growth rate for wild fish in order to validate the age measurements (Weatherley, 1972). One alternative is to note changes in the length of fish which have been captured, tagged, released and subsequently recaptured. Comparison of the growth of cod measured by these two independent methods for a sea loch on the west coast of Scotland indicated that mean length and weight at age may provide an underestimate of growth — probably in this case because the method of capture was size selective (Hawkins *et al.*, 1985). In other circumstances one might expect tagged fish to grow slower, because of the effects of the tag in impairing growth.

If reliable data on mean length and weight at age or on the changes in length and weight of tagged fish with time are available, the specific growth rate of the wild population can be calculated. Taken in conjunction with laboratory data on the growth–ration relationship, this estimate of growth can provide an indirect measure of food consumption in the wild, as mentioned earlier. However, to estimate the energy retained as growth (*P*), which is the main purpose of the measurements, the growth (as length or live weight) must be converted into energy units. The simplest method is to multiply the growth in live weight by a single calorific value derived per unit weight of fish (from bomb calorimetry of homogenized fish carcasses). This method has been used by Healey (1972), examining the sand goby *Gobius minutus* in an estuary. The method assumes that the energy value of the fish remains unchanged during the course of the study, and sometimes throughout an annual cycle. This assumption is most unlikely to be valid for any wild population of fish and has been strongly criticized by Elliott (1979) and others. It is well established that the energy content per unit live weight of fish changes markedly with variations in body size and maturity, feeding rate, season and activity level of the fish (e.g. Brett *et al.*, 1969; Elliott, 1976a; Staples & Nomura, 1976; Craig, 1977; Flowerdew & Grove, 1980; Love, 1980; Soofiani, 1983; Weatherley & Gill, 1983). The changes are most pronounced for those species storing a great deal of lipid (like the salmonids and clupeids), but even with a species like the cod, the calorific content may vary between 21.5 and 24.1 kJ g^{-1} ash-free dry weight, depending on the ration size (Soofiani, 1983) and is also accompanied by changes in water content. Any estimate of growth in terms of energy units must make allowance for these pronounced variations in energy content, preferably by measuring the changes in energy content of fish of differing length and maturity stage, at different times of the year. This requires an elaborate programme of sampling, and careful treatment of the samples. It is especially important to homogenize the whole body of the fish examined because of variations in the energy content of the different tissues, including the viscera. Detailed sampling of this kind has been carried out for the northern pike *Esox lucius* (Diana, 1983) and for a population of benthic fishes (Small, 1975) under field conditions.

Where direct information on these changes in energy content are not available, all is not lost. The length/live weight ratio, or condition factor, of the fish is often a good indicator of changes in energy content, and if the variation of energy content with condition factor is known, from a few samples, or from laboratory studies, it may not be necessary to monitor the changes in energy content directly, from the wild population itself. Such a method has been used by Soofiani (1983) to calculate the energy accumulated as growth in populations of juvenile cod in Loch Torridon.

There are a number of alternative methods which can be applied to estimate the energy accumulated as growth if direct measurements are not available. One is to estimate the food intake, and then subtract appropriate levels of energy for the waste products and metabolism. The excess energy is then considered to be available for body growth and gonad production (Mills and Forney, 1981). The method is strongly dependent upon the accurate estimation of metabolism in the wild, which is usually the least certain component of the budget.

Gonad growth of mature fish in the field is usually determined directly by sampling fish at suitable intervals. It usually starts very slowly but ultimately becomes equal to the somatic growth, and becomes more rapid near the spawning season. At this time the entire energy surplus may be channelled to gonad production, and somatic growth may cease, especially in the female (Chapter 9). In addition, during the later stage of gonadal development, some of the body reserves, especially lipid and to a lesser extent protein, are catabolized to provide the necessary energy requirements for maturation. As a result of this phenomenon, a severe depletion of body reserves occurs after spawning (Beamish *et al.*, 1979). For example, in sockeye salmon, *Oncorhynchus nerka*, examined in a tributary of the Fraser River, British Columbia, the lipid content depleted by over 90% in the period between the onset of the spawning migration and the completion of spawning (Idler & Clemens, 1959).

Estimates of the energy expended in gonad production are usually made by comparing the mean energy content of the gonads between each sampling date. It is important to measure the actual energy content rather than weight because the calorific value of the gonad may change substantially during development.

11.2.3 Measurement of Energy Loss as Waste Products (F+U)

In preparing an energy budget for fish it is also necessary to determine the proportion of the consumed energy which is lost and is therefore not available for either growth (P) or metabolism (R). Primary sources of energy loss through the waste products are: first, energy lost in the faeces (F), including not only undigested food material but also sloughed intestinal epithelial cells, mucus, catabolized digestive enzymes, and bacteria. These faecal energy losses are reported to vary from 2 to 31% of the consumed energy (Elliott, 1979). The second source of loss is nitrogenous excretion (U), which represents the end-product of protein metabolism within the animal (Cowey, 1980). The main nitrogenous product in fishes is ammonia, which is largely formed in the liver and excreted passively at the gills (Wood, 1958; Forster & Goldstein, 1969; Payan & Matty, 1975). Ammonia is reported to form between 80 and 98% of the nitrogenous excretion of freshwater fish (Brett, 1962; Forster & Goldstein, 1969; Solomon & Brafield, 1972; McLean & Fraser, 1974; Brett & Zala, 1975), and between 75 and 85% in marine fish (Wood, 1958; Jobling, 1981a). Urea is the other major nitrogenous product and occasionally forms a high proportion of the total nitrogenous product, especially under hatchery conditions or for starving fish (Fromm, 1963; Brett & Zala, 1975). In general, the nitrogenous excretions (U) are usually the smallest component in an energy budget, accounting for only 7% of the consumed energy in carnivorous fish (Brett & Groves, 1979). For a comprehensive review of the energy losses through the waste products (U), and various methods available for determining U, see Braaten (1979). Brafield (in Chapter 10 of this volume) has also reviewed some of the methods available for measuring nitrogenous excretion under laboratory conditions.

The energy lost from the fish, whether as faeces or dissolved nitrogenous material, can seldom, if ever, be measured directly for fish in the wild. Both the particulate and dissolved matter are rapidly dispersed in the surrounding medium, and cannot be collected for subsequent examination. Inevitably, then, these components of the budget must be estimated from the results of laboratory experiments. For the faeces, estimates of the assimilation or absorption efficiency $(C-F/C)$ are required for the mixture of various food items and ration levels consumed. The estimates must be appropriate to the environmental temperature and other conditions prevailing in the wild, and must be derived from fish of appropriate size and condition, because some workers have found that the percentage of loss in the faeces is greatly influenced by these factors (e.g. Pandian, 1967b; Solomon & Brafield, 1972; Elliott, 1976b; Brett & Groves, 1979; Soofiani, 1983). Similarly, rates of nitrogenous excretion must also be estimated for fish consuming a similar diet to fish in the wild, again with the same reservations regarding the experimental conditions.

These two components $(F+U)$ for the waste products are relatively small, at least in carnivorous fish, and being based on laboratory experiments can be estimated relatively precisely. However, they essentially depend upon the food intake and are usually expressed as a proportion of this. In determining values in the wild it is therefore

especially important to ensure that the estimate of the energy intake
(C) is precise.

11.2.4 Measurement of the Energy Released in Metabolism (R)

A knowledge of metabolism (or the energy lost as heat) is basic to any
energy budget since it reflects the total energy expended by the fish in
basal metabolism and in activities like feeding and swimming, and can
comprise a large fraction of the total energy income. However, there
are problems for the measurement of this energy of metabolism in
field populations of fishes, so that this component is perhaps the
weakest link in the whole budget. Neither direct calorimetric
measurements, nor indirect measurements through the oxygen
consumption, are possible in the field since the fish can rarely if ever
be restrained or confined. Estimates of metabolism under field
conditions can only be approximate, and must inevitably involve a
number of assumptions.

It is convenient to separate the metabolism into a standard or
routine component (R_S), a component associated with feeding (R_F),
and a component associated with active movement (R_A). Standard
metabolism is measured as the minimum metabolic rate for an unfed,
resting fish (Fry, 1971), usually by monitoring oxygen consumption.
It is sometimes quite difficult to measure, because of the difficulty in
eliminating all spontaneous activity (Beamish & Mookherjii, 1964).
An indirect and more common way of estimating R_s is to extrapolate
to zero activity from a curve showing the oxygen consumption at
various levels of swimming activity (Brett, 1964; Fry, 1971; and see
Chapter 2). As an alternative, routine metabolism is the rate of
oxygen consumption observed in unfed fish whose activity is minimal.
It is likely that most experiments reported in the literature represent
measurements of low or intermediate routine metabolism. They are
usually obtained by monitoring the oxygen consumption over a range
of weights and temperatures (see reviews of Fry, 1971; Brett, 1972;
Weatherley, 1972; Brett & Groves, 1979). The relationship between
metabolism and fish weight follows a power law, values of the weight
exponent varying between 0.7 and 0.9, with an extreme range of
0.5–1.0 (Winberg, 1956; Beamish, 1964; Brett, 1964; Fry, 1971;
Kamler, 1976). In general, the weight exponent significantly
decreases as the age of the fish increases (Kamler, 1976) or as the
level of metabolism decreases (Brett & Glass, 1973; Elliott, 1976c;
Staples & Nomura, 1976). Much more complex and varied
relationships exist between temperature and metabolism, but

discussion of these is beyond the scope of this chapter (see Ege & Krogh, 1914; and Elliott, 1976c, for details).

The maximum oxygen consumption measured under laboratory conditions is usually that due to active metabolism. This is defined as the maximum rate of oxygen consumption in fish swimming at their maximum sustainable speed. It is usually measured in a tunnel respirometer (Brett, 1964; Fry, 1971). Of course, it should be noted that the maximum oxygen consumption may not always occur at the maximum sustainable swimming speed but can also be reached under other conditions. Soofiani & Priede (1985) have reported that in juvenile cod the maximum oxygen consumption occurs during the repayment of oxygen debt following exhausting exercise, rather than that at the maximum sustained swimming speed (Figure 11.2). The peak rate is some 40% higher than the peak level observed during sustained swimming. In the wild, most fish operate at neither their standard metabolic rates nor at their active rates, but somewhere in-between, depending on their level of swimming activity and their level of feeding.

Feeding in fish is accompanied by increased heat production and followed by an increase in the rate of oxygen consumption. This energy loss associated with feeding (R_F) is generally assumed to result from several causes, including excited locomotor and other activities, the mastication, digestion and absorption of the food, and the biochemical transformation of the absorbed material. A large element is undoubtedly associated with the deamination of proteins (Borsook, 1936; Beamish *et al.*, 1975), and this part is generally known as the specific dynamic action or SDA (but cf. Chapter 8). Beamish (1974) has referred to the whole increase in metabolic rate following feeding as the 'apparent SDA', making no distinction between the various contributing factors. The increase in metabolic rate following feeding is ration, temperature and weight dependent (see Averett, 1969; Beamish, 1974; Beamish *et al.*, 1975; Brett, 1976; Soofiani & Hawkins, 1982). Brett (1976) has developed a model displaying the metabolic rates as isopleths for all levels of feeding. Thus, given the food intake for fish in the field, the increase in metabolic rate above the standard rate can be estimated, at least for those few fish for which laboratory data are available.

The high energy demand following feeding to satiation can severely reduce the metabolic scope of the fish, leaving little capacity for performing any other activity (Vahl & Davenport, 1979; Jobling, 1981b; Soofiani & Hawkins, 1982). The scope for activity is greatest

Figure 11.2: Rate of Oxygen Consumption of Juvenile Cod Under Different Conditions of Activity and Temperature, Adjusted to an Arbitrary Standard Fish Weight of 150 g. ■, the maximum oxygen consumption recorded after exercise to exhaustion; ●, the minimum resting oxygen consumption in unfed fish; ○, the highest oxygen consumption recorded during sustained swimming in a tunnel respirometer. Redrawn from Soofiani & Priede (1985) with some alterations.

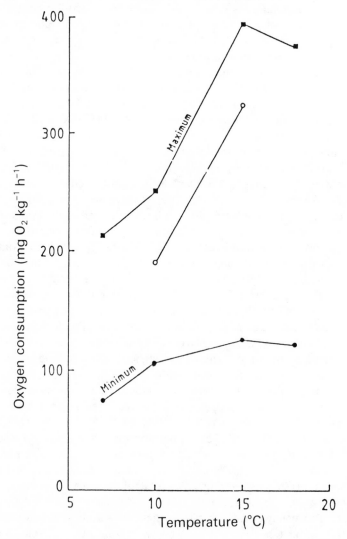

in fasting, non-swimming individuals, and is measured as the difference between the standard and active metabolic rates (Fry,

1971). Our own study of the feeding metabolism of juvenile cod fed to satiation showed that the metabolic rate greatly increased following feeding, resulting in a pronounced reduction in the scope for activity — by between 83 and 97% depending on the temperature. Indeed, the rate of oxygen consumption for cod fed to satiation corresponds closely to the active metabolic rate, suggesting that there is very little remaining scope. For a further discussion see Chapter 2 of this volume.

Estimates of the metabolic expenditure associated with feeding are generally obtained from laboratory experiments. These may then be applied to fish in the field, provided that the food intake of the fish is known. A number of other factors must also be considered, in particular the composition of the diet. The magnitude of increase in oxygen consumption may depend strongly on the biochemical composition of the food, the higher the protein content the higher being the magnitude (Jobling & Davies, 1980). It is also important in measuring R_F in the laboratory to ensure that the diet presented is physically similar to that eaten in the wild. A partly macerated, more highly digestible diet presented in the laboratory may lead to an underestimate of R_F for fish in the field, especially where the fish feed on large, entire, prey organisms. Since the estimates of R_F depend upon the magnitude of the food intake, it is especially important to ensure that the estimate of C is precise.

Perhaps the most difficult component of the metabolism to estimate is that devoted to activity (R_A), comprising that associated with swimming migrations, escape from predators, foraging and capturing prey, reproductive behaviour, depth changes and other aspects of performance. It can include the metabolic adaptations necessary to cope with low oxygen and other adverse environmental conditions, and with disease, injury or stress. In the absence of direct information on the respiration rates of fish in the field, it is generally necessary to assume that swimming is the most important component. The most important source of data in estimating these various components is then obtained from the various power–performance relationships derived for different species, showing the metabolic costs of swimming at different speeds. If the pattern of swimming by the fish is known (in the form of a distribution of the time spent swimming at different speeds), then it is possible to estimate the metabolic costs from the power–performance relationship.

There are few easy ways of examining the time spent swimming at different speeds. Cine film and video tapes of the fish are a possible source of data, but can only be used for relatively sedentary fish, moving within a restricted area. Feldmeth & Jenkins (1973) measured tail-beat frequencies for rainbow trout in a stream, and from a frequency–swimming speed curve obtained in a laboratory calculated actual swimming speeds. Acoustic and radio tags can provide data on movements on a larger scale, but are much less precise in the positional data they provide and can often only yield data at widely spaced intervals. The actual errors in measuring the distance travelled by the fish (reviewed by Hawkins & Urquhart, 1983) can be quite large. In brief, tracking really only provides the appropriate information for a fish which is in regular motion, swimming at a fairly steady speed and in a straight line. Sudden movements, like those involved in jumping, attacking prey, and escaping predators, may take up only a small part of the available time to animals, but may incur a heavy energy expenditure. Moreover, power–performance curves for fish are usually obtained in the laboratory with fish swimming at a constant velocity. Fish rarely swim smoothly in the field, but twist and turn and may suddenly burst into activity to capture prey or escape predators. It has been shown that an attack or escape episode involving 20 s at burst speed can be equal, in terms of energy expended, to 15 min of active metabolism (at the maximum sustained swimming speed), or about 3 h of standard metabolism (Brett & Groves, 1979). Thus, relatively brief episodes in the life of the animal may incur significant metabolic expenditure. These events may easily be missed by the field observer, thus leading to an underestimate of the total metabolic expenditure of the fish. Indeed, we have concluded from an examination of the energy budget of cod in the wild (Soofiani, 1983) that a large discrepancy in the energy expenditure (about 40% of the energy intake) must be attributed to foraging and other activities. The energy expended in activity can vary widely, even within the life cycle of a particular species. Brett's (1983) figures for sockeye salmon migrating up the Fraser River (when the fish are not feeding) suggest that in these fish the metabolic fraction of the energy budget is very large (75%) compared with the same fish swimming offshore (44%).

We have remarked that the energy expended in activity and hence in total metabolism can be only crudely estimated for most fish in the wild. It is possible, with the development of more advanced biotelemetry techniques, that parameters such as heart rate (Wardle & Kanwisher, 1974; Priede & Tytler, 1977; Priede & Young, 1977) or gill ventilation rate (Oswald, 1978) can be used to estimate the metabolic rate of free-living fish indirectly. It is also possible that some independent way of estimating metabolism in the wild can be developed. An example of such a technique applied to other invertebrates is measurement of the loss of particular ions from the animal, on the assumption that this loss is dependent upon the metabolic rate. These methods have the advantage that they take account of all the metabolic energy losses, not just those associated with physical movement through the water. The alternative is to estimate the total energy of metabolism (R) from the other components of the balanced energy equation.

11.3 Compiling the Energy Budget

It is beyond the scope of this chapter to review the many studies there have been of energy budgets for fish in the wild. Much of the data which have been accumulated are in any case specific to the animal under study, and cannot be generalized to other fish living under different circumstances. What we can do is underline the particular problems that have been encountered in compiling these budgets.

Few workers have been able to obtain direct measures of all the various components of the energy budget. The practical problems in doing so are far too great. One approach has been to measure the energy intake and then to subtract the estimated metabolism, faeces and nitrogenous excretion to provide a prediction of the surplus energy available for growth, of both the body and the gonads. Recent examples of this approach are provided by Mills & Forney (1981) and Minton & McLean (1982). An alternative has been to measure the growth rate, to correct for metabolism, and then to predict the energy intake, as carried out by Chekunova (1979) for the Baltic herring — a species for which direct measurement of the food intake presents enormous practical difficulties. Both methods have the disadvantage that they rely on estimates of metabolism for fish in the wild, obtained largely from the results of laboratory investigations. This metabolic component is almost certainly the weakest link in any budget because it is the one which is least amenable to direct measurement, for reasons already discussed. Conditions encountered in the laboratory are likely to reduce the energy expended in the search and capture of prey, in escape from predators, and in maintaining station in the presence of currents. A more sensible approach is to recognize this problem and to try to measure food intake and growth — both of which are amenable to measurement for all except the most difficult species. A prediction of metabolism can then be made for comparison with estimated values. At the very least it is usually possible to set lower and upper limits to the metabolic rate (corresponding to the standard and active rates of metabolism, respectively) which are usually known for the species. Clearly, if the predicted rate falls outside the range of possible values, then there are errors in the measurement of either food intake or growth or both.

It is not uncommon for energy budgets from wild fish to fail to balance. Diana (1983), in a study of the northern pike *Esox lucius*, compared both estimated and predicted food intake rates, and estimated and predicted growth rates. The estimated ration ranged from 30 to 160% of the predicted ration, depending upon the time of year. At the time of maximum growth, in summer, the ration was less than predicted. Diana concluded that the estimation of food intake is particularly prone to error (though, as we have discussed earlier, substantial errors can also arise in estimating growth). Any error in estimating the energy intake (C) will of course also affect the estimation of F, U, and R_F (the component of metabolism devoted to food processing).

Healey (1972), in his study of the sand goby *Gobius minutus*, also measured an energy intake which was less than the energy required to account for observed growth plus standard metabolism. As for the pike (Diana, 1983), the imbalance was especially evident during periods of rapid somatic growth. From an analysis of published feeding and growth studies, Healey concluded that energy imbalances for fish on low rations were commonplace, fish seeming to be able to shunt more energy into growth when on a restricted ration than one would predict from studies of standard metabolism. In our own study of juvenile cod we have concluded, like Diana, that the estimation of food intake is especially prone to error (Hawkins *et al.*, 1985). Healey, however, concluded that the observed energy shortage at low rations was a real phenomenon, the corollary of this being that levels of metabolism for fish in the wild may be extremely low. This conclusion is supported by the study of Feldmuth & Jenkins (1973) on the energy expenditure of rainbow trout *Salmo gairdneri* in a stream, where metabolism was estimated indirectly by counting caudal fin beat frequencies. The tail-beat frequency provided an estimate of the swimming speed. Then, from measured metabolic rates of fish at different speeds in a water tunnel, it proved possible to estimate the metabolic rates of the free-swimming fish. The food intake required to offset the energy expenditure by the fish was subsequently calculated. The actual food intake of fish in the stream, measured from the stomach contents, was less than the predicted requirement, suggesting that metabolism was being overestimated.

The results of these studies emphasize the need for better methods of estimating metabolic rates in the wild, whether by direct measurement of oxygen consumption, through indirect estimation of metabolic rates by the telemetry of heart-beat and ventilation rates or EMG (Priede & Tytler, 1977; Priede, 1983), or by other means. The simple doubling of the standard metabolic rate originally proposed by Winberg (1956) does not now seem appropriate, especially in view of recent measurements of the activity levels of fish in the wild. Young *et al.* (1972) demonstrated that brown trout in a small loch averaged swimming speeds of 0 to 0.14 body lengths per second, and Soofiani & Priede (1985) quote average swimming speeds of less than 0.1 body lengths per second for juvenile cod in a sea loch. At these speeds, the metabolic rates (derived by way of the oxygen consumption–swimming speed relationship for fish in the laboratory) are only a little above the standard rate. Several other authors have concluded that the metabolic costs of locomotion are minimal, and certainly less than suggested by Winberg (Warren & Davis, 1967; Kitchell *et al.*, 1977; Diana, 1979; Adams *et al.*, 1982; Minton & McLean, 1982). It is perhaps a mistake, however, to assume that swimming is the main component of active metabolism. We have already pointed out that brief bursts of activity can be particularly expensive. It is possible that foraging behaviour, which may involve an element of burst activity, forms an important component of active metabolism, as Adams *et al.* (1982) have suggested.

Some workers have reported imbalances in the energy budgets of fish which they have attributed to higher metabolic rates than estimated. Minton & McLean (1982), in their study of the sauger *Stizostedion canadense*, found that in winter the food consumption measured for the fish did not produce the growth expected, and suggested that this discrepancy might be due to increased activity metabolism at this time. An increase in field metabolism of some 3 to 4 times the standard rate would be required to account for the imbalance. Ware (1975) has also indicated that activity metabolism can be high, and suggested rates of about 3 times standard metabolism for juvenile pelagic fish. Large increases in activity metabolism are certainly possible under some circumstances. Brett & Groves (1979) suggested that the metabolic rate of upstream-migrating sockeye salmon was over 8 times the standard rate, but this is probably exceptional.

Not all workers have observed large imbalances in the energy budgets of fish in the wild. Though Minton & McLean (1982) reported large imbalances in the winter budget, in summer the predicted and observed growth rates of fish agreed within 2–10%. Mills & Forney (1981), in their study of yellow perch *Perca flavescens* in Oneida Lake, found good agreement between their predictions of growth and actual measurements.

Several general conclusions can be drawn from the energy budgets obtained from fish in the wild, or under free-ranging conditions in the laboratory. Carnivorous fish show a higher assimilation efficiency and lower growth efficiency than herbivorous fish, as Winberg originally suggested. It appears that more energy is lost through respiration by fish at higher levels in the food chain, both through the increased costs of processing protein, and perhaps also through increased foraging costs. The level of metabolism appears to be particularly high in actively swimming pelagic fish like the herring (Chekunova, 1979).

It has become evident, however, that the energy budget can change markedly over a seasonal cycle, or throughout the lifetime of the fish. Wootton *et al.* (1980, and see Chapter 9), have shown for the small stickleback *Gasterosteus aculeatus* that in winter only a small percentage of the energy income is invested in ovarian or somatic growth. Most of the energy is dissipated in routine metabolism. In spring a much greater proportion of the energy income goes into growth of the body and gonads, and respiration accounts for a smaller proportion of the energy income. Diana (1983) has stressed the differences that exist between the different sexes over the same seasonal cycle for the northern pike. Though the pattern of somatic growth is similar for both sexes, females ingest more energy annually than males, the extra energy being diverted into growth of the ovaries. There are large differences in the timing of somatic and gonadal growth for the two sexes. These results emphasize the need to examine the sexes separately, and to consider the relative changes in the somatic and gonadal tissues at different times of the year.

Brett (1983) has drawn attention to the pronounced changes in the energy budget of the sockeye salmon over the lifetime of the fish. In the freshwater lacustrine period, the salmon fingerling shows retarded growth and restricted ration, a large part of the energy accrued being dissipated in metabolism. Once in the sea, both food intake and growth are high, and though the fish may range widely in its movements, little energy is expended in locomotion. This situation changes again with the return of the adult to fresh water, where the main energy income is derived from breakdown of the somatic tissues, and the main outputs are the growth of the gonads and the massive expenditure of energy in swimming up river (Idler & Clemens, 1959). These studies of the sockeye salmon have been especially valuable in showing the value of energy budgets in promoting an understanding of the major selective forces acting upon an important food fish. We must remember, however, that the success in constructing energy budgets for the Pacific salmon is founded upon several decades of intensive physiological and ecological research. We can only hope that one day comparable bodies of data may exist for a whole range of fish.

References

Adams, S.M., McLean, R.B. and Parrotta, J.A. (1982), Energy partitioning in largemouth bass under conditions of seasonally fluctuating prey availability. *Trans. Am. Fish. Soc. 111*, 549

Averett, R.C. (1969), Influence of temperature on energy and material utilisation by juvenile coho salmon. Ph.D. thesis, Oregon State University, Corvallis

Bajkov, A.D. (1935), How to estimate the daily food consumption of fish under natural conditions. *Trans. Am. Fish. Soc. 65*, 288

Beamish, F.W.H. (1964), Respiration of fishes with special emphasis on standard oxygen consumption. II Influence of weight and temperature on respiration of several species. *Can. J. Zool. 42*, 177

Beamish, F.W.H. (1974), Apparent specific dynamic action of largemouth bass, *Micropterus salmoides. J. Fish. Res. Bd Can. 31*, 1763

Beamish, F.W.H. and Mookherjii, P.S. (1964), Respiration of fishes with special emphasis on standard oxygen consumption. I. Influence of weight and temperature on respiration of goldfish, *Carassius auratus* L. *Can. J. Zool. 42*, 161

Beamish, F.W.H., Niimi, A.J. and Let, P.F.K.P. (1975), Bioenergetics of teleost fishes: environmental influences. In: Bolis, L., Maddrell, H.P. and Schmidt-Nielsen, K. (eds). *Comparative Physiology — Functional Aspects of Structural Materials*, pp. 187–209. Amsterdam, North-Holland

Beamish, F.W.H., Potter, I.C. and Thomas, E. (1979), Proximate composition of the adult anadromous sea lamprey, *Petromyzon marinus*, in relation to feeding, migration and reproduction. *J. Anim. Ecol. 41*, 1

Birkett, L. (1969), The nitrogen balance in plaice, sole and perch. *J. exp. Biol. 50*, 375

Borsook, H. (1936), The specific dynamic action of protein and amino acids in animals. *Biol. Rev. 11*, 147

Braaten, B.R. (1979), Bioenergetics — a review on methodology. In: Halver, J.E. and Tiews, K. (eds). *Proc. World Symp. on Finfish Nutrition and Fishfeed Technology*, vol. 2, pp. 461–97. Berlin, Heenemann

Brafield, A.E. and Llewellyn, M.J. (1982), *Animal Energetics*. Glasgow, Blackie

Brett, J.R. (1962), Some considerations in the study of respiratory metabolism in fish, particularly salmon. *J. Fish. Res. Bd Can. 19*, 1025

Brett, J.R. (1964), The respiratory metabolism and swimming performance of young sockeye salmon. *J. Fish. Res. Bd Can. 21*, 1183

Brett, J.R. (1965), The relation of size to rate of oxygen consumption of sustained swimming speed of sockeye salmon, *Oncorhynchus nerka. J. Fish. Res. Bd Can. 23*, 1491

Brett, J.R. (1972), The metabolic demand for oxygen in fish, particularly salmonids, and a comparison with other vertebrates. *Respir. Physiol. 14*, 151

Brett, J.R. (1976), Feeding metabolic rates of sockeye salmon (*Oncorhynchus nerka*) in relation to ration level and temperature. *Envir. Can., Fish. Mar. Serv. Tech. Rep. No. 675*, 1

Brett, J.R. (1983), Life energetics of sockeye salmon, *Oncorhynchus nerka*. In: Aspey, W.P. and Lustick, S.I. (eds). *Behavioural Energetics: The Cost of Survival in Vertebrates*, pp. 29–63. Columbus, Ohio State University Press

Brett, J.R. and Glass, N.R. (1973), Metabolic rates and critical swimming speeds of sockeye salmon, *Oncorhynchus nerka*, in relation to size and temperature. *J. Fish. Res. Bd Can. 30*, 379

Brett, J.R. and Groves, T.D.D. (1979), Physiological energetics. In: Hoar, W.S., Randall, D.J. and Brett, J.R. (eds). *Fish Physiology.* vol. VIII, pp. 279–352. New York, Academic Press

Brett, J.R. and Zala, C.A. (1975), Daily patterns of nitrogen excretion and oxygen consumption of sockeye salmon, *Oncorhynchus nerka*, under controlled conditions. *J. Fish. Res. Bd Can. 32*, 2479

Brett, J.R., Shelbourn, J.E. and Shoop, C.T. (1969), Growth rate and body composition of fingerling sockeye salmon, *Oncorhynchus nerka*, in relation to temperature and ration size. *J. Fish. Res. Bd Can. 26*, 2363

Brocksen, R.W., Davies, G.E. and Warren, C.E. (1968), Competition, food consumption and production of sculpins and trout in laboratory stream communities. *J. Wildl. Mgmt. 32*, 51

Carline, R.F. and Hall, J.D. (1973), Evaluation of a method for estimating food consumption rates of fish. *J. Fish. Res. Bd Can. 30*, 623

Chekunova, V.I. (1979), Energy requirements of the Baltic herring, *Clupea harengus membras. J. Ichthyol. 19*, 118

Condrey, R.E. (1982), Ingestion-limited growth of aquatic animals: the case for Blackman kinetics. *Can. J. Fish. Aquat. Sci. 39*, 1585

Cowey, C.B. (1980), Protein metabolism in fish. In: Buttery, P.J. and Lindsay, D.B. (eds). *Protein Deposition in Animals*, pp. 271–287. London, Butterworth

Craig, J.F. (1977), The body composition of adult perch, *Perca fluviatilis*, in Windermere, with reference to seasonal changes and reproduction. *J. Anim. Ecol. 46*, 617

Daan, N. (1973), A quantitative analysis of the food intake of North Sea cod, *Gadus morhua* L. *Neth. J. Sea Res. 6*, 479

Davies, G.E. and Warren, C.E. (1971), Estimation of food consumption rates. In: Ricker, W.E. (ed.). *Methods for Assessment of Fish Production in Fresh Waters*, 2nd edn, pp. 227–48. IBP Handbook No. 3. Oxford, Blackwell

Diana, J.S. (1979), The feeding pattern and daily ration of a top carnivore, the northern pike, *Esox lucius. Can. J. Zool. 57*, 2121

Diana, J.S. (1983), An energy budget for northern pike, *Esox lucius. Can. J. Zool. 61*, 1968

Ege, R. and Krogh, A. (1914), On the relation between temperature and the respiratory exchange in fishes. *Int. Rev. Hydrobiol. 7*, 48

Elliott, J.M. (1972), Rate of gastric evacuation in brown trout, *Salmo trutta* L. *Freshwat. Biol. 2*, 1

Elliott, J.M. (1975a), Number of meals in a day, maximum weight of food consumed in a day and maximum rate of feeding for brown trout, *Salmo trutta* L. *Freshwat. Biol.* 5, 287

Elliott, J.M. (1975b), The growth rate of brown trout (*Salmo trutta* L.) fed on maximum rations. *J. Anim. Ecol.* 44, 805

Elliott, J.M. (1976a), Body composition of brown trout (*Salmo trutta* L.) in relation to temperature and ration size. *J. Anim. Ecol.* 45, 273

Elliott, J.M. (1976b), Energy losses in the waste products of brown trout, *Salmo trutta* L. *J. Anim. Ecol.* 45, 561

Elliott, J.M. (1976c), The energetics of feeding, metabolism and growth of brown trout (*Salmo trutta* L.) in relation to body weight, water temperature and ration size. *J. Anim. Ecol.* 45, 923

Elliott, J.M. (1979), Energetics of freshwater teleosts. In: Miller, P.J. (ed.). *Fish Phenology: Anabolic Adaptiveness in Teleosts*, pp. 29–61. *Symp. Zool. Soc. Lond. No. 44*. London, Academic Press

Elliott, J.M. and Persson, L. (1978), The estimation of daily rates of food consumption for fish. *J. Anim. Ecol.* 47, 977

Feldmeth, C.R. and Jenkins, T.M. (1973), An estimate of energy expenditure by rainbow trout (*Salmo gairdneri*) in a small mountain stream. *J. Fish. Res. Bd Can.* 30, 1755

Fischer, Z. (1979), Selected problems of fish bioenergetics. In: Halver, J.E. and Tiews, K. (eds). *Proc. World Symp. on Finfish Nutrition and Fishfeed Technology*, vol. II, pp. 17–44. Berlin, Heenemann

Flowerdew, M.W. and Grove, D.J. (1980), An energy budget for juvenile thick-lipped mullet, *Crenimugil labrosus* (Risso). *J. Fish Biol.* 17, 395

Forster, R.P. and Goldstein, L. (1969), Formation of excretory products. In: Hoar, W.S. and Randall, D.J. (eds). *Fish Physiology*, vol. I, pp. 315–50. New York, Academic Press

Fromm, P.O. (1963), Studies on renal and extra-renal excretion in a freshwater teleost, *Salmo gairdneri. Comp. Biochem. Physiol.* 10, 121

Fry, F.E.J. (1971), The effects of environmental factors on the physiology of fish. In: Hoar, W.S. and Randall, D.J. (eds). *Fish Physiology*, vol. VI, pp. 1–98. New York, Academic Press

Hawkins, A.D. and Urquhart, G.G. (1983), Tracking fish at sea. In: MacDonald, A.G. and Priede, I.G. (eds). *Experimental Biology at Sea*, pp. 103–66. New York, Academic Press

Hawkins, A.D., Soofiani, N.M. and Smith, G.W. (1985), Growth and feeding of juvenile cod, *Gadus morhua* L. *J. Cons. int. Explor. Mer.* (in press)

Healey, M.C. (1972), Bioenergetics of a sand goby, *Gobius minutus*, population. *J. Fish. Res. Bd Can.* 29, 187

Hyslop, E.J. (1980), Stomach contents analysis — a review of methods and their applications. *J. Fish Biol.* 17, 411

Idler, D.R. and Clemens, W.A. (1959), The energy expenditures of Fraser River sockeye salmon during the spawning migration to Chilko and Stuart Lakes. *Int. Pacif. Salm. Fish. Common. Prog. Rep. No. 6*, 80 pp.

Jobling, M. (1981a), Some effects of temperature, feeding and body weight on nitrogenous excretion in young plaice *Pleuronectes platessa* L. *J. Fish Biol.* 18, 87

Jobling, M. (1981b), The influences of feeding on the metabolic rate of fishes: a short review. *J. Fish Biol.* 18, 385

Jobling, M. and Davies, P.S. (1980), Effects of feeding on the metabolic rate and specific dynamic action in plaice, *Pleuronectes platessa* L. *J. Fish Biol.* 16, 629

Jones, R. (1974), The rate of elimination of food from the stomachs of haddock

(*Melanogrammus aeglefinus*), cod (*Gadus morhua*) and whiting (*Merlangius merlangus*). *J. Cons. int. Explor. Mer.* 35, 225

Kamler, E. (1976), Variability of respiration and body composition during early developmental stages of carp. *Polskie Archwm Hydrobiol.* 23, 431

Kerr, S.R. (1971), Prediction of fish growth efficiency in nature. *J. Fish. Res. Bd Can.* 28, 809

Kitchell, J.F., Steward, D.J. and Weininger, D. (1977), Applications of a bioenergetics model to yellow perch (*Perca flavescens*) and walleye (*Stizostedion vitreum vitreum*). *J. Fish. Res. Bd Can.* 34, 1922

Love, R.M. (1980), *The Chemical Biology of Fishes*. London, Academic Press

McLean, W.E. and Fraser, F.J. (1974), Ammonia and urea production of coho salmon under hatchery conditions. *Envir. prot. ser. Pac. Region, Surveillance Rep. Eps* 5-PR-74-5

Mann, K.H. (1978), Estimating the food consumption of fish in nature. In: Gerking, S.D. (ed.). *Ecology of Freshwater Fish Production*, pp. 250–73. Oxford, Blackwell

Mills, E.L. and Forney, J.L. (1981), Energetics, food consumption and growth of young yellow perch on Oneida Lake, New York. *Trans. Am. Fish. Soc.* 110, 479

Minton, J.W. and McLean, R.B. (1982), Measurements of growth and consumption of sauger, *Stizostedion canadense*: Implication for fish energetics studies. *Can. J. Fish. Aquat. Sci.* 39, 1396

Oswald, R.L. (1978), The use of telemetry to study light synchronization with feeding and gill ventilation rates in *Salmo trutta. J. Fish Biol.* 13, 729

Pandian, T.J. (1967a), Intake, digestion, absorption, and conversion of food in the fishes, *Megalops cyprinoides* and *Ophiocephalus striatus. Mar. Biol.* 1, 16

Pandian, T.J. (1967b), Transformation of food in the fish (*Megalops cyprinoides*) I. Influence of quality of food. *Mar. Biol.* 1, 60

Payan, P. and Matty, A.J. (1975), The characteristics of ammonia excretion by a perfused isolated head of trout (*Salmo gairdneri*): effect of temperature and CO_2-free ringer. *J. Comp. Physiol.* 96, 167

Priede, I.G. (1983), Heart rate telemetry from fish in the natural environment. *Comp. Biochem. Physiol.* 76A, 515

Priede, I.G. and Tytler, P. (1977), Heart rate as a measure of metabolic rate in teleost fishes: *Salmo gairdneri*, *Salmo trutta* and *Gadus morhua. J. Fish Biol.* 10, 231

Priede, I.G. and Young, A.H. (1977), The ultrasonic telemetry of cardiac rhythms of wild brown trout (*Salmo trutta* L.) as an indicator of bio-energetics and behaviour. *J. Fish Biol.* 10, 299

Ricker, W.E. (ed.) (1971), *Methods for Assessment of Fish Production in Fresh Waters*, 2nd edn. IBP Handbook No. 3. Oxford, Blackwell

Small, J.W., Jr (1975), Energy dynamics of benthic fishes in a small Kentucky stream. *Ecology*, 56, 827

Solomon, D.J. and Brafield, A.E. (1972), The energetics of feeding, metabolism and growth of Perch, *Perca fluviatilis* L. *J. Anim. Ecol.* 41, 699

Soofiani, N.M. (1983), An energy budget for juvenile cod, *Gadus morhua* L. in a Scottish sea loch. Ph.D. thesis, University of Aberdeen

Soofiani, N.M. and Hawkins, A.D. (1982), Energetic costs at different levels of feeding in juvenile cod, *Gadus morhua* L. *J. Fish Biol.* 21, 577

Soofiani, N.M. and Priede, I.G. (1985), Aerobic metabolic scope and swimming performance in juvenile cod, *Gadus morhua* L. *J. Fish Biol.* (in press)

Staples, J. and Nomura, N. (1976), Influence of body size and food ration on the energy budget of rainbow trout, *Salmo gairdneri* Richardson. *J. Fish Biol.* 9, 29

Steele, J.H. (1974), *The Structure of Marine Ecosystems*. Cambridge, Mass., Harvard University Press

Ursin, E. (1979), Principles of growth in fishes. In: Miller, P.J. (ed.). *Fish Phenology: Anabolic Adaptiveness in Teleosts*, pp. 63–87. *Symp. Zool. Soc. Lond. No. 44.* London, Academic Press

Vahl, O. and Davenport, J. (1979), Apparent specific dynamic action of food in the fish, *Blennius pholis. Mar. Ecol. Prog. Ser. 1*, 109

Wardle, C.S. and Kanwisher, J.W. (1974), the significance of heart rate in free-swimming *Gadus morhua*: some observations with ultrasonic tags. *Mar. Behav. Physiol. 2*, 311

Ware, D.M. (1975), Growth, metabolism, and optimal swimming speed of a pelagic fish. *J. Fish. Res. Bd Can. 32*, 33

Ware, D.M. (1978), Bioenergetics of pelagic fish. Theoretical change in swimming speed and ration with body size. *J. Fish. Res. Bd Can. 35*, 220

Warren, C.E. and Davies, G.E. (1967), Laboratory studies on the feeding, bioenergetics and growth of fish. In: Gerking, S.D. (ed.). *The Biological Basis of Freshwater Fish Production*, pp. 175–214. Oxford, Blackwell

Weatherley, A.H. (1972), *Growth and Ecology of Fish Populations*. New York, Academic Press

Weatherley, A.H. and Gill, H.S. (1983), Protein, lipid, water and caloric content of immature rainbow trout, *Salmo gairdneri* Richardson, growing at different rates. *J. Fish Biol. 23*, 653

Winberg, G.G. (1956), Rate of metabolism and food requirements of fishes. *Beloruss, State University, Minsk* (Fish. Res. Bd Can. Transl. Ser. No. 194, 1960)

Windell, J.T. (1978), Digestion and the daily ration of fishes. In: Gerking, S.D. (ed.). *Ecology of Freshwater Fish Production*, pp. 159–83. Oxford, Blackwell

Wood, J.D. (1958), Nitrogen excretion in some marine teleosts. *Can. J. Biochem. Physiol. 36*, 1237

Wootton, R.J., Allen, J.R.M. and Cole, S.J. (1980), Energetics of the annual reproductive cycle in female sticklebacks, *Gasterosteus aculeatus* L. *J. Fish Biol. 17*, 387

Young, A.H., Tytler, P., Holliday, F.G.T. and MacFarlane, A. (1972), A small sonic tag for measurement of locomotory behaviour in fish. *J. Fish Biol. 4*, 57

12 ENERGETICS AND FISH FARMING

B. Knights

12.1 Introduction

The primary aims of fish farming are to maximize survival and growth at minimal cost. These aims are particularly difficult to attain in intensive culture at high densities where input costs (particularly of food) are high. Information on optimal food quality (Chapter 6) and feeding regimes (Chapters 3 and 4) is essential, as is a knowledge of water quality, stocking density, water velocity, handling and other stresses which can affect growth. However, practical and economic constraints discourage objective studies on farms, particularly the problems of controlling variables in large-scale systems, lack of analytical facilities, labour costs and risks to stock. Thus growth has to be empirically related to water quality, feed regimes, etc. Fish-feed manufacturers often carry out basic research on food quality and feeding but these again usually rely on comparative growth experiments. In contrast, the extent of knowledge in and practical utilization of energetics approaches in the farming of poultry and ruminants, for example, is much more advanced (MAFF, 1974, 1975).

Bioenergetic studies of fish have largely been theoretical or applied to natural populations. Other chapters show that these have followed two main approaches: first, short-term studies of individual quiescent fish in carefully controlled laboratory experiments using natural foods, and, secondly, studies of wild populations taken together with extrapolations from appropriate laboratory studies. The thesis of this chapter is that bioenergetic approaches and the use of predictive energetic models are of potentially greater benefit in aquaculture than has generally been recognized. Realization of these benefits, however, is dependent on better liaison between experimental scientists and practical fish farmers and a better understanding of practical limitations and differences in emphasis, terminology and methodology.

A major benefit of energy budgets is that they can be 'instantaneous', i.e. carried out over relatively short time periods comparable in duration to biological and cultural cycles (e.g. one feeding cycle, one day, etc.). This is of particular benefit in making rapid predictions of effects on growth of a certain diet formulation or feed regime or an acute culture stress, such as handling. Responses to chronic stresses (such as low oxygen or high density) can also be studied in this way provided they are of relatively constant magnitude and adequate prior acclimation is allowed. Such approaches avoid long growth experiments which are prone, especially on farms, to interference from uncontrolled variables (e.g. due to feeder, pump or aeration failure) and take up much space, time and labour. Longer duration 'cumulative' budgets are of value with regard to longer cycles in nature, e.g. covering different phases in growth or seasonal changes in temperature and water quality. These studies are more prone to the problems experienced in growth experiments and an adequate picture can be gained from a series of integrated instantaneous budgets taken at appropriate intervals. Brett and co-workers have used this approach to model the energetics of sockeye salmon, *Oncorhynchus nerka* (Brett, 1983), and Schreck (1981, 1982) has derived models of the effects of stress on the energetics of smoltification; equivalent studies on farmed species have not been attempted. Detailed studies are obviously more easily controlled and managed in laboratory-scale systems but provided they adequately mimic conditions on farms, the information derived can be scaled up. It is particularly important to use comparable population size structures, densities, tank hydrodynamics, water qualities, food and feeding regimes. As many fish as possible should be used to compensate for individual variation and, perhaps, for mortalities. Those budget components which are difficult to measure and/or are known to be of minor importance can be derived from measurements of other parameters, as discussed in other chapters of this volume. The more comprehensive the study, however, the more complete and accurate the picture gained (Jobling, 1983b).

In this chapter, 'stress' refers to the environmental stimulus which induces a 'stress response'; the primary response involves stimulation of the nervous and hormonal systems which induce rapid homeostatic (secondary) changes in physiology, morphology and behaviour (Pickering, 1981; Schreck, 1981). Tertiary effects may result, e.g. reduced viability due to histopathological changes and increased susceptibility to additional stresses and disease. Feeding and absorption may also be affected. Taken together with the energy expenditure involved in stress responses, growth and reproductive potential can also be reduced. The magnitude and duration of these effects, during both the exposure and recovery phases, depend on the magnitude and duration of the original stress. Acclimation may occur with chronic exposure, however, and repeated acute exposures can lead to acclimation and/or habituation (Schreck, 1981). These factors are all of potential importance to fish farmers because of the stresses inherent in culture.

Most studies of stress in fish have, however, concentrated on primary hormonal changes and secondary changes in blood chemistry, haematology, tissue chemistry and histopathology (Wedemeyer & McLeay, 1981). Secondary physiological changes, such as diuresis (Lloyd & Swift, 1976) and changes in gas exchange (Smart, 1978), have also been measured. However, very few unequivocal relationships between these parameters and feeding, growth and survival have been established. The main confusing factors are the naturally high variability of individual fish (Hilge, 1980; Tomasso *et al.*, 1981; Miller *et al.*, 1983) and the stress caused in the very act of sampling fish for analysis. Furthermore, growth and feeding correlates have generally to be extrapolated from other individuals in an experimental population. Because of the significance of energy costs of homeostasis and the methodological advantages in constructing energy budgets, it is surprising that more use has not been made of bioenergetic approaches. Tertiary effects of feeding inhibition, poor conversions, poor growth, abnormal behaviour, high disease incidence and increased mortality have always been used by fish farmers as empirical indices of incipient stress; there is, however, scope for a more rigorous and objective approach. Studies on energy metabolism can be used to predict growth inhibition and suggest ways of avoiding or minimizing stresses and of aiding recovery. Cost–benefit analyses can then be used to make decisions regarding needs for extra aeration, benefits to be gained from changing management practices, etc.

To clarify the similarities and differences in emphasis, terminology and methodology between the 'scientific' bioenergetic and the more empirical farming approaches, this chapter will discuss the various components of the standard energetics equation, $C = F + U + R + P$ (Chapter 9). Areas where the application of bioenergetics could be of particular benefit to fish farmers will be discussed, with special reference to the limited relevant literature and work in the author's laboratory on intensive culture of European eels (*Anguilla anguilla* L.) in warm water.

12.2 Consumption (*C*), Defaecation (*F*) and Absorption (*A*)

The emphasis in aquaculture is to find the minimum amount of food (at the lowest price) required to produce the largest yield of marketable flesh in the shortest period of time. Commercial feeds used in intensive farming need to have an optimal nutrient composition, palatability and digestibility. Optimal feed regimes (meal size and feeding frequency) in relation to body size and temperature also need to be determined to maximize consumption and absorption, and production efficiencies. Relevant methodologies have been discussed in other chapters and in reviews by Smith (1979) and Utne (1979).

From studies on bioenergetics, it is possible to calculate absorption ($A = C - F$), absorption efficiency (A/C x 100%), and, together with growth, net (P/A x 100%) and gross production efficiency (P/C x 100%). The concepts of digestible energy (DE) and metabolizable energy (ME) are used extensively in studies of other farmed animals such as poultry (M.A.F.F., 1974) and ruminants (M.A.F.F., 1975). The former is equivalent to absorption (A), the latter is equal to DE minus energy losses in nitrogen excretion (and methane in ruminants), i.e. it represents the energy extracted from the feed which is actually available for maintenance and production. These parameters have been used in fish growth and nutrition studies but Jobling (1983b) has criticized their frequent misuse. Furthermore, many authors have applied incorrect DE and ME values or have calculated them from incorrect values of digestibility and absorption efficiencies. The main barriers to determining DE and ME accurately for different species under different culture conditions are the problems of measuring C, F and U, as discussed further below.

Absorption (A) is a useful measure because it indicates the energy potentially available from energy consumed (total or gross energy, C) for metabolism and growth. Brett & Groves (1979) and Elliott (1979) point out in their reviews that more information is required on the effects of temperature and body size but conclude that the assumption of Winberg (1956) that absorption efficiencies in fish are about 83–85% is generally too high for herbivores and often too low for carnivores. Instead, they (and Fischer, 1979) quote typical values of about 40–50% for herbivores and 70–90% for carnivores, with values tending to be larger in small fish and tending to increase with temperature and with falling ration size.

Few studies have been carried out with commercial diets, and high values might be expected because the components are finely ground and nutritionally balanced. As compared with studies involving natural foods such as insects and crustaceans, they usually contain a lower proportion of indigestible material. Elvers and eel fingerlings typically show absorption efficiencies in energy terms of 90–95% when fed on paste or dry commercial feeds (Tarr & Hill, 1978; Nielsen & Jurgensen, 1982; Knights, unpublished results).

Measuring food consumption and faecal production in large ponds, tanks, raceways or cages is as difficult as it is in natural water bodies because of the problems of (a) observing individuals feeding, and (b) collecting and separating uneaten food and faeces. Instead, farmers have to find a balance between ensuring enough food is made available while minimizing wastage. They are constrained to using feed conversion ratios (weight of food given : body weight gain) or food conversion efficiency (body weight gain : weight of food given). Some of the methods used to estimate food consumption in the wild might be applicable in culture (see Chapter 5). For example, sampling of stomach contents can be carried out, but accurate interpretation requires laboratory studies of evacuation rates and relationships to appetite, fish size, frequency of feeding and temperature. Non-destructive techniques involve use of radio-opaque labelled foods which tend to be unpalatable, and force-feeding and stomach-pumping are stressful. The captive nature of a farmed population allows sequential sampling for autopsy after feeding. However, this is wasteful of valuable stock, and for eels, commercial dry feeds liquefy rapidly in the stomach and are readily regurgitated (Knights, unpublished results). Stomach evacuation and subsequent absorption in eels are also rapid, making accurate measurement of consumption difficult.

Nitrogen balances can also be used but Ricker (1971) suggests that the method requiring fewest simplifying assumptions and the minimum of facilities is that which derives relationships between food consumption and growth in the laboratory and then extrapolates consumption from measurement of growth in the wild. This, however, assumes comparability of nutritional and metabolic states in the wild and in the laboratory and requires studies at a number of ration levels. Feeding and metabolic conditions are relatively constant for 'captive' farm fish, and laboratory studies can be designed to be more comparable from the outset.

One further indirect means of measuring consumption on farms is that based on measuring R (energy expended in respiration) and P (growth energy) for insertion into the basic energetics model for fish, derived by Winberg (1956), i.e. $pC = R + P$ (Jobling, 1983b). Here p is a correction factor for conversion of ration consumed to ration absorbed and available for growth and metabolism. Feeding metabolic rate is assumed to be twice resting metabolic rate. Winberg (1956) estimated p as 0.8. Other authors disagree with this figure (as discussed below), and ideally, p should be determined in any study using the equation (Jobling, 1983b).

A variant of this approach has been used in an aquaculture context by Kinghorn (1983) in studying genetic variation in food conversion efficiency and growth of rainbow trout (*Salmo gairdneri*) in fry tanks. Instead of the factor p defined above, metabolizable energy concentration (MEC) in the commercial pellet diet was calculated using the method of Phillips & Brockway (1959), i.e. by assigning calorific energy equivalents (corrected for average digestibility) to the dietary constituents determined by proximate analysis. Thus $C = (R + P)/\text{MEC}$; oxygen consumption was continuously measured in a semi-closed system (similar to that used by Staples & Nomura (1976) as discussed later) and oxycalorific coefficients were applied. The energy component of growth (P) was calculated following proximate analysis of fish sampled at the beginning and end of the growth period. It was assumed that energy lost in damaged tissues, mucus, etc. was negligible and that for a fixed rate of growth, factors affecting oxygen consumption (e.g. temperature and activity) had similar effects on food consumption. From earlier discussion and the critical review of Jobling (1983b), it could be suggested that this method of calculating C relies on too many assumptions and requires direct experimental verification. However, Kinghorn (1983, Table III) found calculated production efficiencies were equal to or only a little higher than those determined directly by other authors, suggesting that this method is in fact reasonably accurate.

Simpler open systems have been developed in the author's laboratory for studying eel fingerlings at high density (Figure 12.1).

Figure 12.1: Experimental System Used for Author's Studies on European Eel (*Anguilla anguilla* L.) Showing: A, clean water inlet to fingerling tank, excess water flowing to waste at B, waste food and faeces being collected in trap C; D, temperature control bath flow; E, pumped recirculation/aeration system with air supply via F; G, escape-proof mesh. Inset H: alternative system for collecting waste food and faeces, solids collecting on fine suspended mesh (E.A. Seymour, personal communication).

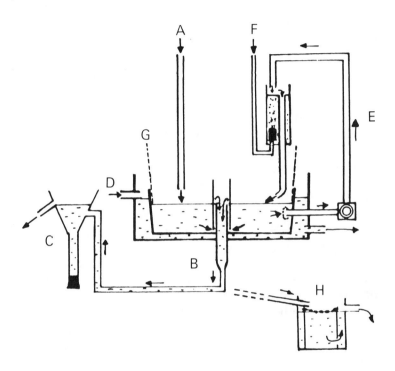

Fish were offered excess commercial paste feed in mesh baskets; these were removed after feeding and uneaten food was collected. Particles lost into suspension (and faeces voided) were swept by the vortical flow to the centre of the round tanks and out via the monk-type overflow. As current speed fell in the mouth of the funnel trap, particles sank to the bottom and were regularly drained off. This system has been further adapted (E.A. Seymour, personal communication) for use with dry granular feeds; as shown in the inset to Figure 12.1, the funnel trap was replaced with a fine nylon mesh suspended just above the water surface. Uneaten granules and faeces collected on the mesh, rather than soaking in water, which helped to minimize break-up and leaching of constituents. Dry particulate feeds have advantages in that they are fairly stable in water and can be sieved into constant size groups and matched to fish size (Knights, 1983). They can also be marked with non-absorbed lipid dyes which produce coloured faeces so that the passage through the gut of single marked meals can be followed. A further advantage is that particles of known size from a discrete batch of feed have almost constant chemical compositions and energy contents. Faecal strings can be easily separated from uneaten food particles although collections and separations have to be made regularly to avoid leaching of constituents (Brett & Groves, 1979).

Open systems and particulate feeds also allow observation of individual fish. For example, eels have been marked with sub-epidermal injections of acrylic paint (Knights, unpublished results) and the number of granules consumed by each fish in a feeding period could be recorded. Even when food was offered to excess, the more aggressive (and usually larger) eels ate more and feeding was inhibited in smaller subordinates. These observations help to explain the marked differential growth rates in farmed eels; social interaction and dominance hierarchies therefore need to be taken into account in any high density situations, particularly at low feeding levels. Jobling & Wandsvik (1983) certainly found this to be a confusing factor in studies of food intake and growth of Arctic charr (*Salvelinus alpinus*), where the larger more dominant fish showed hyperphagia and highest growth rates at low feeding frequencies.

12.2.1 Effects of Stress

The effects of stress on C, F and A have been little studied although it is generally recognized that appetite can be inhibited and absorption and growth possibly affected by them. More work is needed, particularly on chronic and acute effects in farming situations.

As an example of the approach to studying appetite inhibition, M. Collins (unpublished results) acclimated eels to 25°C (close to the optimum growth temperature determined by Sadler, 1979) and fed them regular meals to excess to determine average time to satiation and maximum ration. He then subjected eels to an increase or decrease in temperature of 5 or 10°C for 1 h (such acute stresses are not uncommon in intensive culture of eels in warm-water effluents such as those from power-station cooling processes). Return of appetite during the recovery phase was then measured by offering food at regular intervals and noting the quantity of food consumed per unit time. The results of such studies indicate how feed ration and timing can be adjusted to avoid wastage. Ideally, effects of stresses on absorption need to be assessed too, particularly with chronic stresses or where acute stresses occur just after feeding.

Handling of fish can also affect appetite. For example, variable inhibition patterns occur in salmonids. Pickering *et al.* (1982) found that brown trout, *Salmo trutta*, refused to feed for three days after a standardized handling stress; Wedemeyer (1976) quotes young coho salmon (*Oncorhynchus kisutch*) losing appetite for 4 to 7 days after handling, but rainbow trout (*Salmo gairdneri*), under identical conditions, resumed feeding the next day. It must be emphasized, however, that although appetite may be inhibited in the short term and energy intake reduced, long-term growth effects may be negligible (e.g. Pickering *et al.*, 1982). Physiological adaptations and hyperphagia might occur to maintain growth rates and thus the recovery phase following an acute stress or the acclimation occurring during chronic stress needs to be studied.

12.3 Excretory (Nitrogenous) Energy Losses (*U*)

Loss of energy by this route has generally been found to be minor in comparison to energy used in respiration, as discussed below. Acceptance of Winberg's (1956) assumption that the energy value of ammonia is negligible has, however, led some authors to overestimate potential net energy available from food consumed. Loss of nitrogen in excretion is also important in aquaculture because, together with faecal nitrogen, it represents a potential reduction in protein accumulation (Cowey & Sargent, 1972; Halver, 1972; Steffens, 1981). Increases in ambient ammonia can also affect growth and survival (Wickins, 1980; Colt & Armstrong, 1981; Parker & Davis, 1981; Smart, 1981). It is argued below that there is a direct link between high ambient ammonia and energy loss via hypermetabolism.

Accurate measurement of ammonia at low concentrations is a major problem, and intermittent stop-flow techniques are commonly used to accumulate sufficiently high levels (e.g. by Solomon & Brafield, 1972, and Brett & Zala, 1975). Continuous absorption from tank outflows on a resin column has also been used by Ogino *et al.* (1973). Few workers have analysed urea nitrogen (UN) as well as total ammonia nitrogen (TAN), although it can be an important additional component of nitrogenous excretion in some species under certain conditions. Furthermore, the rate of excretion of TAN and/or UN can change markedly with time (tending to reach maxima some time after feeding) and with changes in ambient ammonia levels. For example, Brett & Zala (1975) found a constant rate of UN excretion of about 2.2 mg kg^{-1} in actively swimming sockeye salmon, but TAN varied between 8.2 mg kg^{-1} in starved fish to a maximum of about 35 mg kg^{-1} 4 to 4.5 h after feeding. At high ambient ammonia levels, rainbow trout were also found to excrete low and relatively constant levels of UN although TAN excretion was inhibited (Olson & Fromm, 1971); in contrast, goldfish in similar experiments increased UN and maintained their rate of total nitrogen (TN) excretion. Even salmonids, however, exhibit variations in UN excretion in culture, as shown by farm studies of fingerling Chinook salmon (Burrows, 1964) and coho salmon (McLean & Fraser, 1974); less UN and more TAN appeared to be excreted at high stocking densities and temperatures and low water exchange rates. These results conflict with those for goldfish (Olson & Fromm, 1971) and eels (Knights, unpublished results) where UN increases with falling water quality and rising ambient ammonia. Experiments on eels (discussed further below) show that salinity may have to be taken into account because UN excretion is higher in fresh water than in sea water (by 120–180% in unfed fish and by 300–330% on one satiation meal of commercial feed per day). This might result from the greater potential for flushing out urea in the higher urine flow rates expected in an hyposmotic medium (Eddy, 1981; Hunn, 1982).

Relationships between high ambient ammonia and energy expenditure are discussed further below. Results of studies on eels (Knights, unpublished results) have shown, however, that U was a relatively minor and constant proportion of the energy budget, i.e. about 2% of energy consumed in sea water and 3% in fresh water (Table 12.1). These figures agree fairly well with those of Nielson & Jurgensen (1982) for eels and those proposed by Winberg (1956). They are , however, lower than the average of 7% for young carnivorous fish calculated by Brett & Zala (1975). The lower losses in eels may be due to the use of commercial diets and to nitrogen balance factors as discussed below.

Nitrogen budgeting is worth further consideration because the protein component of commercial diets is a major cost factor (Steffens, 1981). Although this usually involves complex analyses of food inputs and faecal and excretory outputs of nitrogen, the constant proportion of protein in commercial feeds means that reasonably accurate estimates of nitrogen input can easily be made if the amount of food eaten is known. Furthermore, if energy and nitrogen absorption efficiencies are high, faecal losses can be assumed to be negligible. Thus only measurement of excretory nitrogen is needed.

In farm studies, accurate measurements of protein retention are not usually feasible, and instead efficiency of protein utilization has to be calculated as increase in body weight $\div$ protein intake, i.e. the protein efficiency ration (PER). A more sophisticated and informative measure is increase in body protein $\div$ protein intake, i.e. the productive protein value (PPV) or apparent net protein utilization (NPU_a). Apparent biological value (BV_a) can also be calculated if nitrogen losses are measured, i.e. increase in body protein $\div$ protein intake – faecal and excretory nitrogen. Growth periods of 2 to 3 months are needed to determine PER, PPV and BV_a accurately (Utne, 1979; Steffens, 1981). However, from the arguments above, it can be seen that instantaneous increases in body protein are easily estimated from determinations of food consumption and nitrogen excretion. Thus, if growth is also estimated (or measured directly), PER and PPV can be calculated. Many assumptions and indirect calculations are involved but the final estimates are of value, especially for comparative purposes.

The aim in fish culture should be to produce PER and PPV values as high as possible and so cost–benefit analyses are needed to determine the appropriate balance between nitrogen retention efficiency and growth. For example, Ogino *et al.* (1976) found that when feeding casein to juvenile trout the maximum PPV values were obtained at a dietary protein level of 21% but growth rate was maximal at 55%. Higher protein levels generally lead to greater utilization of protein for energy metabolism but increasing the proportion of carbohydrates or high quality fats as alternative energy sources can have a beneficial protein sparing effect (Ogino *et al.*, 1976). Steffens (1981) quotes the example of an optimum diet for growth of juvenile rainbow trout consisting of 35% protein and 18% lipid. Carbohydrates would be expected to be more beneficial for omnivores and herbivores in protein sparing, as shown for juvenile carp by Takeuchi *et al.* (1979).

Changes in nitrogen budgets occur with age because of the falling protein requirement for growth as fish mature. Temperature is also important because of its effects on digestive and metabolic efficiencies and substrate usage. Atherton & Aitken (1970), for example, found that rainbow trout increased nitrogen retention and growth at higher temperatures and with higher lipid contents in their diet. This correlated with a reduction in nitrogen loss in faeces (from 6–15% to about 3% of intake) and in excretion (from 40–60% to 10–20% of intake). Nitrogen retained relative to weight gained rose from 20–30% to as high as 90%.

12.3.1 Effects of Accumulation of Ammonia

The studies described above all suffer the disadvantages that they entail long periods of observation and there is much to gain from using the instantaneous integrated approach discussed previously. An investigation of the effects of high ambient ammonia on energy and nitrogen budgets in eels (*Anguilla anguilla*) will be used to illustrate this (Knights, unpublished results). The basic aim of the work was to examine the effects of reduced water exchange rates occurring on eel farms due to lack of water, a need to reduce pumping costs or a need to recirculate water to maintain temperature. These are common problems in warm-water culture of eels in power station or industrial process cooling effluents, and are not unknown in other culture systems.

Supplementary aeration can maintain dissolved oxygen but waste ammonia can accumulate. To imitate this situation, exchange rates were reduced in successive experiments in the system shown in Figure 12.1 from over 2 to 0.15 litres min^{-1} kg^{-1}. Experiments were carried out in fresh water and sea water (30‰ at 25°C. Fish were fed once daily to satiation on paste feed over an acclimation period of at least 2 weeks, extra aeration being provided as necessary to maintain dissolved oxygen above 80% saturation. Oxygen consumption and nitrogenous excretion (as well as food consumption, and, over 24 h, faecal output) were regularly monitored (see Figure 12.2 and 12.3 for data from seawater experiments). Ammonia accumulated during the post-prandial period but pH was progressively depressed, presumably due to increases in carbon dioxide and acid metabolite concentrations in the water. This influenced the equilibrium reaction between ionized (NH_4^+) and un-ionized (NH_3) ammonia in fresh water (Trussell, 1972) such as to reduce the relative concentration of the more toxic NH_3 (Smart, 1981). Similar effects were seen in seawater experiments (Figure 12.3) but un-ionized ammonia (UIA) levels were generally lower, as would be expected from the effects of salinity on the equilibrium reaction (Whitfield, 1974; Bower & Bidwell, 1978). Urea excretion showed a marked tendency to increase under

Figure 12.2: Author's Eel Experiments (unpublished results). Oxygen consumption and total ammonia nitrogen (TAN) and urea nitrogen (UN) excretion rates at different water exchange rates in seawater (30°/₀₀) experiments. Mean control values are shown as black dots. See text for further explanation and discussion.

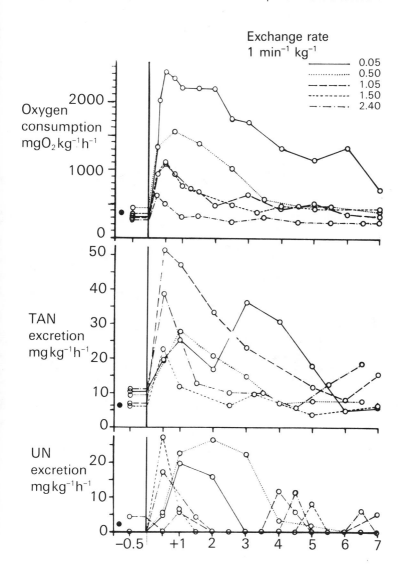

ammonia stress during the post-prandial peaks. This helped to

Figure 12.3: Author's Eel Experiments (unpublished results). Ambient total ammonia nitrogen (TAN), pH and un-ionized ammonia nitrogen (UIAN) at different exchange rates in sea water (30°/₀₀) experiments. Mean control values are shown as black dots. See text for further explanation and discussion.

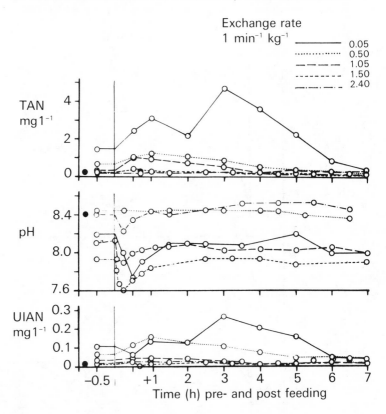

maintain total nitrogen excretion rates in the face of unfavourable gradients for diffusive loss of UIA and $Na^+ : NH_4^+$ active exchange at the gills (Kormanik & Cameron, 1981; Evans *et al.*, 1982). As noted previously, UN was generally more significant in freshwater experiments. Nitrogen intake was calculated by dividing the weight of protein ingested (51% of dry feed) by 6.25, and nitrogen loss was calculated by graphical integration of TN excretion with time. Faecal losses were assumed to be negligible, as discussed earlier. Nitrogen intake was variable but was generally about 10–20% higher in fresh water at all exchange rates (although weight and energy absorption efficiencies were very similar in both media). As ambient ammonia concentrations increased, high variability was seen in UN and TAN excretion rates (e.g. as in Figure 12.2) but TN rates showed a smoother pattern with progressive suppression of peak values and increases in duration. The result was that similar amounts of nitrogen were lost through excretion in the post-prandial period. PPV thus stayed fairly constant at about 60–70% in fresh water and 70–80% in sea water. These results do, however, imply that protein retention is generally lower in fresh water, possibly because more protein has to be metabolized to provide the higher energy demands of osmotic and ionic regulation and/or because of higher Na^+: NH_4^+ exchanges at the gills. As might be expected, growth rates were lower in fresh water than in sea water (by about 30–40% in both media). PER values fell from about 2.2 in fresh water and 2.6 in sea water to about 1.8 and 2.1, respectively. However, since protein ingestion and PPV were relatively unaffected by ambient ammonia, the conclusion must be that only part of the observed inhibitions of growth were due to nitrogen losses. Other explanations must be sought and a correlation appears to exist between growth inhibition and considerable increases in oxygen consumption (Figure 12.2). Furthermore, energy for this wasteful hypermetabolism must come from substrates other than protein. Explanations for these phenomena are discussed below but this example has illustrated how nitrogen budgets can be derived secondarily from studies of energetics and can be of relevance in aquaculture.

12.4 Respiratory Energy Loss (R)

Total energy of metabolism comprises R_S (standard metabolism), R_F (additional energy released during feeding excitation and activity, digestion, absorption and biochemical transformations) and R_A (additional energy used in swimming and metabolic responses to stress). Earlier chapters have shown that most laboratory studies have concentrated on single fish in quiescent states such that they measure largely R_S. Some studies have measured R_F, primarily with reference to apparent or real specific dynamic action (SDA); others have attempted to produce known levels of activity and hence R_A by forced swimming at known velocities. Field studies have involved measurement of swimming speeds and distances covered (e.g. by telemetry), allowing extrapolation from laboratory measurements of R_A to calculate energy expenditure in the wild (Chapter 11).

Relatively few studies have attempted to measure all components in a single system and even fewer have attempted integrated studies of the components ultimately of most significance in fish farming, i.e. R_F and R_A. Not only are these valuable in assessing the effects of culture variables on growth, but measurements of oxygen consumption also yield useful information on oxygen requirements.

12.4.1 Methodology

Direct methods of measuring oxygen consumption commonly rely on measuring the fall in oxygen with time in a sealed vessel or the fall between inlet and outlet (via the Fick principle) in flow-through respirometers and then application of oxycalorific coefficients (Chapter 10). These methods are difficult to apply in systems with high fish densities and flow rates and the need for easy access for feeding, cleaning, etc. The closed respirometer system developed by Brett *et al.* (1971), described earlier, has much to recommend it for studies of energetics, but simpler open-tank systems can be used. For example, Staples & Nomura (1976) measured oxygen consumption of rainbow trout in open rectangular tanks (1 x 1 x 5 m, 5000 litres capacity) by measuring the drop in dissolved oxygen (DO) between inlet and outlet. They assumed from control experiments that oxygen absorption from the air was negligible, but did not take into account the extra absorption that will occur as DO is reduced by the fishes' own respiration. This means that the R values that they calculated are probably underestimates at higher levels of metabolism. To account for oxygen absorption in the eel tanks described earlier (Figure 12.1), the water was deoxygenated before and/or after experiments by nitrogen scrubbing, and the rise in DO against time was plotted at the levels of flow, aeration and turbulence used in each experiment. After appropriate replication, a line of best fit was plotted by eye and the slope at any particular DO represented the rate of reoxygenation (milligrams of oxygen per minute). Oxygen consumption could then be calculated by drawing a tangent to the curve at experimentally determined DOs; i.e. by assuming that the rate of oxygen consumed was equal to the rate of reoxygenation at that DO. Such techniques are used by engineers to assess efficiencies of aeration systems, and have been used on large tanks on an eel farm (Aston *et al.*, personal communication). Sodium sulphite was used to deoxygenate the water. These methods appear to be reliable and reproducible, and adequate for comparative energetic studies; in the eel study, metabolic rates were comparable to published values in the literature.

12.4.2. *Active Metabolic Energy* (R_A) *and Stress*

Standard (R_S) and maintenance metabolic energy expenditure have little direct relevance in fish farming but are useful for comparative purposes in determining metabolic scope in relation to higher levels of metabolism due to feeding activity and to stress. Brett & Groves (1979) quote an average standard metabolic rate for a wide range of fish of about 90 mg O_2 kg^{-1} h^{-1} (range 26–299 mg O_2 kg^{-1} h^{-1}) and a Q_{10} (increase in rate for a 10°C rise in temperature) of about 2.3. At high levels of maintained activity, R_A is typically 8 to 10 times R_S with possibilities of increases up to 1000 times R_S in short, anaerobic bursts (see Chapter 2). Brett and co-workers (see Brett & Groves, 1979, and Brett, 1983) have accumulated a considerable body of data and derived very useful models for relationships between swimming speeds, temperature, body size and metabolism for salmonids. More information and models of this type are badly needed for farmed species.

Swimming activity is a major component of energy expenditure, especially in intensive tank or raceway systems with high water velocities. Temperature is also important because a rise accelerates metabolic rate but reduces oxygen solubility. Temperature preferenda, growth optima and oxygen requirements are generally well researched and understood for most commercial species because they are such fundamental factors. Less is known, however, about the effects of acute temperature shocks. The same is true for acute hypoxia. Critical chronic oxygen levels below which growth is significantly inhibited are often quoted for particular species (e.g. see Smart, 1981) but in most cases, objective proof is still lacking.

Other important factors in fish farming are the sensory disturbances associated with general management (Schreck, 1981) and particularly the sensory and hypoxic stresses due to handling and transport (Wedemeyer, 1976; Schreck, 1981; Pickering *et al.*, 1982). Increasing population density can also be stressful but some acclimation probably occurs (Fagerlund *et al.*, 1981; Schreck, 1981; Feldmeth, 1983; Jobling & Wandsvik, 1983). Problems are exacerbated if agonistic social interactions occur and territories and/or dominance hierarchies are established; subordinate fish may be more stressed and show a relatively higher R_A. Conversely, dominant fish may expend excessive energy in aggressive activities and territoriality (Li & Brocksen, 1977). There is a dearth of knowledge on this topic for fish in culture. A major difficulty is that R_A cannot be directly determined for an individual in a population. This can only be achieved by observing activity levels and extrapolating from controlled studies of energy costs at comparable levels of activity (Feldmeth, 1983). This still does not account for any energy expenditure due to responses to psychological stress which do not involve hyperactivity. Peters *et al.* (1980), for example, demonstrated that social stress in confined pairs of eels led to variable but significant physiological and morphological changes in subordinates, including a decrease in liver glycogen and an increase in blood glucose and lactate. Effects on food consumption and, ultimately, on growth in populations also need to be studied. In behavioural studies of marked eel fingerlings, for example, individuals can be assigned to different levels of apparent dominance/subordinance by observing agonistic behaviour (Knights, unpublished results). Simultaneous observations of activity and feeding success and measurement of growth showed that subordinates (generally smaller fish) tended to spend more time swimming in the water column rather than resting on the bottom, ate less food and showed poorest growth rates. No energetic studies were made but the results suggest a correlation between high R_A and low C and P and links with the marked differential growth rates seen in eels in culture.

For all the stresses discussed above, measurements of the magnitude and duration of increments of R_A (and of negative increments in C if feeding is inhibited) are of potential benefit in assessing relative energy costs inherent in the stress responses, both during initial adjustment, acclimation (if it occurs) and during any recovery phase. Not only will this yield information on growth effects but it will also provide information on how feeding, aeration and general management practices can be adjusted to avoid or minimize stress or at least to promote recovery and avoid imposing further stresses (c.f. Speece, 1981). However, there is a problem in assessing R_A from oxygen consumption and predicting oxygen needs if fish use anaerobic metabolism. An oxygen debt may be accumulated and this is where study of the recovery phase, when the debt is being paid off, is particularly important. Care also needs to be taken in selecting appropriate oxycalorific coefficients because different substrates may be used at different times. Measurement of CO_2 excreted can be used to calculate RQ (respiratory quotient = CO_2 produced/O_2 consumed) to give crude indications of the substrates used (Elliott & Davison, 1975; Kutty, 1968, 1981). Ammonia quotients (AQ = TN excreted/O_2 consumed) are also supposed to indicate the proportion of energy being derived from protein metabolism (Kutty, 1978, 1981).

12.4.3 Feeding Metabolic Energy (R_F)

The distinctions between feeding excitation and activity and the metabolic processing (specific dynamic action, SDA) components of R_F (apparent SDA) are often difficult to make. They could be of great significance in culture because of possible relations to growth and post-prandial oxygen demands. For a range of species, Brett & Groves (1979) quote average increases in feeding metabolic rate over standard and routine rates of 3.7 and 1.7, respectively; Jobling (1981) quotes increases of about twice low routine rates. Winberg's (1956) estimate of an increase of twice the standard rate appears to be an underestimate and more applicable to the increase above routine rate. However, in studies of *Anguilla anguilla* at 25° (Knights, unpublished results), feeding metabolic rates were only about 1.6–1.7 times standard and 1.2–1.3 times pre-feed 'routine' rates. Durations of apparent SDA were also short (about 2 h), with the result that the magnitudes of energy expenditure per meal were low, R_F costs being about 2% of ingested energy in sea water and 5% in fresh water. After subtracting energy expended in feeding excitation and activity, true SDA costs would be expected to be very low. The eels were feeding to satiation on one meal per day, consuming about 2.2–2.7% of body weight in dry food, which is about average for eels of similar size (2.5 g) (E.A. Seymour, personal communication).

Other studies in the literature have obtained much higher R_F costs of about 9–20% of ingested energy per meal for a variety of fish species feeding on natural foods (Brett & Groves, 1979; Jobling, 1981). Magnitudes of R_F may be affected by interspecific body size or temperature differences, but Staples & Nomura (1976) suggest that, because of the ease of finding, consuming and digesting high-energy commercial diets, less energy has to be expended in feeding and metabolic processing than is the case with natural foods. Furthermore, they contend that the energy costs of processing food relative to energy intake are probably not affected much by body size, a view supported by Brett & Groves (1979).

Jobling has argued that SDA represents the high energy costs entailed in elevated rates of protein synthesis and turnover following feeding, i.e. that R_F is an inescapable cost of growth (Jobling, 1981, 1983a). He expands his arguments in Chapter 8 and suggests SDA and growth might be interactive (i.e. that high rates of growth will be reflected in high rates of metabolism and high SDA) rather than competitive (i.e. that a diet inducing a high SDA will reduce the amount of food energy available for growth). The relationships between growth efficiency and R_F costs in fish may in fact be more easily explained on the basis of levels of nitrogen turnover and excretion. Thus, if the balance of amino acids in the diet is poor, and if much protein has to be catabolized to produce energy, high levels of transamination and deamination will occur with concomitant increases in metabolic rate. Furthermore, ammonia in the body will rise, leading to progressive hypermetabolism if it cannot be excreted quickly enough. Evidence to support these ideas comes from the studies of the effects of ammonia accumulation on *Anguilla anguilla* (Knights, unpublished results). It was pointed out earlier that nitrogen intake and the magnitude of nitrogen excretion were relatively unaffected by high ambient ammonia in these studies; the time taken to excrete comparable amounts of nitrogen took progressively longer, however, and this was accompanied by hypermetabolism (e.g. see Figure 12.2). Peak post-prandial oxygen demands increased by up to twice and six times the pre-feed levels in fresh water and sea water, respectively, and durations of the SDA increased by up to seven and four times, respectively. This resulted in marked increases in the magnitude of R_F as a percentage of energy ingested per meal, R_F values rising as high as 18%. Blood ammonia is known to rise in the post-prandial period due to deamination and metabolic processing of assimilated amino acids (e.g. Plakas *et al.*, 1980). As ammonia, particularly UIA, accumulates in the surrounding water, passive UIA excretion at the gills would be inhibited and blood levels would rise (Fromm & Gillette, 1968; Maetz, 1972; Kormanik & Cameron, 1981). Acute ammonia stress is known to increase $Na^+:NH_4^+$ exchange and cause internal alkalosis (with compensatory changes in $Na^+:H^+$ exchange); plasma Na^+ tends to rise in fresh water, and urine flow rate (and, in some cases, UN excretion) increases (Alabaster & Lloyd, 1980; Wickins, 1980; Eddy, 1981; Kormanik & Cameron, 1981; Smart, 1981). Smart (1978, 1981) has argued, however, that the initial effects of ammonia stress are exerted by the direct effects of elevated blood UIA on metabolism of the central nervous system. This view is supported by Arillo *et al.* (1981), who showed that ambient UIAN concentrations as low as $20\mu g$ litre^{-1} increased UIA in the brain of trout, which interfered with cerebral amino acid metabolism and energy production and caused a fall in the neurotransmitter GABA. The result was neuronal hyperexcitability, increased muscle tone and hence gross hypermetabolism. Other physiological and morphological responses to high ammonia are therefore possibly secondary to these effects or to the primary hormonal changes in response to generalized stress (Mazeaud & Mazeaud, 1981). Hypermetabolism may be energy costly but the concomitant cardiovascular responses and hyperventilation would be adaptive in helping to excrete more ammonia at the gills; these responses, particularly hyperventilation, are, however, also energy costly (Jones & Schwarzfeld, 1974). To return to the eel example, oxygen consumption increased due to feeding activity and excitement but was dropping at the time of the post-prandial peaks in nitrogen excretion when ambient ammonia was low. Similar results were found by Brett & Zala (1975) for sockeye salmon (*Oncorhynchus nerka*) fed on Oregon Moist Pellets, although their fish showed an anticipatory increase in metabolism before feeding. The low true SDA energy cost in each case may be due to the ease of excreting ammonia because hyperventilation was already occurring; the relatively high surface area:volume ratio and thin skins may also have been a help to the eels. Furthermore, the diets contained easily digested and absorbed and balanced mixtures of amino acids, together with high levels of protein-sparing nutrients. Thus the needs for amino acid interconversions and for catabolism for energy production were relatively low. This in turn would lead to a relatively low level of deamination and hence ammonia to be excreted, which was little affected by increased energy needs during the hypermetabolism at high ambient ammonia levels. Arillo *et al.* (1981) certainly found that ammonia-induced hypermetabolism in trout (*Salmo gairdneri*) fed commercial diets correlated with depletion of carbohydrate stores in the liver, but that amino acid metabolism was little affected. Peters *et al.* (1980) have also demonstrated that social stress in eels causes a reduction in liver glycogen and increases in blood glucose and lactate.

Objective proof is still required, but these results suggest that many apparent and real SDA energy costs quoted in the literature may be distorted relative to those expected in culture. Most other studies have been on quiescent individual fish where feeding excitation and activity may produce exceptionally high initial rises in metabolism. Then, at the peak of ammonia excretion, the low level of activity and low ventilation rates inhibit ammonia excretion until stimulated by increases in blood concentration. Furthermore, the amount of ammonia to be excreted may be relatively large because of a less favourable balance of amino acids in natural (largely invertebrate) foods used and a greater reliance on protein metabolism for energy production. Finally, and possibly most significantly, many studies may have allowed excessive ammonia accumulation in closed systems, without any allowance for acclimation, to achieve easily measurable levels. This in turn could have led to hypermetabolism, as seen in the eels at high ambient ammonia levels.

In conclusion, it can be seen that there is a need for more information on relationships between apparent and real SDA, R_F, diet and ammonia excretion/accumulation in laboratory studies as well as in culture. However, the data imply that real SDA may be a less important source of energy loss in culture than laboratory studies suggest. Use of nutritionally appropriate diets appears important but energy expended in feeding excitation and activity may be the more significant component, providing ammonia accumulation is avoided. Matching of feeding practices and food particle sizes to a species' feeding behaviour and structures are important here (Knights, 1983). Knowledge of the duration and peak levels of oxygen demand and ammonia excretion can be used to ensure that water exchange rates, supplementary aeration and perhaps biological treatment are adequate to meet demands (e.g. see Speece, 1981). Changes in management practices may be of value (e.g. reduce meal size but increase feeding frequency or offer food continuously via automatic feeders). The actual growth inhibitory effects of high ambient ammonia/feeding metabolism are discussed further in the next section.

12.5 Growth Energy (P) and Energy Budgeting

Fish farmers have to measure growth as increments in wet weight (or perhaps length), expressed either as percentage body weight gain per day or (by exponential transformation) as a more instantaneous estimate, i.e. specific growth rate (SGR):

$$SGR = \frac{(\text{Log}_e \text{ final weight} - \log_e \text{ initial weight}) \times 100}{\text{Time in days}} \%$$

Because food is a major cost factor, at least in intensive farming, growth efficiencies are generally expressed as feed conversion ratio (weight of food given/weight gain) or the converse relationship, feed conversion efficiency, often expressed as a percentage. These estimates are open to many errors because, as discussed earlier, relationships between food given and food actually consumed are generally unknown. Confusion can also occur because weights are sometimes expressed as dry and sometimes wet weights.

Growth can be more objectively expressed via construction of nitrogen budgets but these pose methodological problems and are prone to errors because of the protein-sparing roles of carbohydrates and lipids and possible interconversions with ingested protein (Braaten, 1979; Fischer, 1979). The use of instantaneous energy budgeting to estimate growth is a particularly useful tool for applied research because it allows the calculation of gross production efficiency (P/C x 100%). This is of more relevance than food conversion efficiency because it takes actual food consumption into account and, by considering energy contents, avoids problems of variations in water content if using wet weights and of high energy and fat content if using dry weights. Brett & Groves (1979) quote typical gross efficiencies in carnivorous fish feeding on natural foodstuffs (mainly invertebrates) as ranging from 10 to 25% in juveniles, rising to 60% or more in adults. Herbivores generally show low efficiencies (10–20%) because of the low digestibility of plant material, although this figure can be raised by compensatory reductions in metabolic rate. Higher efficiencies would be expected for carnivores in culture feeding on high quality commercial diets; for example, in the studies on eel discussed above, maximum estimated efficiences were 58% in fresh water and 66% in sea water on 1 feed per day at 25°C. Larger eels (average 11.45 g) on maximum commercial rations at comparable temperature show gross efficiencies of about 46% (Nielsen & Jurgensen, 1982). Further studies of relationships between gross conversion efficiencies, feed formulation, ration size and feeding frequency (particularly in relation to body size and temperature) would be of great benefit to feed manufacturers and fish farmers.

Besides optimizing feeds and conversions, studies of energetics can be used to predict levels of culture stresses which inhibit growth. Furthermore, they can show whether inhibition is exerted via effects on C, A or R (individually or in combination) and suggest means of avoiding stress or, at least, aiding recovery.

To illustrate this approach, it is instructive to compare Winberg's (1956) theoretical budget for a typical carnivorous fish with the averages derived by Brett & Groves (1979, Table VI) from the

Table 12.1: Relative Energy Budgets for Various Fish Species Compared with Those for Eels, *Anguilla anguilla* L., Showing Average Percentage Losses of Energy· via Different Routes (for further explanation and discussion, see text).

	C	F	U	R	P	Source
Winberg						
Theoretical budget	100 =	16 +	4 +	60 +	20	Winberg, 1956
Young carnivores	100 =	20 +	7 +	44 +	29	Brett & Groves, 1979
Young herbivores	100 =	41 +	2 +	37 +	20	Brett & Groves, 1979
Young eels in freshwater at:						
Low ambient ammonia	100 =	3 +	3 +	36 +	58	Knights, unpublished results
High ambient ammonia	100 =	3 +	3 +	48 +	46	Knights, unpublished results
Young eels in seawater at:						
Low ambient ammonia	100 =	3 +	2 +	29 +	66	Knights, unpublished results
High ambient ammonia	100 =	3 +	2 +	42 +	53	Knights, unpublished results
Young eels in fresh water	100 =	11 +	5 +	35 +	46	Nielsen & Jurgensen, 1982

literature for young carnivores and herbivores that fed and grew well on natural foods. These in turn can be compared with average budgets for eels (*Anguilla anguilla*) fed on commercial diets, derived by Nielsen & Jurgensen (1982) and Knights (unpublished results). Budgets are expressed in Table 12.1 to show partitioning of energy as a percentage of ingested energy.

Data for actual energy consumed (C) are difficult to compare because different times were used and amounts consumed were expressed differently. All fish were feeding at or near maximum rations, however, except for the eels in Knights' study which were only fed once a day to satiation and were not achieving maximum potential growth rate, a point returned to later. Winberg (1956) underestimated the value for F at 15–17% and U at 3–5% (averaged in Table 12.1) compared with Brett & Groves (1979), but F appears much lower in the eels. This correlates with the high absorption efficiencies found when using commercial feeds, as discussed earlier. The higher F values found by Nielsen & Jurgensen (1982) are probably due to the fact that their eels were fed on trout pellets rather than diets specifically designed for eels. Herbivores show low absorption efficiencies because of the high content of indigestible cellulose in natural plant diets. Efficiencies might be increased in culture if better plant diets could be found or if supplementary formulated feeds were supplied. Excretory nitrogen (U) losses in herbivores appear low because of the low protein content in their diets but they were comparable to losses in the eel studies. It has been argued above that the latter result is due to the ideal amino acid profile of the eel diets and to the protein-sparing effects of other nutrients in energy production. Excretory losses were probably higher in fresh water because of enhanced $Na^+ : NH_4^+$ exchange at the gills to maintain hydromineral balance.

Winberg (1956) overestimated the energy costs of feeding and hence R; the average R for carnivores calculated by Brett & Groves (1979) is in turn some 1.2 and 1.5 times greater than that for eel fingerlings in fresh water and sea water, respectively (Table 12.1). They estimated from data on single satiation feeds that R was comprised of ($7R_S + 23R_A + 14R_D$), whereas the appropriate equivalents for eels studied by Knights (unpublished results) were ($10R_S + 17R_A + 2R_F$) in sea water and ($12R_S + 19R_A + 5R_F$) in fresh water. The larger values for R_S are explained by the relatively high R_S of small fish (average 2.5 g) and high temperature (25°C). The relative energy losses in active metabolism appear low, despite the possible stresses due to high density and the need to swim against water currents. Metabolic energy costs of feeding were particularly low, possibly relating to the relatively small amounts of waste ammonia produced and the ease of excreting it, as discussed earlier. It must be noted, however, that this cost is lower than might be expected if the eels were fed more often. Eels of similar size under comparable conditions in fresh water were studied by E.A. Seymour (personal communication) who found that growth rates increased with increasing feeding frequency to a plateau at about four feeds per day. Total daily food consumption only increased by about 35% but there was little change in absorption efficiency and therefore probably little change in the relative value of F or U. There is no reason to assume that R_S and R_A would have increased but there would have been four apparent SDA periods in one day. Assuming R_F costs were of the same magnitude for each feed, the estimated relative proportions of energy lost in metabolism in fresh water would have been ($qR_S + 14R_A + 15R_F$). Retention of energy for growth (gross production efficiency) would have consequently fallen to about 56%. These calculations, although involving many assumptions, do help to emphasize the importance of losses due to R_F in culture, and that although increasing feeding frequency may accelerate growth rates, a plateau is reached as R_F costs increase. Decreases in absorption efficiencies might also occur at higher feeding rates, although Nielsen & Jurgensen (1982) found no significant correlation with increasing rations in their eels.

The estimated gross production efficiencies for the eels appear large, even at higher feeding levels, reflecting the high potential growth rates of small fish under warmwater culture conditions. Ideally, any indirect estimates from energetic parameters should be checked against actual growth in the experiment or against comparable experiments measuring growth directly. On one meal per day, highest mean population SGRs were estimated from energetic conversions to be about 1.8 and 1.2% per day in sea water and fresh water, respectively. On four feeds per day in fresh water, a relative P of 56% would represent an SGR of about 1.5%. These do in fact compare favourably with SGRs measured directly in eels of comparable size and at similar temperatures and ration levels in fresh water (Sadler, 1979, 1981; Nielsen & Jurgensen, 1982; E.A. Seymour, unpublished results). They are also comparable to those found in UK eel farms (Eel Producers Association, personal communication). However, in the experiments, estimated P, SGR and gross production efficiency tended to decrease appreciably as ambient UIAN rose about 0.1 mg litre^{-1}. This corresponds well with the growth inhibitory level of 0.12 mg litre^{-1} determined by Sadler (1981) in direct ammonia dosing experiments on eels; similar levels are critical in inhibiting growth in other species (Alabaster & Lloyd, 1980; Wickins, 1980; Colt & Armstrong, 1981; Parker & Davis, 1981; Smart, 1981). The energy budget approach has not only predicted correctly the critical UIA concentration but, as discussed earlier, has also suggested how its effect is exerted. The ammonia-induced hypermetabolism caused a 69% increase in R energy costs in sea water and a 75% increase in fresh water, leading to the decreases in P and SGR. Minimum water exchange rates can now be derived, i.e. about 1.0 litre min^{-1} kg^{-1} in both media. Adjusted appropriately for stocking density, pH, salinity, temperature and ambient ammonia in the water supply (Trussell, 1972; Whitfield, 1974; Bower & Bidwell, 1978). Lower exchange rates can only be used if adequate biological denitrification can be arranged. If water supplies are interrupted in an emergency, feeding should be stopped and ambient pH decreased and/or salinity increased appropriately if possible. The data also indicate the relative needs for supplementary aeration during feeding and the post-prandial period. Water exchange rates could also be altered accordingly, if possible, but it must be noted that to meet all oxygen needs as opposed to merely minimizing ammonia accumulation would require increases of about six orders of magnitude.

More studies of fish energetics are needed under conditions comparable to those on fish farms but it appears likely that, for a certain feed type, ration size and feeding frequency, F and U are fairly constant and predictable proportions of C and that they form a relatively small part of the energy budget. Thus studies of the energetics of fish in culture, other than those designed specifically to examine C, F and U relationships, can probably be simplified by only measuring C and R; slight variations in $(F + U)$ from values determined in initial studies or derived from the literature would only introduce minor errors. It is interesting to note the parallels between this approach and that used to derive rational design criteria for salmonid hatcheries and farms based on largely empirical data for fish metabolism. Haskell (1955) was the first worker to propose that carrying capacity and water requirements were directly related to oxygen consumption and accumulation of metabolic products which in turn are both proportional to the ration fed. Other workers have expanded and refined Haskell's theory; Willoughby *et al.* (1972), for example, proposed that average oxygen required and average ammonia produced were 250 g and 32 g per kilogram of feed respectively over a temperature range of 5–18°C. Westers & Pratt (1977) used these figures to derive a theoretical model for the rational design and production potential of salmonid hatcheries. The disadvantages in these approaches are that although they implicitly recognize the importance of C and R and relationships between them, they have relied on empirical data from farm and hatchery studies. Furthermore, they have not measured actual growth, which can be affected by chronic stresses such as crowding, intra- and interspecific interactions, etc., as discussed earlier. Studies on energetics would be useful in assessing relative inhibitory effects of such factors on growth. Equally, if not more important, could be the use of energetics in studies of acute stresses such as handling. As an example of the value of such a study, consider the situation found in eel culture, where average growth rates in a population decline (e.g. because of differential growth and increasing aggression); size grading helps to minimize this but the stresses inherent in grading itself produce growth checks. Periodic monitoring of C and R in a growing population (or instantaneous studies of acclimated populations of different size-frequency distributions) can be used to estimate declines in P and hence allow determination of critical densities, growth periods and/or size structures. The magnitude of growth checks produced by standardized handling procedures could then be assessed via their effects on R during handling and on R and C in the recovery phase. It would then be possible, after estimating P in new population structures and densities, to carry out cost–benefit analyses on management techniques to optimize growth. Further information would be gained on how to adjust feeding, water exchange rates and aeration to minimize respiratory and perhaps ammonia stresses.

12.6 Summary

This chapter has tried to demonstrate the value of bioenergetic approaches to fish culture. The major problems facing the farmer are the lack of experimental and analytical facilities, lack of time and expertise and high labour costs. For the laboratory scientist, the major problems are related to scaling-up facilities to work on high densities of fish under experimental conditions comparable to those on the farm. These are not insurmountable and perhaps a more significant barrier is the tendency towards an overemphasis on the theoretical aspects of energetics and applications to fish in the wild. The latter may well be useful in extensive pond culture but studies of intensive culture require a different approach.

Methodologies and control of variables may not be so rigorous as those possible in small-scale single-animal studies but potential errors are reduced by using standard feeds and working on the larger scale, e.g. with larger numbers of fish, relatively large amounts of faeces, high total oxygen demands, etc. In any case, relative differences in metabolism, growth and efficiencies and the development of predictive models are generally of greater importance in aquaculture than absolutely accurate data. Eventually, data and models must be applied to the large-scale farm situation for full commercial verification. It is here that reasoned predictions can be used to minimize costs and risks to stock while any 'fine tuning' necessary is carried out in the context of the unique set of conditions prevailing on individual farms.

12.7 Acknowledgements

The author would like to thank co-workers at the Polytechnic of Central London for their work and support in preparation of this chapter. Particular thanks are due to Alan Brafield, Andy Seymour and Honor Roberts for their constructive criticisms. Much of the work on eels has been supported and encouraged by the Eel Producers Association and it is hoped the views expressed in this chapter will convince other researchers of the need for a greater applied emphasis in fish bioenergetics.

References

Alabaster, J.S. and Lloyd, R. (1980), *Water Quality Criteria for Freshwater Fish*: London, Butterworth

Arillo, A., Margiocco, C., Melodia, F., Mensi, P. and Scenone, G. (1981), Ammonia toxicity mechanism in fish: studies on rainbow trout (*Salmo gairdneri* Rich.). *Ecotoxicol. Env. Safety* 5, 316

Atherton, W.D. and Aitken, A. (1970), Growth, nitrogen metabolism and fat metabolism in *Salmo gairdneri* Rich. *Comp. Biochem. Physiol. 36*, 719

Bower, C.E. and Bidwell, J.P. (1978), Ionization of ammonia in seawater: effects of temperature, pH and salinity. *J. Fish. Res. Bd. Can. 35*, 1012

Braaten, B.R. (1979), Bioenergetics — a review of methodology. In : Halver, J.E. and Tiews, K. (eds). *Finfish Nutrition and Fishfeed Technology*, vol. 2, pp. 461–504. Berlin, Heenemann

Brett, J.R. (1983), Life energetics of sockeye salmon, *Oncorhynchus nerka*. In : Aspey, W.P. and Lustick, S.I. (eds). *Behavioural Energetics: The Cost of Survival in Vertebrates*, pp. 29–63. Columbus, Ohio State University Press

Brett, J.R. and Groves, T.D.D. (1979), Physiological energetics. In: Hoar, W.S., Randall, D.J. and Brett, J.R. (eds). *Fish Physiology. Bioenergetics and Growth* vol. VIII, pp. 279–352. New York, Academic Press

Brett, J.R. and Zala, C.A. (1975), Daily patterns of nitrogen excretion and oxygen consumption of sockeye salmon (*Oncorhynchus nerka*) under controlled conditions. *J. Fish. Res. Bd. Can. 32*, 2479

Brett, J.R., Sutherland, D.B. and Heritage, G.D. (1971), An environmental control tank for the synchronous study of growth and metabolism of young salmon. *J. Fish. Res. Bd. Can. Tech. Rep. 283*, 11 pp

Burrows, R.E. (1964), Effects of accumulated excretory products on hatchery-reared salmonids. *Fish Wildl. Serv. U.S. Res. Rep. 66*, 1

Colt, J.E. and Armstrong, D.A. (1981), Nitrogen toxicity to crustacea, fish and molluscs. In : Allen, L.J. and Kinney, E.C. (eds). *Proc. Bioengineering Symposium for Fish Culture* (F.C.S. Publ. 1), pp. 34–47, Bethesda, American Fisheries Society

Cowey, C.B. and Sargent, J.R. (1972), Fish nutrition. *Adv. Mar. Biol. 10*, 383

Eddy, F.B. (1981), Effects of stress on osmotic and ionic regulation in fish. In: Pickering, A.D. (ed.). *Stress and Fish*, pp. 77–102, New York, Academic Press

Elliott, J.M. (1979), Energetics of freshwater teleosts. *Symp. zool. Soc. Lond. 44*, 29

Elliott, J.M. and Davison, W. (1975), Energy equivalents of oxygen consumption in animal energetics. *Oecologia 19*, 195

Evans, D.H., Claiborne, J.B., Farmer, L., Mallery, C. and Krasny, E.J. (1982), Fish gill ionic transport methods and models. *Biol. Bull. 163*, 108

Fagerlund, U.H.M., McBride, J.R. and Stone, E.T. (1981), Stress-related effects of hatchery rearing density in coho salmon. *Trans. Am. Fish. Soc. 110*, 644

Feldmeth, C.R. (1983), Costs of aggression in trout and pupfish. In: Aspey, W.P. and Lustic, S.I. (eds). *Behavioral Energetics: The Cost of Survival in Vertebrates*, pp. 117–138. Columbus, Ohio State University Press

Fischer, Z. (1979), Selected problems in fish bioenergetics. In: Halver, J.E. and Tiews, K. (eds). *Finfish Nutrition and Fishfeed Technology*, vol. 1, pp. 17–44, Berlin, Heenemann

Fromm, P.O. and Gillette, J.R. (1968), Effect of ambient ammonia on blood ammonia and nitrogen excretion of rainbow trout (*Salmo gairdneri*). *Comp. Biochem. Physiol. 26*, 887

Halver, J.E. (ed.) (1972), *Fish Nutrition*. New York, Academic Press

Haskell, D.C. (1955), Weight of fish per cubic foot of water in hatchery troughs and ponds. *Prog. Fish Cult. 17*, 117

Hilge, V. (1980), Long-term observations on blood parameters of adult mirror carp (*Cyprinus carpio* L.) held in a closed warmwater system. *Arch. FischWiss. 31*, 41

Hunn, J.B. (1982), Urine flow rate in freshwater salmonids. *Prog. Fish Cult. 44*, 119

Jobling, M. (1981), The influence of feeding on the metabolic rate of fishes : a short review. *J. Fish Biol. 18*, 385

Jobling, M. (1983a), Towards an explanation of specific dynamic action (SDA). *J. Fish Biol. 23*, 549

Jobling, M. (1983b), A short review and critique of methodology used in fish growth and nutrition studies. *J. Fish Biol. 23*, 685

Jobling, M. and Wandsvik, A. (1983), Effect of social interactions on growth rate and conversion efficiency of Arctic charr, *Salvelinus alpinus* L. *J. Fish Biol. 22*, 577

Jones, D.R. and Schwarzfeld, T. (1974), The oxygen cost to the metabolism and efficiency of breathing in trout (*Salmo gairdneri*), *Resp. Physiol. 21*, 241

Kinghorn, B.P. (1983), Genetic variaton in food conversion efficiency and growth in rainbow trout. *Aquaculture 32*, 141

Knights, B. (1983), Food particle-size preferences and feeding behaviour in warmwater aquaculture of European eel (*Anguilla anguilla* L.) *Aquaculture 30*, 173

Kormanik, G.A. and Cameron, J.N. (1981), Ammonia excretion in animals that breathe water: a review. *Mar. Biol. Letters 2*, 11

Kutty, M.N. (1968), Respiratory quotients in goldfish and rainbow trout. *J. Fish. Res. Bd Can. 25*, 1689

Kutty, M.N. (1978), Ammonia quotient in sockeye salmon (*Oncorhynchus nerka*). *J. Fish. Res. Bd Can. 35*, 1003

Kutty, M.N. (1981), Energy metabolism of mullets. In: Oren, O.H. (ed.). *Aquaculture of Grey Mullets* (IBP Publ. 26), pp. 219–64. Cambridge, Cambridge University Press

Li, H.W. and Brocksen, R.W. (1977), Approaches to the analysis of energetic costs of intraspecific competition for space by rainbow trout (*Salmo gairdneri*). *J. Fish Biol. 11*, 329

Lloyd, R. and Swift, D.J. (1976), Some physiological responses by freshwater fish to low dissolved oxygen, high carbon dioxide, ammonia and phenol with particular reference to water balance. In: Lockwood, A.P.H. (ed.) *Effects of Pollution on*

Aquatic Organisms, pp. 47–69. Cambridge, Cambridge University Press

McLean, W.E. and Fraser, F.J. (1974), Ammonia and urea production of coho salmon under hatchery conditions. *Environm. Prot. Ser. Pac. Region Surveillance Rep. EPS-5-PR-74-5.* Washington D.C., Environmental Protection Agency

Maetz, J. (1972), Branchial sodium exchange and ammonia excretion in the goldfish *Carassius auratus*. Effects of ammonia loading and temperature changes, *J. exp. Biol*, **56**, 601

M.A.F.F. (1974). Poultry Nutrition. *Ministry of Agriculture, Fisheries and Food; Reference Book 174*, London, HMSO

M.A.F.F. (1975). Energy allowances and feeding for ruminants. *Ministry of Agriculture, Fisheries and Food; Techn. Bull. 33*. London, HMSO

Mazeaud, M.M. and Mazeaud, F. (1981), Adrenergic responses to stress in fish. In: Pickering, A.D. (ed.). *Stress and Fish*, pp. 49–75. New York, Academic Press

Miller, W.R., Hendricks, A.C. and Cairns, J. (1983), Normal ranges for diagnostically important haematological and blood chemistry characteristics of rainbow trout (*Salmo gairdneri*). *Can. J. Fish. Aquat. Sci.* **40**, 420

Nielsen, L.H. and Jurgensen, E.J. (1982), Al i intensiv akvakultur. *Rapport fra Vandkvalitetsinstituttet, Hoersholm, Denmark*. 139 pp

Ogino, C., Kakimo, J. and Chen, H-S. (1973), Protein nutrition in fish. II Determination of metabolic faecal nitrogen and endogenous nitrogen excretion in carp. *Bull. Jpn. Soc. Sci. Fish.* **39**, 519

Ogino, C., Chiou, J.Y. and Takeuchi, T. (1976), Protein nutrition in fish. IV Effects of dietary energy sources on utilization of protein by rainbow trout and carp. *Bull. Jpn. Soc. Sci. Fish.* **42**, 213

Olson, K.R. and Fromm, P.O. (1971), Excretion of urea by two teleosts exposed to different concentrations of ambient ammonia. *Comp. Biochem. Physiol.* **40A**, 999

Parker, N.C. and Davis, K.B. (1981), Requirements of warmwater fish. In: Allen, L.J. and Kinney, E.C. (eds). *Proc. Bioengineering Symposium for Fish Culture* (F.C.S. Publ. 1), pp. 21–28. Bethesda, American Fisheries Society

Peters, G., Delventhal, H. and Klinger, H. (1980), Physiological and morphological effects of social stress in the eel (*Anguilla anguilla* L.) *Arch. FishWiss.* **30**, 157

Phillips, A.M. and Brockway, D.R. (1959), Dietary calories and the production of trout in hatcheries. *Prog. Fish Cult. 21*, 3

Pickering, A.D. (1981), Introduction: The concept of biological stress. In: Pickering, A.D. (ed.). *Stress and Fish*, pp. 1–9. New York, Academic Press

Pickering, A.D., Pottinger, T.G. and Christie, P. (1982), Recovery of brown trout, *Salmo trutta* L., from acute handling stress: a time-course study. *J. Fish Biol.* **20**, 229

Plakas, S.M., Katayama, T., Tanaka, Y. and Deshimaru, O. (1980), Changes in levels of circulating plasma free amino acids of carp (*Cyprinus carpio*) after feeding a protein and an amino acid diet of similar composition. *Aquaculture 21*, 307

Ricker, W.E. (ed.) (1971), *Methods for Assessment of Fish Production in Freshwaters* (IBP Handbook No. 3). Oxford, Blackwell

Sadler, K. (1979), Effects of temperature on growth and survival of European eel, *Anguilla anguilla* L. *J. Fish Biol.* **15**, 499

Sadler, K. (1981), The toxicity of ammonia to the European eel (*Anguilla anguilla* L.) *Aquaculture 26*, 173

Schreck, C.B. (1981), Stress and compensation in teleostean fishes: responses to social and physical factors. In: Pickering, A.D. (ed.). *Stress and Fish*, pp. 295–321. New York, Academic Press

Schreck, C.B. (1982), Stress and rearing of salmonids. *Aquaculture 28*, 241

Smart, G.R. (1978), Investigations of the toxic mechanisms of ammonia to fish — gas

exchange in rainbow trout (*Salmo gairdneri*) exposed to acutely lethal con-
cc centrations. *J. Fish Biol. 12*, 93

Smart, G.R. (1981), Aspects of water quality producing stress in intensive fish culture. In: Pickering, A.D. (ed.). *Stress and Fish*, pp. 277–93, New York, Academic Press

Smith, R.R. (1979), Methods for determining digestibility and metabolizable energy of feedstuffs. In: Halver, J.E. and Tiews, K. (eds). *Finfish Nutrition and Fishfeed Technology*, vol. II, pp. 453–60. Berlin, Heenemann

Solomon, D.J. and Brafield, A.E. (1972), The energetics of feeding, metabolism and growth of perch (*Perca fluviatilis* L.). *J. Anim. Eco. 41*, 699

Speece, R.E. (1981), Management of dissolved oxygen and nitrogen in fish hatchery waters. In: Allen, L.J. and Kinney, E.C. (eds). *Proc. Bioengineering Symposium for Fish Culture* (F.C.S. Publ. 1), pp. 53–62, Bethesda, American Fisheries Society

Staples, D.J. and Nomura, M. (1976), Influence of body size and food ration on the energy budget of rainbow trout *Salmo gairdneri* Richardson. *J. Fish Biol. 9*, 29

Steffens, W. (1981), Protein utilization by rainbow trout (*Salmo gairdneri*) and carp (*Cyprinus carpio*): A brief review. *Aquaculture 23*, 337

Takeuchi, T., Watanabe, T. and Ogino, C. (1979), Availability of carbohydrates and lipids as dietary energy sources for carp. *Bull. Jpn. Soc. Sci. Fish. 45*, 977

Tarr, R.J.Q. and Hill, B.J. (1978), Oxygen consumption, food assimilation and energy content of southern African elvers (*Anguilla* sp.). *Aquaculture 15*, 141

Tomasso, J.R., Davis, K.B. and Simco, B.A. (1981), Plasma corticosteroid dynamics in channel catfish exposed to ammonia and nitrite. *Can. J. Fish Aquat. Sci. 38*, 1106

Trussell, R.P. (1972), The percent un-ionized ammonia in aqueous ammonia solutions at different pH levels and temperatures *J. Fish. Res. Bd. Can. 29*, 1505

Utne, F. (1979), Standard methods and terminology in finfish nutrition. In: Halver, J.E. and Tiews, K. (eds). *Finfish Nutrition and Fishfeed Technology*, vol. 2, pp. 437–44. Berlin, Heenemann

Wedemeyer, G.A. (1976), Physiological responses of juvenile coho salmon (*Oncorhynchus Kisutch*) and rainbow trout (*Salmo gairdneri*) to handling and crowding stress in intensive fish culture. *J. Fish. Res. Bd Can. 33*, 2699

Wedemeyer, G.A. and McLeay, D.J. (1981), Methods for determining tolerance of fishes to environmental stressors. In: Pickering, A.D. (ed.). *Stress and Fish*, pp. 247–75. New York, Academic Press

Westers, H. and Pratt, K.M. (1977), Rational design of hatcheries for intensive salmonid culture, based on metabolic characteristics. *Prog. Fish. Cult. 39*, 157

Whitfield, M. (1974), The hydrolysis of ammonium ions in seawater — a theoretical study. *J. mar. Biol. Assoc., UK 54*, 565

Wickins, J.F. (1980), Water quality requirements for intensive aquaculture: a review, *EIFAC Symposium, Stavanger, Norway, May 1980*, EIFAC/80/Symp: R/2, mimeographed. Rome, F.A.O. Publications

Willoughby, H., Larsen, N. and Bowen, J.T. (1972), The pollutional effects of fish hatcheries. *Amer. Fishes U.S. Trout News 17*, 6

Winberg, G.G. (1956), [Rate of metabolism and food requirements of fishes] Belorussian State Univ. Minsk. *Fish Res. Bd Can Transl. Ser. No. 194*, 1960, pp.1–253

SUBJECT INDEX

341

SYSTEMATIC INDEX

347